METHODS IN MOLECULAR BIOLOGY™

For further volumes:
http://www.springer.com/series/7651

Cellular and Subcellular Nanotechnology

Methods and Protocols

Edited by

Volkmar Weissig, Tamer Elbayoumi, and Mark Olsen

Department of Pharmaceutical Sciences, Midwestern University College of Pharmacy, Glendale, AZ, USA

Editors
Volkmar Weissig
Department of Pharmaceutical Sciences
Midwestern University College of Pharmacy
Glendale, AZ, USA

Tamer Elbayoumi
Department of Pharmaceutical Sciences
Midwestern University College of Pharmacy
Glendale, AZ, USA

Mark Olsen
Department of Pharmaceutical Sciences
Midwestern University College of Pharmacy
Glendale, AZ, USA

ISSN 1064-3745 ISSN 1940-6029 (electronic)
ISBN 978-1-62703-335-0 ISBN 978-1-62703-336-7 (eBook)
DOI 10.1007/978-1-62703-336-7
Springer New York Heidelberg Dordrecht London

Library of Congress Control Number: 2013933992

Printed on acid-free paper

Humana Press is a brand of Springer
Springer is part of Springer Science+Business Media (www.springer.com)

Preface

"While early ideas about the impact of nanotechnology on healthcare focused on fanciful ideas involving small submarines and cancer-zapping robots, current advances have been enabled by advances in imaging, control over materials and an increased understanding of how biology works at the nanoscale" (Tim Harper, CEO Cientifica).

This book is dedicated to showcase the most recent advances that have been made in utilizing the enormous potential of nanotechnology for probing, imaging, and manipulating life on a cellular and subcellular level. All chapters were written by leading experts in their particular fields. Daniel Moyano and Vincent Rotello describe a novel "chemical nose" approach, i.e., nanoparticle-based sensor arrays for the differentiation of biomolecules through pattern recognition that utilizes functionalized gold nanoparticles as receptors and Green Fluorescent Protein as transducer. This new strategy allows the identification of cellular signatures in early stages of cancer without previous knowledge of specific receptors or ligands. Sunaina Surana and Yamuna Krishnan demonstrate the utility of an externally introduced, pH-triggered DNA nanomachine inside the multicellular eukaryote *Caenorhabditis elegans*. This nanomachine uses FRET to effectively map spatiotemporal pH changes associated with endocytosis in coelomocytes of wild type as well as mutant worms. Syed K. Sohaebuddin and Liping Tang describe a method which allows the assessment of lysosomal membrane integrity upon exposure to various nanoparticles. The electron microscopic visualization of 1–2 nm gold nanoparticles, which are used as nano markers, allows Valeriy Lukyanenko and Vadim Salnikov to determine the precise localization of a variety of nano-objects within a cell. The same author also describes a saponin-based method for membrane permeabilization allowing the delivery of particles up to 20 nm in size to the perinuclear and perimitochondrial space of cardiomyocytes.

Howard Gendelman's laboratory provides in two chapters protocols for the isolation of subcellular compartments containing sequestered nanoparticles. Indriati Pfeiffer and Michael Zäch describe the use of nanostructured SiO_2 surfaces prepared by the colloidal lithography technique to scrutinize the formation of suspended lipid bilayers from a solution of nano liposomes. These authors employ atomic force microscopy (AFM) and quartz crystal microbalance with dissipation monitoring (QCM-D) to characterize nanostructure fabrication and lipid bilayer assembly on the nanostructured surface. QCM-D is also being utilized by Rickard Frost and Sofia Svedhem to monitor the interaction of nanoparticles with lipid membranes in real time. The authors demonstrate how the outcome of such analysis provides information on the adsorption process (importantly kinetics and adsorbed amounts) as well as on the integrity of both the nanoparticles and the lipid membrane upon interaction. A protocol for studying the interactions of nanoparticles with proteins is provided by Lennart Treuel and Marcelina Malissek. These authors describe a procedure to study the adsorption of proteins onto nanoparticle surfaces based on circular dichroism (CD) spectroscopy. Jerry Chang and Sandra Rosenthal describe the principles, methodologies, and experimental protocols for quantum dot-based single-molecule imaging.

Ben Zhong Tang and his colleagues describe the fabrication of fluorescent silica nanoparticles (FSNPs) containing aggregation-induced emission (AIE) luminogens. By employing surfactant-free sol–gel reaction the authors are able to generate FSNPs with uniform size and high surface charge and colloidal stability. Simon C.W. Richardson group applies single cell imaging technology for studying the intracellular trafficking of both biological and synthetic macromolecules and they demonstrate the possibility of temporally dissecting novel and default trafficking of both macromolecular "drugs" and macromolecular drug delivery systems. Irene Canton and Giuseppe Battaglia describe a polymersomes-mediated delivery of fluorescent probes for targeted and long-term imaging in live cell microscopy. Junghae Suh and colleagues explain in their chapter one of the most complicated aspects of real-time particle tracking, i.e., the mean square displacement (MSD) calculation, in a simple manner designed for the novice particle tracker. By providing comprehensive instructions needed to perform particle tracking experiments, their chapter will enable researchers to gain new insight into the intracellular dynamics of nanocarriers, potentially leading to the development of more effective and intelligent therapeutic delivery vectors. Mi-Sook and Song Her provide a direct method for quantifying cellular transduction of PTD in vitro and in vivo using bioluminescence imaging. Their methodology exploits noninvasive techniques to create an environment suitable for the real-time imaging of PTD transduction and appears therefore as a promising tool for studying the mechanism of PTD transduction and the in vivo application of new therapeutic candidates. Achim Göpferich group describes a procedure for monitoring the intracellular route of polyplexes based on the use of labeling PEI and pLL with a reduction-sensitive fluorescent dye. Katye M. Fichter and Tania Q. Vu describe the use of single nanoparticle quantum dot (QD) probes to quantitatively investigate the complex endocytic trafficking pathways that receptors undergo following ligand activation. The use of cell-penetrating peptides (CPPs) to facilitate the cellular internalization of quantum dots (QDs) is described by Yue-Wern Huang and colleagues. Their approach is based on simple noncovalent interactions between CPPs and QDs. Lo and Wang describe the use of peptide-based carriers for the intracellular delivery of biologically active proteins as well as methods for the qualitative and quantitative evaluation of their delivery efficiency. Jan van Hest's laboratory presents a novel strategy for the preparation of gold nanoparticles exhibiting a stimuli-responsive behavior, which is based on the use of a ligand consisting of only a single repeat of the elastin-based pentapeptide VPGVG. The authors provide protocols for the solid-phase peptide synthesis of thiol-terminated VPGVG ligand and for the preparation of gold nanoparticles covered with the pentapeptide through a ligand-exchange reaction. Jae Sam Lee and Ching-Hsuam Tung have developed an improved CPP-based cellular delivery vector, named lipo-oligoarginine peptide (LOAP), by conjugating an oligoarginine peptide with a fatty acid moiety. The prepared LOAPs were further stabilized by introducing different combinations of D-Arg residues into the peptide backbone, and were systematically evaluated for their membrane penetrating properties and metabolic stabilities in cells. Andrea Alessandrini and Paolo Facci describe the use of electrochemical scanning tunneling microscopy (ECSTM) and spectroscopy (ECSTS) for studying the electron transport through single redox molecules with the aim of understanding the transport mechanisms ruling the flow of electrons via a single molecule placed in a nanometer-sized gap between two electrodes, while elucidating the role of the redox density of states brought about by the molecule. Yamuna Krishnan's group has constructed an icosahedron from DNA using a modular self-assembly strategy. They describe a method to determine the functionality of DNA polyhedra as nanocapsules by encapsulating different cargo such as gold nanoparticles and functional biomolecules like FITC dextran from solution within DNA icosahedra. The use of polymer-gold nanorods assemblies for the delivery of plasmid DNA into mammalian cells is described by Kaushal Rege's laboratory. Puiyan Lee and Kenneth K.Y. Wong describe a technique for the synthesis of a novel lipophilic nano carrier for the incorporation of hydrophobic and toxic potent cancer drugs, such as gold (III) porphyrin. Tamer Elbayoumi's laboratory provides protocols for preparing mitochondria-targeted nanoemulsions loaded with tocopherol and Cyclosporine A which are able to protect cardiac muscle mitochondria from doxorubicin-induced oxidative stress. Achim Weber and colleagues describe the production of uniform protein-binding biofunctional fluorescent spherical silica core-shell nanoparticles. The authors characterize their novel nanoparticle system including its surface functionalization via microelectrophoresis, dynamic light scattering (DLS) and a colorimetric detection of the amount of nanoparticle-attached protein via a bicinchoninic acid (BCA) assay. Such fluorescently spiked nanoparticle cores with biofunctional shells for molecular recognition reactions may be used as imaging tools or reporter systems. Neskovic and her colleagues describe the assessment of genotoxic properties of purified single wall carbon nanotubes (SWCNT), multiwall carbon nanotubes (MWCNT), and amide functionalized purified SWCNT using cultured human lymphocytes and human fibroblasts. Dusica Maysinger's group has developed a suitable fractionation method for field flow fractionation, an analytical technique that allows the separation of nano and microparticles over a wide size range. The authors present asymmetrical flow field-flow fractionation (AF4) conditions that have proven their reliability for the analysis of quantum dots and other nanoparticles in the 5–50 nm size range. Maxwell B. Zeigler and Daniel T. Chiu give detailed steps necessary to perform laser surgery upon single adherent mammalian cells, where individual organelles are extracted from the cells by optical tweezers and the cells are monitored post-surgery to check their viability. Yaron R. Silberberg and Andrew E. Pelling describe a method to quantify the intracellular mechanical response to an extracellular mechanical perturbation, specifically the displacement of mitochondria. A combined fluorescent-atomic force microscope (AFM) was used to simultaneously produce well-defined nanomechanical stimulation to a living cell while optically recording the real-time displacement of fluorescently labeled mitochondria.

We are extremely grateful to all authors for having spent parts of their valuable time to contribute to this book. It is our hopes that together we have succeeded in providing an essential source of know-how and at the same time a source of inspiration to all investigators who are as fascinated as we are about the potential of applying nanotechnology to all areas of biomedical sciences. Last but not least we would like to thank John Walker, the series editor of "Methods in Molecular Biology" for having invited us to assemble this book and above all for his unlimited guidance and help throughout the whole process.

Glendale, AZ, USA

Volkmar Weissig
Tamer Elbayoumi
Mark Olsen

Contents

Contributors

P. Hans H.M. Adams • *Department of Organic Chemistry, Institute for Molecules and Materials, Radboud University Nijmegen, Nijmegen, The Netherlands*
Andrea Alessandrini • *CNR-NANO-S3, and Physics Department, University of Modena and Reggio Emilia, Modena, Italy*
Giuseppe Battaglia • *Department of Biomedical Science, The Krebs Institute, The University of Sheffield, Sheffield, UK*
Dhiraj Bhatia • *National Centre for Biological Sciences, Tata Institute of Fundamental Research, Bangalore, India*
Miriam Breunig • *Lehrstuhl für Pharmazeutische Technologie, Universität Regensburg, Regensburg, Germany*
Irene Canton • *Department of Biomedical Science, The Krebs Institute, The University of Sheffield, Sheffield, UK*
Saikat Chakraborty • *National Centre for Biological Sciences, Tata Institute of Fundamental Research, Bangalore, India*
Jerry C. Chang • *Department of Chemistry, Vanderbilt University, Nashville, TN, USA*
Sijie Chen • *Division of Biomedical Engineering, Hong Kong University of Science and Technology, Kowloon, Hong Kong, China*
Huey-Jenn Chiang • *Institute of Biotechnology, National Dong Hwa University, Hualien, Taiwan*
Daniel T. Chiu • *Department of Chemistry, University of Washington, Seattle, WA, USA*
Jelena Cvetićanin • *Institute of Nuclear Sciences "Vinča", University of Belgrade, Belgrade, Serbia*
Christopher Dempsey • *Department of Bioengineering, Rice University, Houston, TX, USA*
Paul D.R. Dyer • *School of Science, University of Greenwich, Kent, UK*
Tamer Elbayoumi • *Department of Pharmaceutical Sciences, Midwestern University College of Pharmacy, Glendale, AZ, USA*
Paolo Facci • *CNR-NANO-S3, Modena, Italy*
Amy Faulk • *Department of Pharmaceutical Sciences, Midwestern University College of Pharmacy, Glendale, AZ, USA*
Katye M. Fichter • *Department of Biomedical Engineering, Oregon Health and Science University, Portland, OR, USA*
Rickard Frost • *Department of Applied Physics, Chalmers University of Technology, Göteborg, Sweden*
Howard E. Gendelman • *Department of Pharmacology and Experimental Neuroscience, University of Nebraska Medical Center, Omaha, NE, USA*
Achim Göpferich • *Lehrstuhl für Pharmazeutische Technologie, Universität Regensburg, Regensburg, Germany*

SONG HER • *Division of Bio-Imaging, Chuncheon Center, Korea Basic Science Institute, Chuncheon, Republic of Korea*
MARION HERZ • *Fraunhofer Institute for Interfacial Engineering and Biotechnology IGB, Stuttgart, Germany*
CONSTANTIN HOZSA • *Lehrstuhl für Pharmazeutische Technologie, Universität Regensburg, Regensburg, Germany*
FEIRAN HUANG • *Department of Bioengineering, Rice University, Houston, TX, USA*
HUANG-CHIAO HUANG • *Wellman Center for Photomedicine, Massachusetts General Hospital and Harvard Medical School, Boston, MA, USA*
YUE-WERN HUANG • *Department of Biological Sciences, Missouri University of Science and Technology, Rolla, MO, USA*
GORDANA JOKSIĆ • *Institute of Nuclear Sciences "Vinča", University of Belgrade, Belgrade, Serbia*
IRENA KADIU • *Department of Pharmacology and Experimental Neuroscience, University of Nebraska Medical Center, Omaha, NE, USA*
ARUN K. KOTHA • *School of Science, University of Greenwich, Kent, UK*
YAMUNA KRISHNAN • *National Centre for Biological Sciences, Tata Institute of Fundamental Research, Bangalore, India*
JACKY W.Y. LAM • *Department of Chemistry, Hong Kong University of Science and Technology, Kowloon, Hong Kong, China*
HAN-JUNG LEE • *Department of Natural Resources and Environmental Studies, National Dong Hwa University, Hualian, Taiwan*
JAE SAM LEE • *Department of Radiology, Methodist Hospital Research Institute, Weill Cornell Medical College, Houston, TX, USA*
MI-SOOK LEE • *Division of Bio-Imaging, Chuncheon Center, Korea Basic Science Institute, Chuncheon, Republic of Korea*
PUIYAN LEE • *Department of Surgery, Li Ka Shing Faculty of Medicine, The University of Hong Kong, Pokfulam, Hong Kong*
VINCENT LEMIEUX • *Department of Organic Chemistry, Institute for Molecules and Materials, Radboud University Nijmegen, Nijmegen, The Netherlands*
ANDREJA LESKOVAC • *Institute of Nuclear Sciences "Vinča", University of Belgrade, Belgrade, Serbia*
BETTY REVON LIU • *Department of Natural Resources and Environmental Studies, National Dong Hwa University, Hualian, Taiwan*
SEONG LOONG LO • *Department of Biological Sciences, National University of Singapore, Singapore*
VALERIY LUKYANENKO • *Department of Medicine, Johns Hopkins School of Medicine, Baltimore, MD, USA*
MARCELINA MALISSEK • *Physical Chemistry, University of Duisburg-Essen, Essen, Germany*
DUSICA MAYSINGER • *Department of Pharmacology and Therapeutics, Faculty of Medicine, McGill University, Montreal, Canada*
JOELLYN MCMILLAN • *Department of Pharmacology and Experimental Neuroscience, University of Nebraska Medical Center, Omaha, NE, USA*
SHABANA MEHTAB • *National Centre for Biological Sciences, Tata Institute of Fundamental Research, Bangalore, India*

ALEXANDRE MOQUIN • *Faculty of Pharmacy, and Department of Pharmacology & Therapeutics, Faculty of Medicine, Université de Montréal and McGill University, Montreal, QC, Canada*
DANIEL F. MOYANOA • *Department of Chemistry, University of Massachusetts, Amherst, MA, USA*
OLIVERA NEŠKOVIĆ • *Institute of Nuclear Sciences "Vinča", University of Belgrade, Belgrade, Serbia*
ARI NOWACEK • *Department of Pharmacology and Experimental Neuroscience, University of Nebraska Medical Center, Omaha, NE, USA*
ANDREW E. PELLING • *Department of Physics, Department of Biology, Institute for Science, Society and Policy, University of Ottawa, Ottawa, Canada*
SANDRA PETROVIĆ • *Institute of Nuclear Sciences "Vinča", University of Belgrade, Belgrade, Serbia*
MARIE W. PETTIT • *School of Science, University of Greenwich, Kent, UK*
INDRIATI PFEIFFER • *Department of Cell biology and Genetics, Erasmus Medical Center, Rotterdam, Netherlands*
JAMES RAMOS • *School of Biological and Health Systems & Chemical Engineering, Center for the Convergence of Physical Science and Cancer Biology, Arizona State University, Tempe, AZ, USA*
KAUSHAL REGE • *School of Biological and Health Systems & Chemical Engineering, Center for the Convergence of Physical Science and Cancer Biology, Arizona State University, Tempe, AZ, USA*
SIMON C.W. RICHARDSON • *University of Greenwich, School of Science, Kent, UK*
SANDRA J. ROSENTHAL • *Department of Chemistry, Department of Pharmacology, Department of Chemical and Biomolecular Engineering, Department of Physics and Astronomy, Institute of Nanoscale Science and Engineering, Vanderbilt University, Nashville, TN, USA*
VINCENT M. ROTELLO • *Department of Chemistry, University of Massachusetts, Amherst, MA, USA*
VADIM SALNIKOV • *Kazan Institute of Biochemistry and Biophysics, Kazan Scientific Centre Russian Academy of Sciences, Kazan, Russia*
YARON R. SILBERBERG • *Biomedical Research Institute (BMRI), National Institute of Advanced Industrial Science and Technology (AIST), Kyoto University, Kyoto, Japan*
SYED K. SOHAEBUDDIN • *Department of Bioengineering, University of Texas at Arlington, Arlington, TX, USA*
JUNGHAE SUH • *Department of Bioengineering, Rice University, Houston, TX, USA*
SUNAINA SURANA • *National Centre for Biological Sciences, Tata Institute of Fundamental Research, Bangalore, India*
SOFIA SVEDHEM • *Department of Applied Physics, Chalmers University of Technology, Göteborg, Sweden*
BEN ZHONG TANG • *Department of Chemistry, Hong Kong University of Science and Technology, Kowloon, Hong Kong, China*
LIPING TANG • *Department of Bioengineering, University of Texas at Arlington, Arlington, TX, USA*

GÜNTER E.M. TOVAR • *Institute for Interfacial Engineering IGVT, University Stuttgart, Stuttgart, Germany; Fraunhofer Institute for Interfacial Engineering and Biotechnology IGB, Stuttgart, Germany*
LENNART TREUEL • *Institute of Applied Physics and Center for Functional Nanostructures (CFN), Karlsruhe Institute of Technology (KIT), Karlsruhe, Germany; Physical Chemistry, University of Duisburg-Essen, Essen, Germany*
DJORDJE TRPKOV • *Institute of Nuclear Sciences "Vinča", University of Belgrade, Belgrade, Serbia*
CHING-HSUAN TUNG • *Department of Radiology, Methodist Hospital Research Institute, Weill Cornell Medical College, Houston, TX, USA*
ANA VALENTA-ŠOBOT • *Institute of Nuclear Sciences "Vinča", University of Belgrade, Belgrade, Serbia*
JAN C.M. VAN HEST • *Department of Organic Chemistry, Institute for Molecules and Materials, Radboud University Nijmegen, Nijmegen, The Netherlands*
TANIA Q. VU • *Department of Biomedical Engineering, Oregon Health and Science University, Portland, OR, USA*
SHU WANG • *Institute of Bioengineering and Nanotechnology, Singapore; Department of Biological Sciences, National University of Singapore, Singapore*
ERIN WATSON • *Department of Bioengineering, Rice University, Houston, TX, USA*
ACHIM WEBER • *Fraunhofer Institute for Interfacial Engineering and Biotechnology IGB, Stuttgart, Germany; Institute for Interfacial Engineering IGVT, University Stuttgart, Stuttgart, Germany*
VOLKMAR WEISSIG • *Department of Pharmaceutical Sciences, Midwestern University College of Pharmacy, Glendale, AZ, USA*
FRANÇOISE M. WINNIK • *Faculty of Pharmacy and Department of Chemistry, Université de Montréal, Montreal, QC, Canada*
KENNETH K.Y. WONG • *Department of Surgery, Li Ka Shing Faculty of Medicine, The University of Hong Kong, Pokfulam, Hong Kong*
CHI-HENG WU • *Department of Biological Sciences, Missouri University of Science and Technology, Rolla, MO, USA*
MICHAEL ZÄCH • *Department of Applied Physics, Chalmers University of Technology, Gothenburg, Sweden*
MAXWELL B. ZEIGLER • *Department of Chemistry, University of Washington, Seattle, WA, USA*

Chapter 1

Nanoparticle-GFP "Chemical Nose" Sensor for Cancer Cell Identification

Daniel F. Moyano and Vincent M. Rotello

Abstract

Nanoparticle-based sensor arrays have been used to distinguish a wide range of bio-related molecules through pattern recognition. This "chemical nose" approach uses nanoparticles as receptors to *selectively* identify the analytes, while a transducer reports the binding through a readable signal (fluorescence). Here we describe a procedure that uses functionalized gold nanoparticles as receptors and green fluorescent protein (GFP) as the transducer to identify and differentiate cell state (normal, cancerous, and metastatic), an important tool in early diagnosis and treatment of tumors.

Key words Sensor, Chemical nose, Gold nanoparticle, GFP, Fluorescence, LDA

1 Introduction

Antibody-based sensing techniques are an important tool in the early detection of cancer (1). These techniques employ *specific* recognition to identify the analytes, targeting different biomarkers of each cell state (2). However this approach also limits the applicability of this method due to constrains in the availability of specific markers (3). As an alternative to this methodology, the "chemical nose" approach utilizes multiple *selective* receptors that generate a unique response pattern for each analyte, allowing its classification (4). The identification is achieved by taking advantage of differential interactions between the analytes and the receptors. In the sensor approach described here, gold nanoparticles (AuNPs) are used as the receptors, controlling the nature of the interaction by tuning the chemical properties at the nanoparticle surface (5). AuNP–analyte interactions are then transduced by a fluorescent probe (GFP), initially quenched when bound to the AuNP, and then displaced from the AuNP surface upon the addition of the analyte, with concomitant restoration of fluorescence. This strategy has been successfully applied to identify and differentiate a variety of bio-constructs, from proteins (6) to bacteria (7) and cancer cells (8).

Volkmar Weissig et al. (eds.), *Cellular and Subcellular Nanotechnology: Methods and Protocols*, Methods in Molecular Biology, vol. 991, DOI 10.1007/978-1-62703-336-7_1, © Springer Science+Business Media New York 2013

Here we report the procedure to apply this methodology to the identification of cancer cells, independent of their origin (isogenic), and differentiating among a diversity of cell states (normal, cancerous, and metastatic). This strategy allows a versatile identification of cellular signatures in early stages of cancer, a major hurdle in cancer therapy, without previous knowledge of specific receptors or ligands.

2 Materials

1. Phosphate buffer (PB) solution: 5 mM phosphate, pH 7.4. Mix 3.87 mL of 1.0 M Na_2HPO_4 with 1.13 mL of 1.0 M NaH_2PO_4 solution. Complete to a final volume of 1.0 L using type I ultrapure water (see Note 1). Adjust pH if necessary and store at room temperature (25°C).
2. Functionalized gold nanoparticles solution: three cationic and one neutral gold nanoparticles (Fig. 1) are synthesized by place exchange reaction (9), from pentanethiol-capped 2 nm diameter gold nanoclusters, and using the desired ligands (synthesized as shown in (10)). Prepare 200 μL of a 40 μM stock solution for each nanoparticle in type I ultrapure water, calculating the AuNP concentration by its absorbance at 506 nm ($\varepsilon_{506} = 4.9 \times 10^5\ M^{-1}cm^{-1}$ for 2 nm diameter gold core) according to the reported methodology (11). Store at 4°C (see Note 2).
3. Green fluorescent protein solution: GFP is expressed according to reported procedure (12). Prepare 400 μL of a GFP stock solution of 250 μM in 5 mM PB, calculating the protein concentration by its absorbance at 488 nm ($\varepsilon_{488} = 5.6 \times 10^4\ M^{-1}cm^{-1}$) (see Note 3).

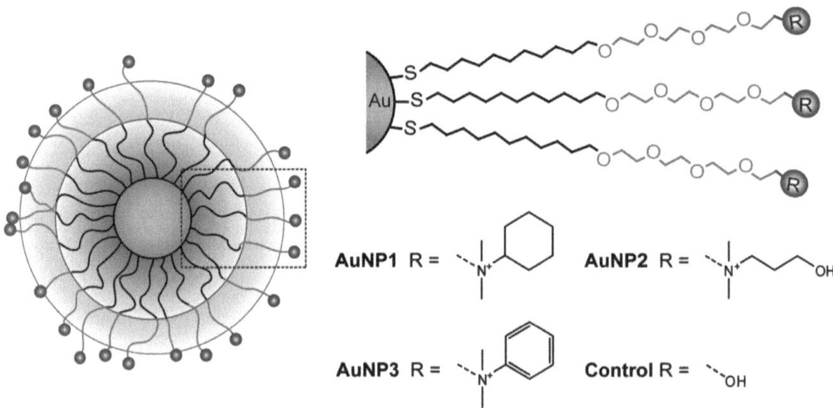

Fig. 1 Chemical structure of the gold nanoparticles, featuring h-bond, π-stacking, and hydrophobic groups at the surface

4. Cell suspension: Grow cancer, metastatic, and normal state cells (analytes) according to the provider procedures in T75/T25 flasks. Remove the media, wash the cells with DBPS buffer, add 2 mL of trypsin EDTA 1× (per container), and collect the cells in serum-containing media. Centrifuge to obtain cell pellets and suspend them in DMEM non-serum medium. Count the cells and dilute the suspension with DMEM to have a final concentration of ~200 cells/μL (see Note 4).
5. 96-well 300 μL black polystyrene plates with dark flat bottom.
6. 50 mL PVC pipette basins, 50 mL PP Falcon tubes, and 7 mL glass vials.

3 Methods

All procedures are done at room temperature.

3.1 Determination of Optimal Sensor Conditions

1. Prepare 40 mL of 150 nM GFP solution in a Falcon tube using 5 mM PB, starting from the initial GFP stock solution (see Note 5).
2. In a 7 mL glass vial, pour 10 μL of one of the nanoparticles stock solutions. Complete the solution to 2 mL using the 150 nM GFP solution. Mix gently and allow this AuNP–GFP complex solution to reach equilibrium for 15 min (see Note 6).
3. During the equilibration time, add sequentially decreasing values of the 150 nM GFP solution into a 96-well plate, according to the scheme (Fig. 2).

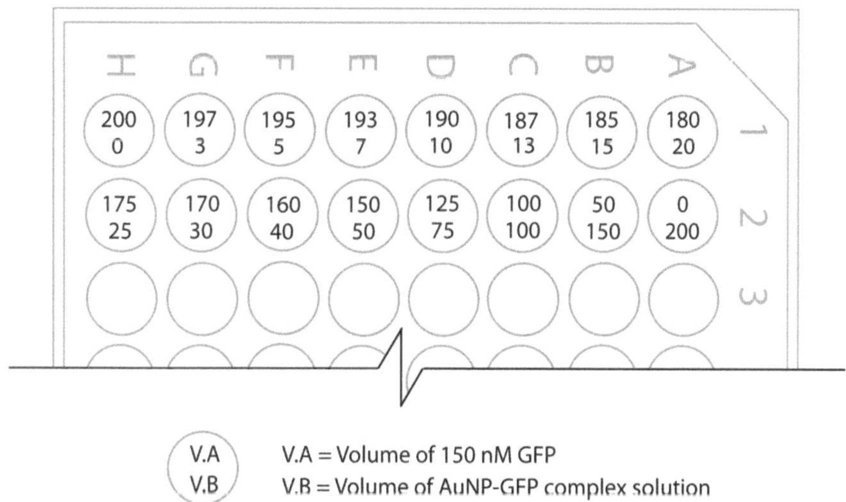

Fig. 2 Sequential volumes of the GFP and AuNP–GFP solutions to be used in the 96-well plate for titration purposes

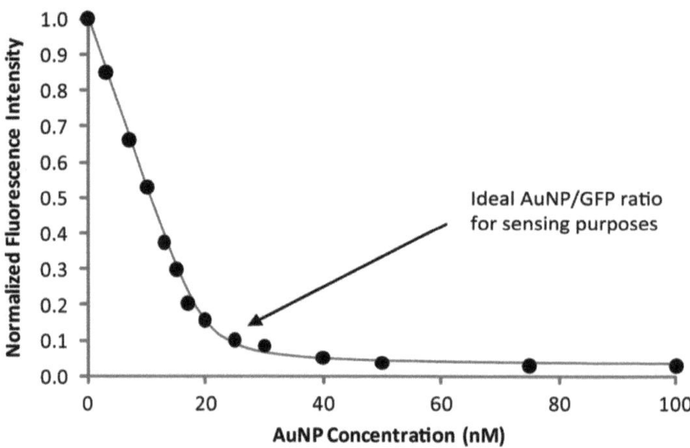

Fig. 3 Cationic-AuNP titration curve featuring the position of the optimal AuNP/GFP ratio to use in the sensing procedure

4. After the 15 min, add sequentially increasing values of the AuNP–GFP complex solution to each well according to the scheme (Fig. 2), to complete a total volume of 200 μL per well. Each nanoparticle has to be titrated by triplicate (see Note 7).
5. Mix gently with the pipette tip to avoid bubbles and leave the system reach equilibrium for another 15 min.
6. After the final equilibration time, take fluorescent intensity measurements at 510 nm, using an excitation wavelength of 475 nm.
7. Plot the normalized fluorescence intensity against the nanoparticle concentration in each well (see Note 8).
8. Select the ideal AuNP/GFP ratio for sensing purposes (Fig. 3). This is the ratio to be used in Subheading 3.2 (see Note 9).
9. Repeat the same procedure using the four different nanoparticles. The AuNP/GFP optimal ratio may differ among all nanoparticles (see Note 10).

3.2 Sensing Procedure

1. Prepare 40 mL of 150 nM GFP solution in Falcon tube using 5 mM PB and starting from the initial GFP stock solution (see Note 5).
2. In a 7 mL vial, pour the amount of the nanoparticle stock solution necessary to reach the optimal AuNP/GFP ratio (150 nM of GFP), completing to 2 mL using the 150 nM GFP solution. Mix gently and wait for 15 min. Repeat this step for the other nanoparticles (see Notes 6 and 10).
3. After the equilibrium time, add 200 μL of each AuNP/GFP solution in the 96-well plate according to the scheme in Fig. 4 (see Note 11). Six replicates are done for each specific nanoparticle.

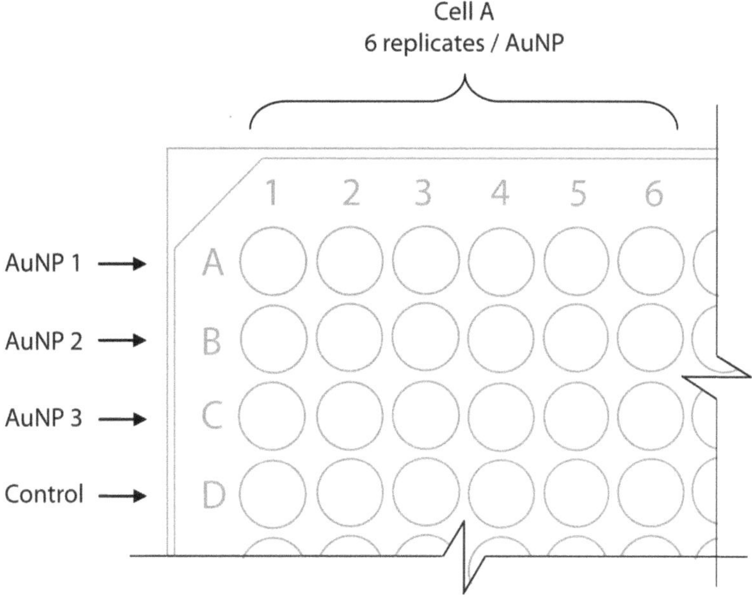

Fig. 4 Input scheme to use the different AuNP–GFP complexes against one cell analyte. The output of this experiment generates the fingerprint for the given analyte

4. Record the fluorescent intensity at 510 nm, using an excitation wavelength of 475 nm. This is the initial intensity (I_i).
5. Add 25 μL of the cell suspension (~5,000 cells) in each well (see Note 12). Mix with the pipette tip gently and leave 30 min for incubation time (see Note 6).
6. After the incubation time, record the fluorescent intensity at 510 nm, using an excitation wavelength of 475 nm. This is the final intensity (I_f).
7. Calculate the $\log(I_f/I_i)$ for each point, generating the average and the standard deviation with the six replicates, to build up the fingerprint of each analyte (see Note 13).
8. Using linear discriminant analysis software (like SYSTAT 11), use all $\log(I_f/I_i)$ values (analytes x replicates x nanoparticles) as an input to generate the canonical score plots and the jackknife classification percentage for cell identification (see Notes 14 and 15).

4 Notes

1. The 1.0 M solution of NaH_2PO_4 has to be fresh each time that the buffer is prepared to avoid later crystallization problems of this solution that can alter the final concentration.

2. After place exchange reaction, nanoparticles are purified by dialysis using a 10,000 MWCO tubing membrane. The nanoparticles are then lyophilized, dispersed in type I ultrapure water, and passed through a 0.22 μm membrane filter. TEM, MS, NMR, and DLS techniques are used to control the quality of the nanoparticles according to literature (10).
3. Continuous freezing cycles of GFP can denature the protein, affecting the reproducibility. To avoid this problem, prepare the stock solution and divide it in different containers, store them at −78°C in dark (aluminum foil), and take out one container each time that the solution is needed. No glycerol or DTT is used given the stability of GFP in PB at low concentrations.
4. If no recommendations are given, grow human- and mouse-type cells in Dulbecco's Modified Eagle's Medium (DMEM, 4.0 g/L glucose) supplemented with 10% fetal bovine serum and 1% antibiotics (100 U/mL penicillin and 100 μg/mL streptomycin) in T75 or T25 flasks and under humidified atmosphere of 5% CO_2 at 37°C. The cells should be maintained in the mentioned conditions and subcultured once every 4 days.
5. Maintain the 150 nM GFP solution in an ice bath and cover the Falcon tube with aluminum foil while performing the experiments to avoid photobleaching and denaturation. Homogenize the solution by mixing gently.
6. During the equilibrium times, the system should be maintained in dark conditions (covering the plate/vial with aluminum foil).
7. Given that each nanoparticle can be titrated (with its three replicates) in 48 wells, 2 different nanoparticles can be analyzed per plate, optimizing the time performance.
8. The normalized fluorescence is obtained by

$$||I|| = \frac{I_c / I_n}{I_c^o / I_n^o},$$

where I_c is the intensity of a given point of a cationic nanoparticle, I_n is the intensity of the same point of the neutral (control) nanoparticle, and I_c^o and I_n^o are the measurements in the absence of nanoparticles in the solution (first point). This is done to eliminate the effect of the gold core natural absorption.
9. Choose a point in the curve at the bottom of the plot, after the principal quenching (initial slope) but before reaching the horizontal tendency. Points in this zone have the biggest ΔI after the addition of the analyte due to its sensitive nature in the competitive binding (13). The control nanoparticle does not have a titration point; in this case, use a concentration similar to the cationic nanoparticles.

10. The total time used to analyze each nanoparticle should be consistent in all the cases to avoid reproducibility problems. Use the same amount of time when preparing as well as mixing the solutions for each one of the nanoparticles.
11. We find that the minimum amount of cationic nanoparticles as predictors to have a good percentage of identification (>90%) is three, featuring functional groups for h-bond, π-stacking, and hydrophobic interactions (14). However more nanoparticles can be used to increase identification accuracy.
12. Four different cells can be run simultaneously in the same plate, optimizing the analysis process (Fig. 4).
13. We find that six is a reasonable number of replicates to have good representation in the canonical graphs with a 95% of confidence (no overlapping).
14. LDA generates canonical factors to minimize the training matrix. The best two canonical factors are the ones used to generate the 2D score plots.
15. In the case of an unknown test sample, *see* Subheading 3.2 with the unidentified cell line. Using Mahalanobis distance analysis, compare the results with the previously generated matrix for proper identification. This analysis is done using SYSTAT as described above.

Acknowledgment

This work was supported by the NIH (GM077173).

References

1. Haab BB (2006) Applications of antibody array platforms. Curr Opin Biotechnol 17:415–421
2. Kingsmore SF (2006) Multiplexed protein measurement: technologies and applications of protein and antibody arrays. Nat Rev Drug Discov 5:310–321
3. Klee EW (2008) Data mining for biomarker development: a review of tissue specificity analysis. Clin Lab Med 28:127–143
4. Albert KJ, Lewis NS, Schauer CL et al (2010) Cross-reactive chemical sensor arrays. Chem Rev 100:2595–2626
5. Moyano DF, Rotello VM (2011) Nano meets biology: structure and function at the nanoparticle interface. Langmuir 27:10376–10385
6. You CC, Miranda OR, Gider B et al (2007) Detection and identification of proteins using nanoparticle–fluorescent polymer 'chemical nose' sensors. Nat Nanotechnol 2:318–323
7. Phillips RL, Miranda OR, You CC et al (2008) Rapid and efficient identification of bacteria using gold-nanoparticle-poly(para-phenylene-ethynylene) constructs. Angew Chem Int Ed 47:2590–2594
8. Bajaj A, Rana S, Miranda OR et al (2010) Cell surface-based differentiation of cell types and cancer states using a gold nanoparticle-GFP based sensing array. Chem Sci 1:134–138
9. Templeton AC, Wuelfing WP, Murray RW (2000) Monolayer-protected cluster molecules. Acc Chem Res 33:27–36
10. De M, Rana S, Akpinar H et al (2009) Sensing of proteins in human serum using conjugates of nanoparticles and green fluorescent protein. Nat Chem 1:461–465
11. Liu X, Atwater M, Wang J et al (2007) Extinction coefficient of gold nanoparticles with different sizes and different capping

ligands. Colloid Surface B Biointerfaces 58:3–7
12. De M, Rana S, Rotello VM (2009) Nickel-ion-mediated control of the stoichiometry of his-tagged protein/nanoparticle interactions. Macromol Biosci 9:174–178
13. Moyano DF, Rana S, Bunz UHF et al (2011) Gold nanoparticle-polymer/biopolymer complexes for protein sensing. Faraday Discuss 152:33–42
14. Bajaj A, Miranda OR, Kim IB et al (2009) Detection and differentiation of normal, cancerous, and metastatic cells using nanoparticle-polymer sensor arrays. Proc Natl Acad Sci USA 106:10912–10916

Chapter 2

A Method to Map Spatiotemporal pH Changes in a Multicellular Living Organism Using a DNA Nanosensor

Sunaina Surana and Yamuna Krishnan

Abstract

Environmental pH has a determining role in the structure of biomolecules, thus playing an important role in regulating cellular activities. Eukaryotic cells must, therefore, strive to stringently regulate pH in various intracellular organelles so as to confer normal functioning at the level of whole organism. Several pH-sensitive probes have been reported, each of which can be used to map the pH dependence of an in vivo process. However, these probes suffer from some inherent drawbacks. Here we demonstrate the utility of an externally introduced, pH-triggered DNA nanomachine inside the multicellular eukaryote *Caenorhabditis elegans*. The nanomachine uses FRET to effectively map spatiotemporal pH changes associated with endocytosis in coelomocytes of wild type as well as mutant worms, in a variety of genetic backgrounds. It shows highest dynamic range in the pH regime 5.3–6.6 and has a half-life of ~8 h, thus positioning it well to interrogate a variety of pH-correlated biological phenomena in vivo.

Key words I-switch, pH sensor, Caenorhabditis elegans, Coelomocytes

1 Introduction

Protons have a determining role in the charge and, in turn, structure of biomolecules. Hence, proton concentration plays an important role in regulating cellular and, in turn, organismal activities. Eukaryotic cells, thus, have stringently controlled pH in their various intracellular organelles (1). Perturbation of cytoplasmic and organellar pH has been shown to lead to defects in receptor-mediated endocytosis, intracellular targeting of newly synthesized lysosomal proteins, calcium homeostasis, protein processing and sorting, and degradation of neurotransmitters (2). In vivo, these cellular defects are manifested in terms of tumor metastasis (1), growth defects (2), neurodegeneration (3), lysosomal storage disorders (4), defects in embryogenesis, spermatogenesis, and excretion (5), to name a few. Hence, pH is an important correlate of biological processes in vivo.

Volkmar Weissig et al. (eds.), *Cellular and Subcellular Nanotechnology: Methods and Protocols*, Methods in Molecular Biology, vol. 991, DOI 10.1007/978-1-62703-336-7_2, © Springer Science+Business Media New York 2013

Due to the molecular and cellular complexity encountered in the environment of a whole organism, there are only few reports of pH-responsive probes in vivo. Most pH measurements in organisms have relied on genetically encodable, pH-sensitive GFP variants termed pHluorins, which can be fused to a specific protein and, thus, used to mark a pathway of interest (6). pHluorins enable ratiometric imaging, which helps to cull differences in protein expression. These properties have led to its use in a variety of systems, right from bacteria to mice. In bacterial systems like *Escherichia coli* and *Listeria*, this probe has been used to measure stress response to optical tweezers (7). On the other hand, Dittman et al. have elegantly used pHluorins to identify genes regulating abundance of vesicular SNAREs (soluble NSF attachment protein receptors) at *Caenorhabditis elegans* cholinergic motor neuron synapses (8). In another report, Poskanzer et al. have used synaptopHluorins to study the dynamics of resting pools of synaptic vesicles in the *Drosophila* neuromuscular junction (9). Despite such promising applications, pHluorins suffer from some issues such as (1) fixed wavelength, which limits its use in the background of GFP-expressing transgenics; (2) simultaneous tracking of two pathways is not possible; (3) quenching of its fluorescence at acidic pH values that leads to difficulties in visualization in highly acidic compartments; and (4) its fixed pH sensitive regime given that many physiological processes often operate at pH regimes beyond the pH sensitivity of pHluorins.

Less widely used pH-sensitive probes in vivo are fluorescein and its derivatives fluorescein isothiocyanate (FITC), 2′,7′ bis (2-carboxyethyl)-5-(and-6)-carboxyfluorescein (BCECF) and carboxyfluorescein (10). Rare examples are the use of BCECF to measure the pH of interstitial space in neoplastic tumors of the rabbit ear (11) and carboxyfluorescein to study intracellular pH in the bacterial species *Lactobacillus delbrueckii* and *Listeria innocua* (12). However, their inability to be targeted specifically to a pathway of interest, fixed wavelength, and suboptimal photophysical properties has precluded their wider applicability in whole organisms.

Thus, there is a need for a pH probe that combines robust fluorescence properties, targeting, and the ability to manipulate wavelengths as well as pH regimes. Here, we describe the functioning of a rationally designed pH-sensitive DNA nanomachine, called the I-switch (13), in vivo. This FRET-based pH sensor utilizes bright and stable fluorophores and can be used at different wavelengths, positioning it well in the context of fluorescent-protein-expressing transgenics. It functions autonomously and reversibly in the organismal milieu, making it amenable to a variety of mutant backgrounds. The structure and charge of the DNA backbone also makes it facile for targeting. It has highest dynamic range in the pH regime 5.3–6.6, enabling it to map spatiotemporal pH changes occurring during endosomal maturation (14).

Importantly, due to its rational design, it is possible to change its dynamic range as well as pH-sensitive regime, positioning it well to track pH changes in other physiological processes in vivo.

2 Materials

2.1 Oligonucleotides and Sample Preparation

1. All oligonucleotides (obtained from MWG Biotech, Germany, or IBA GmbH, Germany) are high-performance liquid chromatography (HPLC) purified and lyophilized (Table 1). Oligonucleotides are dissolved in Milli-Q (MQ) water to prepare 200 μM stocks, aliquoted, and all aliquots stored at −20°C, until further use. Fluorescently labeled oligonucleotides are subjected to ethanol precipitation prior to use to remove any traces of free dye.
2. Ethanol, absolute.
3. 3.0 M potassium acetate solution: 2.94 g CH_3COOK dissolved in 10 mL MQ water and pH adjusted to 5.2.
4. Phosphate buffer – 100 mM KH_2PO_4: 1.36 g KH_2PO_4 dissolved in 10 mL MQ water. 100 mM K_2HPO_4: 1.74 g K_2HPO_4 dissolved in 10 mL MQ water. Each potassium phosphate solution is diluted to 10 mM. Add 10 mM KH_2PO_4 and 10 mM K_2HPO_4 to obtain a buffer of pH 5.5.
5. 3.0 M potassium chloride solution: 2.23 g KCl dissolved in 10 mL MQ water.
6. Heat block and water bath.

2.2 Fluorescence Spectroscopy

1. FluoroMax-4 instrument (Horiba Jobin Yvon, Japan), equipped with a mercury–Xe lamp as the light source.
2. Clamping buffer – 20× salt solution: 3.57 g KCl, 0.116 g NaCl, 0.044 g $CaCl_2$, and 0.081 g $MgCl_2$ dissolved in 20 mL MQ water. 20× HEPES solution: 1.9 g HEPES dissolved in 20 mL MQ water.

Table 1
Oligonucleotide sequences used for the I-switch

Sequence
5′-CCCCAACCCCAA*TACATTTTACGCCTGGTGCC*-3′
5′-*CCGACCGCAGGATCCTATAA*AACCCCAACCCC-3′
5′-*TTATAGGATCCTGCGGTCGG*A*GGCACCAGGCGTAAAATGTA*-3′
5′-Alexa 488-CCCCAACCCCAA*TACATTTTACGCCTGGTGCC*-3′
5′-*CCGACCGCAGGATCCTATAA*AACCCCAACCCC-Alexa 647-3′

3. 10 mL of 1 N HCl and 1 N NaOH solutions.
4. Prior to use, clamping buffers of desired pH values (5.0, 5.5, 6.0, 6.5, and 7.0) are made by mixing the salt and HEPES solutions and diluting to make 1× solution. The pH is then adjusted by using 1 N HCl or 1 N NaOH.
5. Samples are diluted to 100 nM using clamping buffers of various pH.
6. 1 cm quartz cuvette.

2.3 *C. elegans* Maintenance and Strains

1. *C. elegans* is grown at 22°C on nematode growth medium (NGM) containing a lawn of OP50 bacteria.
2. All strains have been obtained from Caenorhabditis Genetics Center (University of Minnesota, USA).
3. Wild-type strain: *C. elegans* isolate from Bristol (strain N2).
4. Mutant strains: *rme-1(b1045)*, *rme-4(b1001)*, *rme-5(b1013)*, and *rme-6(b1014)*.
5. Transgenic strains: *arIs37* [p*myo-3::ssGFP*], *cdIs131* [p*cc1::GFP::RAB-5* + *unc-119(+)* + *myo-2::GFP*], *cdIs66* [p*cc1::GFP::RAB-7* + *unc-119(+)* + *myo-2::GFP*], and *pwIs50* [*lmp-1::GFP* + *Cb-unc-119(+)*].
6. BOD incubator at 22°C for maintenance of nematode stocks.

2.4 Coelomocyte Labeling and pH Clamping

1. 5 μM I-switch sample containing acceptor label only (I_{A647}) and both donor and acceptor labels ($I_{A488/A647}$) is diluted to the desired concentration (see Subheading 3.1) in 1× medium 1.
2. 10× medium 1: 4.37 g NaCl, 0.18 g KCl, 0.055 g $CaCl_2$, 0.1 g $MgCl_2$, and 0.95 g HEPES dissolved in 50 mL MQ water and pH adjusted to 7.3. This is filter-sterilized using 0.22 μm membrane filter.
3. TE2000-S inverted microscope, equipped with a 40×, 0.75 NA objective (Nikon, Japan), and microinjection setup (Narishige, Japan).
4. Borosilicate glass capillaries.
5. 2.0% agarose pads, made on 22 × 50 mm glass coverslips.
6. 2.0% agarose: 0.2 g agarose dissolved in 10 mL MQ water.
7. Halocarbon oil.
8. 10 mM nigericin: 1.0 mg nigericin (Sigma-Aldrich, USA) is dissolved in 133 μL of absolute ethanol. 10 μL aliquots are made and stored at −20°C.
9. 10 mM monensin: 1.0 mg monensin (Sigma-Aldrich, USA) is dissolved in 120 μL of absolute ethanol. 10 μL aliquots are made and stored at −20°C.

10. Clamping buffers: See Subheading 2.2 above. To each of these buffers, 10 mM monensin and 10 mM nigericin are added to obtain final concentration of 100 μM each.

2.5 Competition Experiments

1. Maleic anhydride.
2. Bovine serum albumin (BSA).
3. BSA solution: 6 mg/mL solution in 0.1 M sodium carbonate bicarbonate buffer (pH 9).
4. BSA solution: 6 mg/mL solution in 1× phosphate-buffered saline (PBS) of pH 7.4.
5. 1× PBS: For 1 L, dissolve 8 g NaCl, 0.2 g KCl, 1.44 g Na_2HPO_4, and 0.24 g KH_2PO_4; adjust pH to 7.4; and bring to 1 L with MQ water.
6. Dextran, 10 kDa MW.
7. Dextran solution: 3.2 mg/mL solution in 1× PBS of pH 7.4.
8. Dextran sulfate, 9–20 kDa MW.
9. Dextran sulfate solution: 4.4 mg/mL solution in 1× PBS of pH 7.4.
10. Heparan sulfate, 30 kDa MW.
11. Heparan sulfate solution: 2.0 mg/mL solution in 1× PBS of pH 7.4.
12. Unlabeled I-switch, diluted to 100 nM using 1× medium 1.

2.6 Fluorescence Microscopy and Image Analysis

1. Axiovert Apotome microscope (Zeiss, Bulgaria), equipped with 40×, 1.3 NA objective, metal halide lamp (Zeiss, Bulgaria), and filters suitable for each fluorophore.
2. TE2000-U inverted microscope (Nikon, Japan), equipped with 60×, 1.4 NA objective, mercury arc illuminator (Nikon, Japan), filters suitable for each fluorophore (Chroma, USA), and Cascade II CCD camera (Photometrics, USA).
3. Image acquisition software: MetaMorph (Universal Imaging, USA).
4. Fluoview 1000 confocal microscope (Olympus, Japan), equipped with argon ion laser (Spectra-Physics, USA) for 488 nm excitation and He–Ne laser (Spectra-Physics, USA) for 633 nm excitation and a set of excitation, emission, and dichroic filters suitable for each fluorophore (Olympus, Japan).
5. Image analysis software: ImageJ ver. 1.42 (NIH, freely available from website: http://rsbweb.nih.gov/ij/).
6. 40 mM sodium azide: 13 mg NaN_3 dissolved in 5 mL M9 buffer.

7. 1× M9 buffer: 0.6 g Na_2HPO_4, 0.3 g KH_2PO_4, 0.5 g NaCl, and 0.1 g NH_4Cl dissolved in 100 mL MQ water and pH is adjusted to 7.3. The buffer is filter-sterilized through a 0.22 μm membrane filter.

3 Methods

We describe the use of a rationally designed synthetic DNA assembly called the I-switch that functions as a pH sensor to map spatiotemporal pH changes in a multicellular living organism. The I-switch consists of two DNA duplexes connected to each other by a flexible hinge and bearing cytosine-rich single-stranded overhangs at the duplex termini. Upon protonation, these C-rich overhangs base pair with each other to form an I-motif, causing a structural transition at acidic pH (13, 14). At neutral pH, the I-motif dissociates and entropic forces as well as electrostatic repulsion between the duplex arms cause the reversal of the structural transition (Fig. 1a). This forms the molecular basis of a FRET-based pH sensor using the I-switch.

Endocytosis is known to be accompanied by changes in pH as vesicles mature from the early endosomes to the lysosomes, via the late endosomes (15). These pH changes, measured in cultured cells, are known to be in the regime of 6.0–6.2 in early endosomes to ~5.5 in late endosomes and ~5.0 in lysosomes (16). Given that this shows a good match with the pH-sensitive regime of the I-switch, we describe its use in mapping spatiotemporal pH changes associated with endosomal maturation in vivo. As an example, we use it to map the same along the anionic ligand-binding receptor (ALBR) pathway in the coelomocytes of wild-type *C. elegans*. We show that this pH sensor can map variations in pH in different genetic backgrounds such as mutants and RNAi knockdowns that perturb pH homeostasis. In a model organism like *C. elegans*, where pathways are elucidated primarily by genetics, this is of special importance, since it demonstrates the utility and non-perturbative nature of the I-switch in this model organism.

3.1 Sample Preparation

1. The I-switch is generally prepared at 5 μM concentration in a volume of 50 μL. 1.25 μL of O1, O2, and O3 (each from a 200 μM stock) with 1.65 μL of 3 M KCl are mixed. The volume is made up to 50 μL by adding 10 mM potassium phosphate buffer of pH 5.5 (see Note 1). The solution is briefly vortexed.
2. The solution is heated at 90°C for 5 min in a heat block and then cooled to room temperature at a rate of 1°C per 3 min. Samples are then equilibrated at 4°C overnight. Fluorescently labeled I-switch is prepared in a similar manner with fluorophore-labeled oligonucleotides. Samples are used within 7 days of annealing.

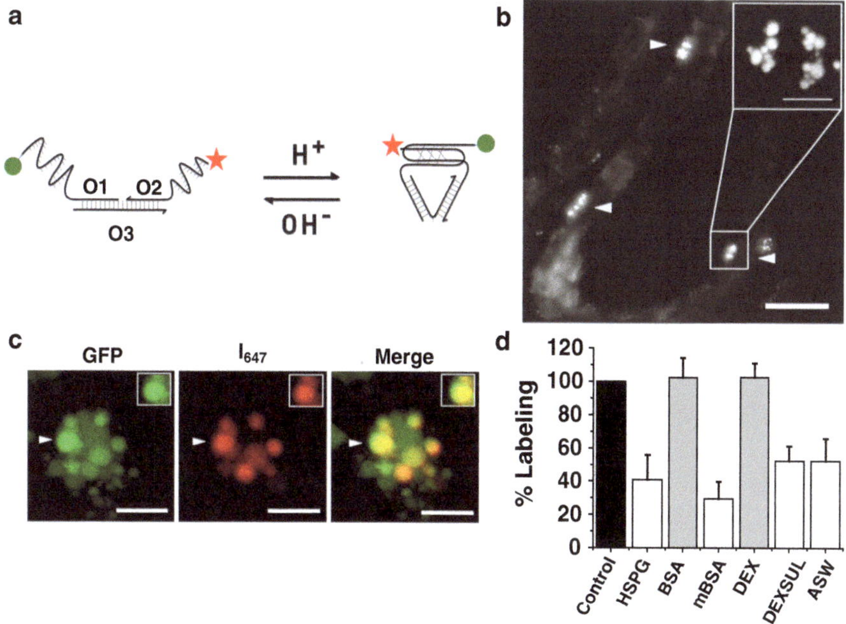

Fig. 1 I-switch marks endosomes of the anionic ligand-binding receptor (ALBR) pathway in coelomocytes of *Caenorhabditis elegans*. (**a**) Schematic showing the principle of the I-switch. (**b**) Epifluorescence image of wild-type *C. elegans* hermaphrodite microinjected with I_{A647}. Arrowheads indicate labeled coelomocytes. Scale bar: 50 μm. Inset: confocal image of I_{A647}-labeled coelomocyte. Scale bar: 5 μm. (**c**) I-switch specifically marks coelomocytes upon injection in the pseudocoelom of *arIs37* hermaphrodites. Scale bar: 10 μm. Inset shows a typical image of one such endosome, showing co-localization between GFP and I_{A647}. (**d**) Competition assay where I_{A647} is co-injected with 300 equivalents of various endocytic markers to establish mode of uptake of I-switch in coelomocytes (HSPG: heparan sulfate proteoglycan, BSA: bovine serum albumin, mBSA: maleylated bovine serum albumin, DEX: dextran, DEXSUL: dextran sulfate, ASW: Competing DNA). Error bars indicate s.e.m. ($n = 20$ worms)

3. Prior to use, samples are diluted in 1× medium 1 and vortexed to enable mixing. They are centrifuged at 9300 rcf for 20 min.

3.2 Microinjections and Coelomocyte Labeling

1. For coelomocyte labeling, I switch made from Alexa 647-functionalized O2 (I_{A647}) is used. 5 μM stock solution of I-switch sample is diluted to 100 nM using 1× medium 1.
2. One-day old hermaphrodites grown on NGM plates (+OP50) are mounted on a 2% agarose pad containing a droplet of halocarbon oil.
3. Injections are performed, using borosilicate capillaries, at 50–55 psi in the dorsal side of the pseudocoelom, just opposite the vulva.
4. Injected worms are released using 1× M9 buffer and transferred to NGM plates (+OP50). Plates are incubated at 22°C for 1 h.

5. After 1 h, injected worms are mounted on a glass slide containing a 2.0% agarose pad, anesthetized using 40 mM NaN_3 in M9 buffer, and imaged.
6. Wild-type hermaphrodites, when injected with I_{A647} and imaged on an Axiovert Apotome microscope, show bright puncta in coelomocytes (Fig. 1b). Uptake is quantified by percentage of coelomocytes labeled postinjection.
7. Confirmation that the I-switch marks coelomocytes in *C. elegans* is obtained by injecting I_{A647} in the strain *arIs37*. Co-localization between GFP and I_{A647} is performed by merging images taken on an Olympus Fluoview confocal microscope (Fig. 1c).
8. For confirming the mode of endocytosis, competition experiments are performed with an excess of anionic ligands, which are known to bind ALBRs with high affinity. I_{A647} is mixed with competitor ligands mBSA, dextran sulfate, heparan sulfate, and unlabeled DNA in a 1:300 ratio, such that the final concentration of I_{A647} is 100 nM. As a control, injections with the neutral molecules BSA and dextran at the same molar ratios are also performed. Imaging is performed on an Axiovert Apotome microscope, equipped with a 40×, 1.3 NA objective.
9. Uptake is quantified by percentage of coelomocytes labeled postinjection; all values are normalized to uptake in hermaphrodites injected with I_{A647} alone (Fig. 1d).

3.3 pH Clamping

1. For pH measurements, doubly labeled I-switch ($I_{A488/A647}$) is used.
2. First, an in vitro pH calibration curve is generated by diluting 5 μM fluorescently labeled I-switch to 100 nM in clamping buffer of the desired pH, ranging from pH 5.0 to 7.0. All samples are vortexed and equilibrated for 30 min at room temperature. The samples are excited at 488 nm, and emission is collected between 500 and 700 nm with a bandwidth of 1 nm (for excitation) and 10 nm (for emission) and spectral scan speed of 1 nm/s.
3. Fluorescent intensities at 520 nm (D) and 665 nm (A) are obtained, and then D is divided by A for every pH value to generate an in vitro pH response curve.
4. For in vivo pH measurements, 5 μM stock solution of $I_{A488/A647}$ is diluted to 500 nM using 1× medium 1.
5. The functionality of a sensor in vivo is determined by assessing the fold change of its donor/acceptor (D/A) ratio in the dynamic regime and comparing this to the in vitro fold change. Good correspondences of the fold change values indicate the sensor integrity in the given environment.

6. One-day old wild-type hermaphrodites are injected with $I_{A488/A647}$, incubated at 22°C for 1 h, and then kept in a dish containing 1 mL clamping buffer of the desired pH for 75 min. This buffer contains the ionophores nigericin and monensin at a final concentration of 100 μM each, which equilibrate the intra-coelomocyte pH to that of the external buffer. Using this method, the pH of coelomocytes is clamped at pH 5.0 in ten worms and at pH 7.0 on ten different worms. Prior to soaking in clamping buffer, the cuticle of the worms is perforated in three regions (anterior, middle, and posterior) with a microinjection needle. After 75 min in the clamping buffer, the worms are mounted on a glass slide using the same clamping buffer (with nigericin and monensin) and the coelomocytes imaged (see Note 2). This is done three times independently, on ten worms each (see Note 3).
7. For each endocytic vesicle, fluorescence intensity at 520 nm (D) is divided by the intensity at 665 nm (A). This gives the D/A ratio for that vesicle. Cells clamped at pH 5.0 show a low D/A ratio, while those clamped at pH 7.0 show elevated values (Fig. 2a, b).
8. Fold change is calculated by dividing the D/A ratio at the higher end of the dynamic regime to that obtained at the lower end (pH 7.0 and pH 5.0, respectively, in this case). This fold change is then compared to the in vitro fold change in order to assess the performance of the sensor (Fig. 2c).
9. pH is then clamped at intermediate values (5.0, 5.5, 6.0, 6.5, and 7.0) to obtain the standard calibration curve (Fig. 2d). This will now be used to calculate pH values from D/A ratios in the system under study (see Note 4).

3.4 Spatiotemporal pH Mapping

1. In order to assess the pH changes occurring during endocytic maturation as a function of time, temporal regimes of the residence times of the I-switch in different populations of vesicles are determined. This is done by performing co-localization studies, as a function of time, of I_{A647} in the GFP-expressing transgenics *cdIs131*, *cdIs66*, and *pwIs50*. These strains express GFP fusions of Rab-5 (early endosomal marker), Rab-7 (late endosomal/lysosomal marker), and Lmp-1 (lysosomal marker).
2. I_{A647} is diluted to 500 nM using 1× medium 1.
3. I_{A647} is injected in wild-type hermaphrodites and, after the requisite time (from 5 to 60 min, at increments of 5 min for the first 30 min), transferred to chilled NGM plates (+OP50), and kept on ice. This method efficiently stops endocytosis and trafficking of vesicles (see Note 5).

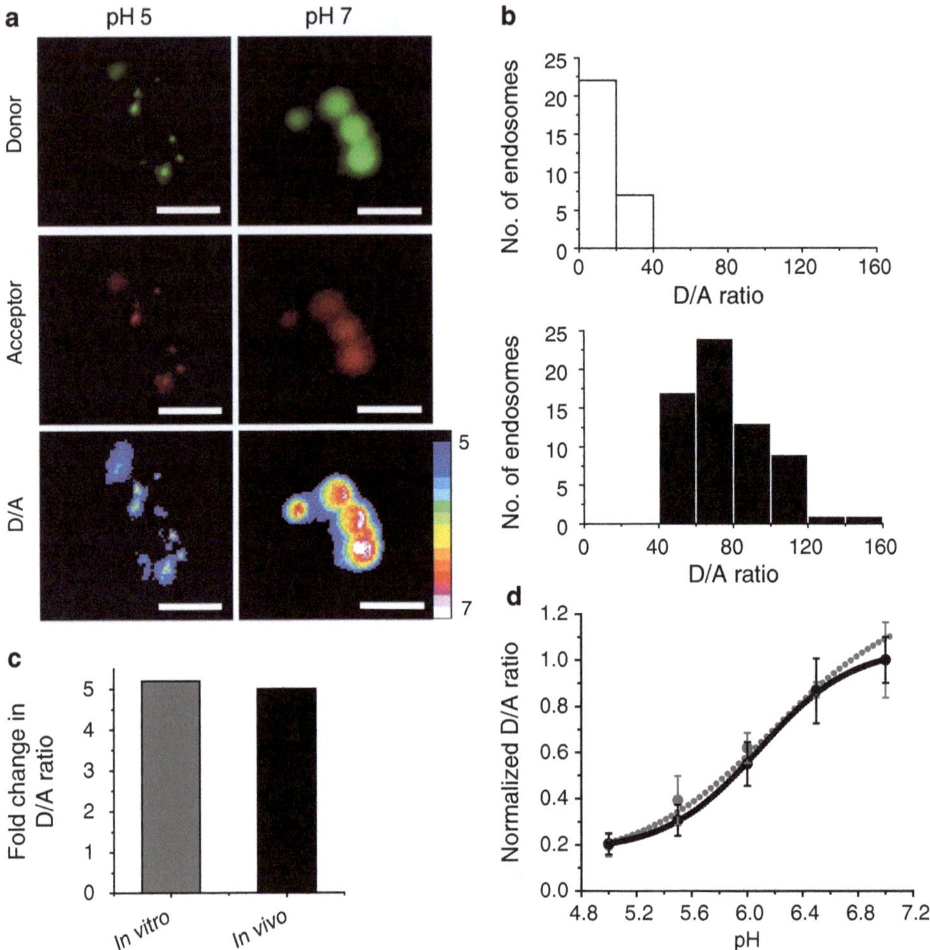

Fig. 2 In vivo characterization of the I-switch. (**a**) Donor channel (D), acceptor channel (A), and respective pseudocolor D/A images of coelomocytes labeled with $I_{A488/A647}$ and clamped at pH 5.0 and 7.0. Scale bar: 10 μm. (**b**) Histograms showing typical spread of D/A ratios of endosomes clamped at pH 5.0 (*white bars*) and pH 7.0 (*black bars*) (n ³ 25 endosomes). (**c**) In vitro and in vivo fold change in D/A ratios of $I_{A488/A647}$ between pH 7.0 and pH 5.0. (**d**) pH calibration curve of $I_{A488/A647}$ in vivo (*black trace*) and in vitro (*gray trace*) showing normalized D/A ratios versus pH. *Error bars* indicate s.e.m. (n ³ 50 endosomes)

4. Coelomocytes are imaged on the Fluoview 1000 confocal microscope. Co-localization of GFP and I_{A647} is determined by counting the numbers of I_{A647}-positive puncta that co-localize with GFP-positive puncta and expressing them as a percentage of the total number of I_{A647}-positive puncta. The time point where the I-switch shows maximal co-localization with an endocytic marker is chosen for pH measurements in that particular endocytic vesicle (Fig. 3a–d).

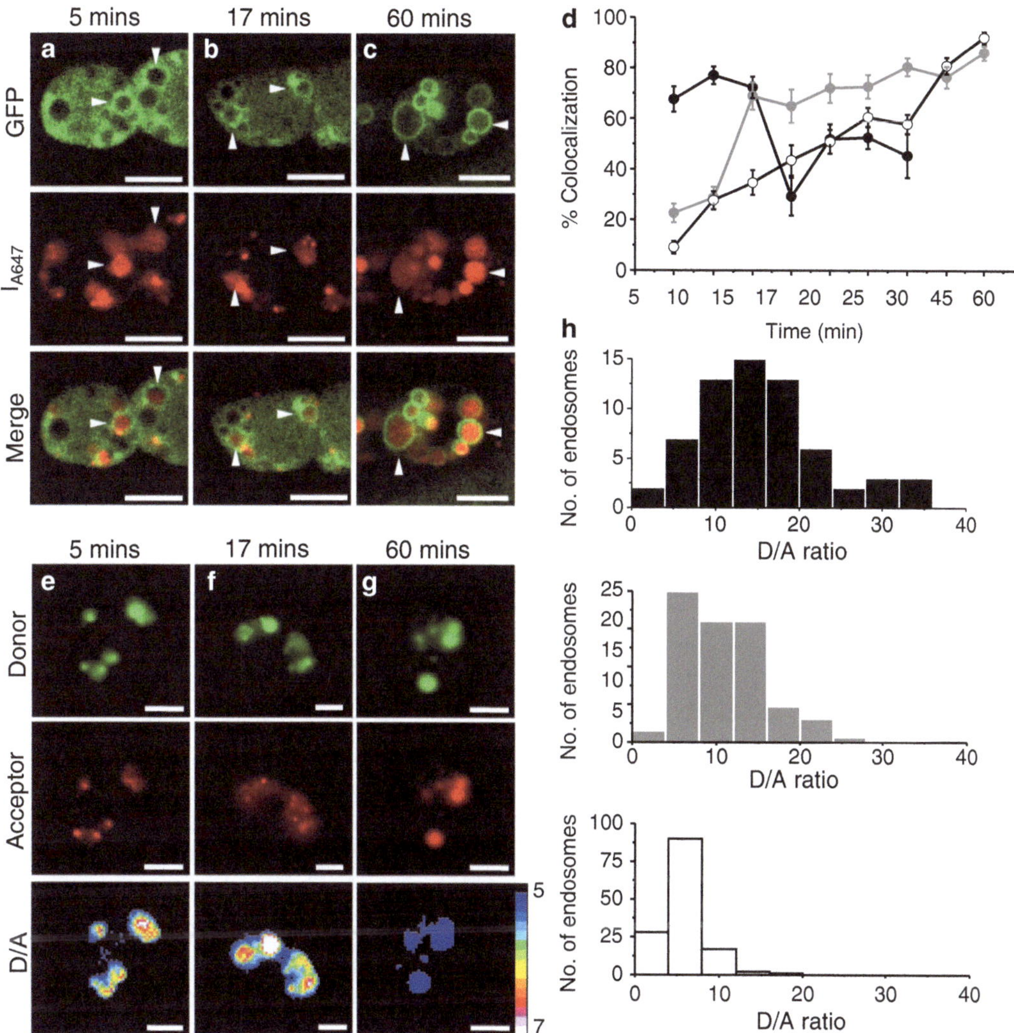

Fig. 3 Spatiotemporal mapping of pH in coelomocytes. Confocal images showing co-localization of I_{A647} with (**a**) GFP::RAB-5 at 5 min. (**b**) GFP::RAB-17 at 17 min. (**c**) LMP-1::GFP at 60 min. Scale bar: 5 μm. (**d**) Trafficking of endocytosed I_{A647}. Percentage co-localization of I_{A647} with GFP-tagged endosomal markers (RAB-5, *black circles*; RAB-7, *gray circles*; LMP-1, *black open circles*) at indicated times ($n \sim 75$ endosomes). (**e–g**) Representative pseudocolor D/A images of $I_{A488/A647}$-labeled coelomocytes in wild-type hermaphrodites at indicated times. Scale bar: 5 μm. (**h**) Histograms of D/A ratios of maturing endosomes; early endosomes at 5 min (*black bars*), late endosomes at 17 min (*gray bars*), and lysosomes at 60 min (*white bars*) ($n \sim 100$ endosomes)

5. Now, to map the pH of a particular population of vesicles, 500 nM $I_{A488/A647}$ is injected in wild-type hermaphrodites. After the requisite time (chosen according to the co-localization studies), the worms are transferred to chilled NGM (+OP50) plates. The worms are anesthetized using 40 mM NaN_3 in M9 buffer, and the coelomocytes are imaged.

3.5 Ratiometric Microscopy and Image Analysis

1. All ratiometric images are collected using a Nikon TE2000-U epifluorescence microscope. Coelomocytes are located and imaged so as to focus maximal number of puncta. Fluorescence images of the cells are obtained by exciting Alexa 488 and collecting emission using the 530±15 nm emission filter. This yields a donor image (D). The cells are then re-excited at 488 nm, and emission of the acceptor is acquired using a 710±40 nm filter. This is the FRET image (A). A third image is obtained by directly exciting the acceptor and collecting emission at acceptor emission wavelength. This is the acceptor image (I).
2. Autofluorescence of each image (D, A, and I) is calculated by measuring mean pixel intensity over an adjacent cell-free area in that image. This autofluorescence is subtracted from the corresponding image, prior to all image processing.
3. Each endosome in the donor is selected by the ROI plug-in within ImageJ program, and total and mean intensities in each endosome are measured and recorded.
4. Each saved ROI is recalled, and total and mean intensities of the corresponding vesicles in the FRET image are measured.
5. Dividing the mean intensity of each endosome in the donor image by the corresponding intensity in the FRET image provides a donor/acceptor (D/A) ratio for that endosome. This is repeated for all the cells imaged (each reading is obtained from coelomocytes of ten worms) to obtain a spread of D/A ratios for that time point (Fig. 3e–h). These values are then used to calculate a mean D/A ratio for the corresponding time point.
6. The standard in vivo calibration curve is then used to convert this D/A ratio to its corresponding pH value (Table 1).

3.6 pH Mapping in Mutants

1. The I-switch is now used to measure pH in genetic backgrounds that perturb the endocytic pathway (Table 2). Two methods are chosen to induce this perturbation: one, a genetic knockout in the *rme-1* gene (this gene functions to recycle internalized receptors to the plasma membrane) and, the other, a knockdown of the VHA-8 protein (this protein is a compo-

Table 2
Mean endosomal pH (±s.e.m.) at various time points postinjection

Strain	5 min	17 min	60 min
Wild type	6.4±0.12	6.0±0.09	5.4±0.03
rme-1	6.1±0.06	6.0±0.1	5.7±0.05
vha-8 (RNAi)	5.7±0.1	5.2±0.07	6.0±0.04

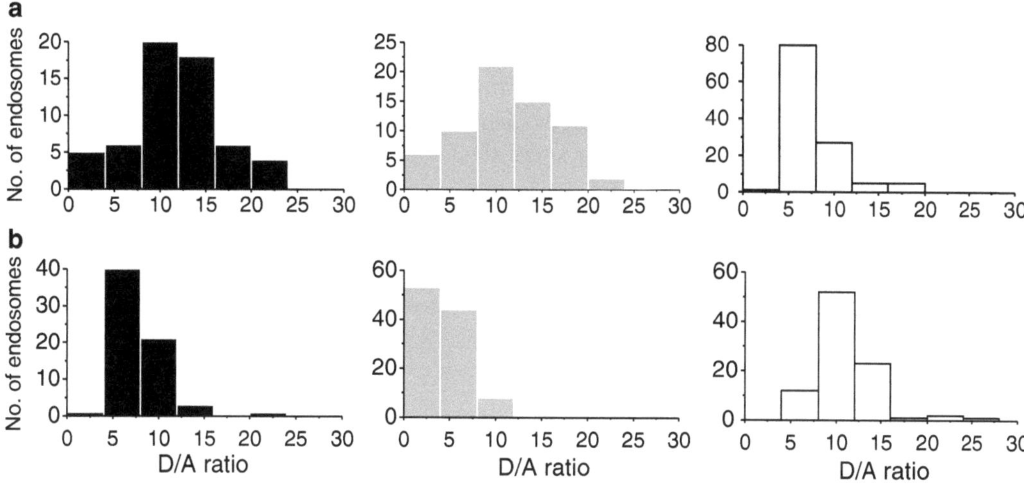

Fig. 4 pH measurements in various genetic backgrounds using the I-switch. (**a**, **b**) Histograms of D/A ratios of endosomes undergoing progressive maturation, from early endosomes at 5 min (*black bars*) to late endosomes at 17 min (*gray bars*) to lysosomes at 60 min (*white bars*) in *rme-1(b1045)* and *vha-8* (RNAi) hermaphrodites, respectively ($n=10$ cells, ^{3} 60 endosomes)

nent of the V-ATPase complex that maintains pH of organelles).

2. Ten hermaphrodites of each strain are injected with 500 nM $I_{A488/A647}$. After the requisite time (according to the co-localization studies in wild-type hermaphrodites), the worms are transferred to chilled NGM plates. The worms are anesthetized using 40 mM NaN_3 in M9 buffer and the coelomocytes are imaged (Fig. 4a, b).
3. Imaging and analysis is performed as outlined in Subheading 3.5, and pH values are calculated from the standard calibration curve (Table 1).

3.7 Targeting the I-Switch to Other Pathways

1. The native I-switch enters coelomocytes via the ALBR pathway. It can be induced to enter other endocytic pathways by saturating the ALBRs with mBSA and tagging the I-switch with the appropriate endocytic ligand. This can be done by injecting the I-switch with mBSA at molar ratios greater than 1:500.

4 Notes

1. It is imperative that the I-switch sample is always annealed in phosphate buffer of pH 5.5 to minimize variability.
2. Coelomocytes are present in the pseudocoelom, and accessibility of the cells to the clamping buffer is sometimes an issue.

Hence, there may be a few cells whose pH may not be efficiently clamped. These cells are discarded from the final analysis.

3. Coelomocytes clamped at pH 5.0 are very distinctive and can immediately be distinguished from those clamped at pH 7.0 due to their small sizes.
4. The in vitro and in vivo calibration curve is normalized by dividing the D/A values at every pH by the D/A value at pH 7.0.
5. Worms should be grown and maintained at 22°C. Maintenance at lower temperatures slows down all physiological processes, including endocytosis, which may alter temporal regimes of trafficking.

Acknowledgments

We thank Sandhya P. Koushika for inputs on experiments, Souvik Modi for technical input, Central Imaging Facility at NCBS and the Caenorhabditis Genetics Center (funded by NIH-NCRR) for nematode strains, and DBT and the Nanoscience and Technology Initiative of DST for funding. S.S. acknowledges the CSIR, and Y.K. acknowledges the Innovative Young Biotechnologist Award and Wellcome Trust–DBT India Alliance for fellowships.

References

1. Casey JR, Grinstein S, Orlowski J (2010) Sensors and regulators of intracellular pH. Nat Rev Mol Cell Biol 11:50–61
2. Stevens TH, Forgac M (1997) Structure, function and regulation of the vacuolar (H^+) ATPase. Annu Rev Cell Dev Biol 13:779–808
3. Syntichaki P, Samara C, Tavernarakis N (2005) The vacuolar H^+-ATPase mediates intracellular acidification required for neurodegeneration in *C. elegans*. Curr Biol 15:1249–1254
4. de Voer G, Peters D, Taschner PEM (2008) *Caenorhabditis elegans* as a model for lysosomal storage disorders. Biochim Biophys Acta 1782:433–446
5. Lee S-K, Li W, Ryu S-E, Rhim TY, Ahnn J (2010) Vacuolar (H+)-ATPases in *Caenorhabditis elegans*: What can we learn about giant H^+ pumps from tiny worms? Biochim Biophys Acta 1797:1687–1695
6. Miesenbock G, De Angelis DA, Rothman JE (1998) Visualizing secretion and synaptic transmission with pH-sensitive green fluorescent proteins. Nature 394:192–195
7. Rasmussen MB, Oddershede LB, Siegumfeldt H (2008) Optical tweezers cause physiological damage to *Escherichia coli* and *Listeria* bacteria. Appl Environ Microbiol 74:2441–2446
8. Dittman JS, Kaplan JM (2006) Factors regulating the abundance and localization of synaptobrevin in the plasma membrane. Proc Natl Acad Sci USA 103:11399–11404
9. Poskanzer KE, Davis GW (2004) Mobilization and fusion of a non-recycling pool of synaptic vesicles under conditions of endocytic blockade. Neuropharmacology 47:714–723
10. Lanz E, Gregor M, Slavik J, Kotyk A (1997) Use of FITC as a fluorescent probe for intracellular pH measurement. J Fluoresc 7: 317–319
11. Martin GR, Jain RK (1994) Noninvasive measurement of interstitial pH profiles in normal and neoplastic tissue using fluorescence ratio imaging microscopy. Cancer Res 4: 5670–5674
12. Siegumfeldt H, Rechinger KB, Jakobsen M (1999) Use of fluorescence ratio imaging for intracellular pH determination of individual bacterial cells in mixed cultures. Microbiology 145:1703–1709
13. Modi S, Swetha MG, Goswami D, Gupta GD, Mayor S, Krishnan Y (2009) A DNA nanomachine that maps spatial and temporal pH changes inside living cells. Nat Nanotechnol 4(325–330)

14. Surana S, Bhat JM, Koushika SP, Krishnan Y (2011) A DNA nanomachine maps spatiotemporal pH changes in a multicellular living organism. Nat Commun 2:340, doi:10.1038/ncomms1340
15. Mukherjee S, Ghosh RN, Maxfield FR (1997) Endocytosis. Physiol Rev 77:759–803
16. Overly CC, Lee KD, Berthiaumet E, Hollenbeck PJ (1995) Quantitative measurement of intraorganelle pH in the endosomal-lysosomal pathway in neurons by using ratiometric imaging with pyranine. Proc Natl Acad Sci USA 92:3156–3160

Chapter 3

A Simple Method to Visualize and Assess the Integrity of Lysosomal Membrane in Mammalian Cells Using a Fluorescent Dye

Syed K. Sohaebuddin and Liping Tang

Abstract

Fluorescent dyes have been used as "nanosensors" for visualization and determination of various processes occurring inside a cell, or intracellular events, such as cell cycle progression and intracellular trafficking. Here, we describe a novel use of acridine orange to visualize lysosomes and discriminate cells with healthy lysosomes from cells with damaged lysosomes in two different types of mammalian cells: fibroblasts and macrophages. This method allows assessment of lysosomal membrane integrity upon exposure to various foreign particles, i.e., engineered nanoparticles. The uniqueness of this method enables investigators to acquire fluorescent images with a dye that is susceptible to photo-bleaching under UV light. These acquired images bolster the quantitative data, providing a visual representation of the cell morphology as well as assess its nucleus and lysosomes.

Key words Lysosomal membrane permeability, Lysosomes, Acridine orange, Nanoparticles, Carbon nanotubes, Lysosomal membrane damage

1 Introduction

The use of fluorescent dyes or enzymes with specificity to a particular target, such as a cellular receptor or an intracellular structure, are invaluable in investigating etiology of diseases and thus developing promising treatments for those diseases (1). Fluorescent dyes and enzymes have also been used to examine the mechanisms of various cellular processes such as endocytosis, exocytosis, and cell death (2). Furthermore, the morphology of intracellular structures such as the nucleus, mitochondria, lysosomes, and actin can be examined using fluorescent dyes (3). These dyes are able to fluoresce due to the presence of a functional group which will absorb energy at a specific wavelength and reemit energy at a different wavelength (4). This causes that molecule to fluoresce, enabling us to detect the presence/absence of that molecule in the experimental samples (4, 5). The amount of fluorescence can be

Volkmar Weissig et al. (eds.), *Cellular and Subcellular Nanotechnology: Methods and Protocols*, Methods in Molecular Biology, vol. 991, DOI 10.1007/978-1-62703-336-7_3,

detected and quantified through the magnitude of fluorescence. Such quantitative analysis is performed using either fluorescent plate readers or flow cytometry (6). To bolster the quantitative data, qualitative assessment can be performed which allows us to observe the physical location(s) of our target in the experimental sample as well as quantify their presence at a particular location(s) in the same sample. The use of both quantitative and qualitative analysis leads to lucid understanding of experimental results and observed phenomenon.

There are, however, some dyes which are not compatible for qualitative assessment mainly because irradiation of the dyes to certain wavelengths of light causes damage to the cell's internal organelles. Damaged organelles may leak the dye to other cellular components such as the cytoplasm, leading to false qualitative results (7). Acridine orange is one of these dyes which, when irradiated with intense blue light, causes damage to the lysosome's membrane (8). Acridine orange is a weak base metachromatic dye capable of crossing plasma membranes and staining nucleic acids and lysosomes. At low concentrations, it can differentiate lysosomes (reddish-orange granules) from other cellular components (diffuse green) (9). Acridine orange molecules become protonated under acidic conditions and hence get trapped within lysosomes. Accumulation of acridine orange molecules in lysosomes leads to a shift in excitation from green = 530 nm to red = 620 nm (10). When pH of the lysosomes rises or if their membranes are damaged, acridine orange molecules become deprotonated, and these molecules can then cross back into the cytoplasm. This shifts the emission back from red to green (11). Therefore, extended exposure (>1 min) of acridine orange loaded cells to blue light leads to lysosomal membrane damage and a shift in the lysosomes' color from red to green.

Here, we have established a methodology to obtain visual representation of cells loaded with acridine orange before acridine orange molecules cause any disruption to the membrane of the lysosomes.

2 Materials

Prepare culture media and PBS using distilled water and autoclave the final solution to sterilize it. Prepare all nanomaterial solutions fresh under sterile conditions and store overnight at 4°C before performing the experiment. Prepare acridine orange solution fresh under sterile conditions and store it overnight at 4°C before performing the experiment. Follow waste disposal regulations when disposing waste materials.

2.1 Cell Culture Components

1. Dulbecco's Modified Eagle's Medium.
2. Calf serum.

3. Penicillin/streptomycin: 10,000 U/mL penicillin (base), 10,000 μg/mL streptomycin (base).
4. Trypsin solution: 2.5 g/mL trypsin, 0.38 g/L EDTA·4Na in Hank's balanced salt solution without calcium and magnesium salts (0.25% EDTA 1 mM).
5. 1× PBS: 137 mM NaCl, 2.7 mM KCl, 4.3 mM Na_2HPO_4, 1.47 mM KH_2PO_4, pH 7.4. Weigh 8 g of NaCl, 0.2 g of KCl, 1.44 g Na_2HPO_4, 0.24 g KH_2PO_4. Dissolve in 800 mL of distilled H_2O. Adjust pH to 7.4. Adjust volume to 1 L with additional distilled H_2O. Sterilize by autoclaving.
6. Sterile tubes, 75 mL cell culture flasks.
7. 6-well cell culture plate and tissue culture treated for cell culture of anchorage-dependent cells.
8. Hemacytometer and glass coverslips.
9. 6-well plates.

2.2 Mammalian Cells

1. 3T3 fibroblasts (American Type Cell Culture, Manassas, VA, USA).
2. RAW 264.7 (macrophages) (American Type Cell Culture, Manassas, VA, USA).

2.3 Nanomaterials (Sun Innovations, Fremont, CA, USA)

1. TiO_2 (anatase, 5–10 nm in diameter).
2. SiO_2 (30 nm in diameter).
3. Multiwalled carbon nanotube (MWCNT) (<8 nm in diameter, 0.5–2 μm length). Carbon nanotubes are allotropes of carbon with a cylindrical structure, and MWCNT are aligned individual nanotubes held together by van der Waals forces.
4. Multiwalled carbon nanotube (MWCNT) (<20–30 nm in diameter, 0.5–2 μm length).
5. Multiwalled carbon nanotube (MWCNT) (>50 nm in diameter, 0.5–2 μm length).

2.4 Fluorescence Microscopy Components

1. Acridine orange dye stock solution: Prepare stock acridine orange stain solution by weighing 5 mg of acridine orange powder. Dissolve this in 5 mL of cell culture medium to obtain a concentration of 1 mg/mL.
2. Leica DMLP microscope (40× lens magnification and FITC-Texas Red dual excitation band fluorescence filter, required for best imaging).
3. Nikon E500 camera (indicate minimum exposure time of 1/8 s required for image acquisition).

3 Methods

3.1 Cell Culture Medium Preparation

1. Warm DMEM, calf serum, and antibiotics to 37°C in a warm bath before use.
2. Sterilize the interior of laminar flow hood by spraying it with 70% ethanol (see Note 1).
3. Once they are warmed up, spray the containers with 70% ethanol and transfer them to a laminar flow hood along with sterile pipettes, pipette tips, sterile tubes, cell culture flasks, 6-well plates, and glass coverslips.
4. Sterilize all items mentioned in the above steps under UV light in the laminar flow hood for 15 min.
5. Prepare the cell culture medium by mixing 1 mL of calf serum and 100 μL of antibiotics for every 9 mL of DMEM (see Note 2).

3.2 Culturing Mammalian Cells

1. Culture the mammalian cells until they are sub-confluent in 75 mL cell culture flasks.
2. Remove the culture medium from the flasks, and rinse the cells three times with approximately 5 mL of 1× PBS.
3. Add 3 mL of trypsin to each flask, enough to cover the surface of the 75 mL cell culture flasks.
4. Place these flasks in 37°C incubator for 3 min.
5. Visualize the cells under microscope to determine the percentage of detached cells. Gently tap the cell culture flasks on your palm to detach any loosely attached cells.
6. Spray these flasks with 70% ethanol and place them under the laminar flow hood.
7. Add 3 mL of calf serum to each flask to deactivate trypsin. Rinse the cells 2–3 times to collect all the cells. Pool the flasks containing the same mammalian cells.
8. After transferring them to a sterile tube, centrifuge the cells at 200–400 × *g* for 10 min.
9. Discard the supernatant and resuspend the cell pellet in 1 mL of cell culture medium. Place a glass coverslip on top of a hemacytometer and pipette 10 μL of this solution into a slot of hemacytometer.
10. Count the cells and calculate the density of the cell solution.
11. Place glass coverslips in the wells of 6-well plates, and seed 25,000 cells on each of the glass coverslips (Fig. 1). Add 3 mL of cell culture medium to each well (see Note 3). Incubate the well plates overnight in a 37°C incubator.

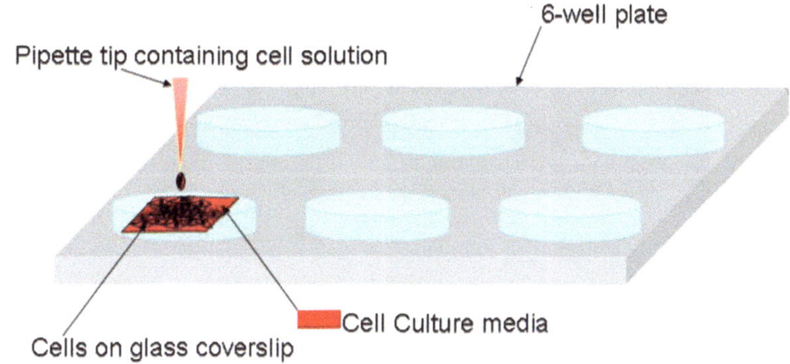

Fig. 1 Illustration represents the seeding of cells on a coverslip in a well of a 6-well plate

3.3 Acridine Orange Stain

1. Prepare working acridine orange stain solution by mixing 5 μL of acridine orange stock solution with 995 μL of cell culture medium to obtain a final concentration of 5 μg/mL (see Note 4).
2. Remove the cell culture medium from the 6-well plates. Rinse the cells three times with 1 mL of 1× PBS.
3. Add 2 mL of acridine orange working solution to the cells in the well plate and incubate this in a 37°C incubator for 15 min.
4. After 15 min, rinse the cells three times with 1× PBS to remove any excess acridine orange stain.

3.4 Treatment with Foreign Particles

1. Prepare nanoparticle solutions at a concentration of 1 mg/mL. Sonicate this solution to disperse the nanoparticles homogeneously in the solution.
2. Add 3 mL of cell culture medium in one of the 6-well plates; this will be the control of the experiment.
3. Add 2.7 mL of cell culture medium to the rest of the wells in the 6-well plate and add 300 μL of nanoparticle solution to their respective wells to obtain a nanoparticle exposure concentration of 100 μg/mL. Incubate the cells in a 37°C incubator for 4 h.
4. After 4 h, remove the cell culture medium from the wells, rinse the cells three times with 1 mL of 1× PBS, and add 3 mL of 1× PBS to each well.

3.5 Imaging the Cells

1. Place the well plate under a microscope and using one area of the control well, switch to the UV light (see Note 5). Under dual FITC/TX-Red filter, bring the cells into focus.
2. Switch to a separate area in the same well and acquire an image. If done properly, one should clearly see the cytoplasm stained as diffuse green and lysosomes as reddish orange (Fig. 2).

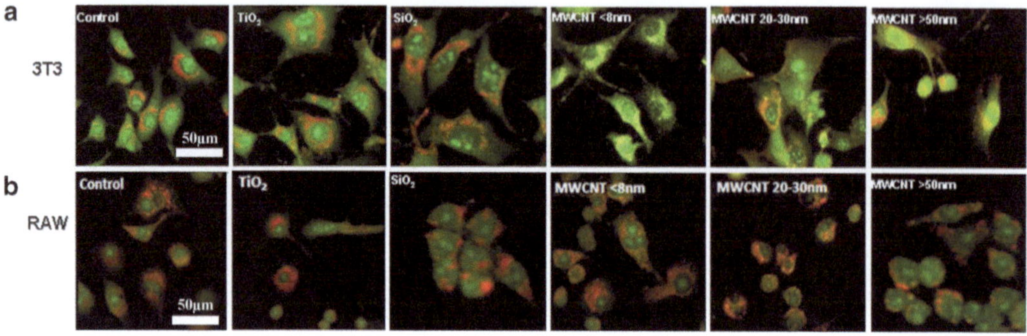

Fig. 2 Visualization of the lysosomes in (**a**) 3T3 fibroblasts and (**b**) RAW macrophages with and without exposure to 100 μg/mL of nanomaterials for 4 h. The lysosomes (*reddish orange*) and cytoplasm (*green*) can be clearly seen in control cells. 3T3 cells exposed to the TiO_2 and SiO_2 nanoparticles exhibit very little to no harm to lysosomes. 3T3 cells exposed to MWCNT <8 nm exhibit severe damage to lysosomes shown by *enhanced green* (cytoplasm) intensity and very low to *none orange* (lysosome) intensity. 3T3 cells exposed to MWCNT 20–30 nm and MWCNT >50 nm exhibit moderate damage to lysosomes. Exposure of nanomaterials to RAW did not lead to any lysosomal membrane damage

3. Follow the same procedure to acquire images of cells in the other wells of the 6-well plate. Cells with lysosomal damage have enhanced green fluorescence in their cytoplasm, and their lysosomes have yellowish to green fluorescence, depending on the extent of lysosomal damage (Fig. 2).

4 Notes

1. To sterilize the interior of the laminar flow hood, cover the work area surface with 70% ethanol and wipe it dry using tissue paper.
2. Prepare 10 mL of cell culture medium for every 75 mL cell culture flask.
3. Be careful not to allow cell solution to leave the glass coverslip so that cells only attach and grow on the glass coverslip. Initially, add only enough medium needed to keep the cells on the glass coverslips. After 2 h of incubation in a 37°C incubator, the cells should have adhered to the glass coverslip. At this time additional medium can be added.
4. Protect acridine orange solution from direct light.
5. Since acridine orange causes photooxidation of lysosomes under blue light, it is always better to use one area of the sample to adjust parameters and settings that will allow acquisition of high-quality images. Once these settings are established, switch over to another area and acquire images immediately and then move over to other areas and acquire more samples.

Acknowledgment

This work was supported by NIH grant EB007271.

References

1. Brugger W, Mocklin W, Heinfeld S et al (1993) Ex vivo expansion of enriched peripheral blood CD34+ progenitor cells by stem cell factor, interleukin-1 beta (IL-1 beta), IL-6, IL-3, interferon gamma, and erythropoietin. Blood 81:2579–2584
2. Li N, Zheng Y, Chen W et al (2007) Adaptor protein LAPF recruits phosphorylated p53 to lysosomes and triggers lysosomal destabilization in apoptosis. Cancer Res 67:11176–11185
3. Moseley JB, Goode BL (2006) The yeast actin cytoskeleton: from cellular function to biochemical mechanism. Microbiol Mol Biol Rev 70:605–645
4. Tsien RY, Waggoner A (1995) Fluorophores for confocal microscopy. In: Pauley JB (ed) Handbook of biological confocal microscopy. Pleum, New York, pp 267–274
5. Rietdrof J (2005) Microscopic techniques. In: Rietdorf J (ed) Advances in biochemical engineering/biotechnology. Springer, Berlin, pp 246–249
6. Antunes F, Cadonas E, Brunk UT (2001) Apoptosis induced by exposure to a low steady-state concentration of H_2O_2 is a consequence of lysosomal rupture. Biochem J 356: 549–555
7. Zdolsek JM, Olsson MG, Brunk UT (1990) Photooxidative damage to lysosomes of cultured macrophages by acridine orange. Photochem Photobiol 51:67–76
8. Zdolsek JM (1993) Acridine orange-mediated photodamage to cultured cells. APMIS 101: 127–132
9. Kobayashi Y, Vohimoto T, Nohara H et al (1999) Mechanism of apoptosis induced by a lysosomotropic agent L-Leucyl-L-Leucine methyl ester. Apoptosis 4:357–362
10. Lovelace MD, Cahill DM (2007) A rapid cell counting method utilizing acridine orange as a novel discriminating marker for both cultured astrocytes and microglia. J Neurosci Methods 165:223–229
11. Zareba M, Raciti MW, Henry MM et al (2006) Oxidative stress in ARPE-19 cultures: do melanosomes confer cryoprotection? Free Radic Biol Med 40:87–100

Chapter 4

Gold Nanoparticle as a Marker for Precise Localization of Nano-objects Within Intracellular Sub-domains

Valeriy Lukyanenko and Vadim Salnikov

Abstract

Delivery of nano-objects to certain intracellular sub-domains is crucial for nanomedicine. Therefore delivery of nano-object to desirable cellular compartment has to be confirmed. The most valuable confirmation of the delivery comes from direct visualization of the nano-object. This visualization usually requires use of microscope and corresponding probe which has to be conjugated with the nano-object. There are two most popular methods of the visualization: confocal and electron microscopy. The former has significant limitations due to diffraction limited resolution of confocal systems and three-dimensional convolution of fluorescence. The latter should be significantly modified for needs of the visualization. Here we describe the method for precise localization of nano-object within the cell using electron microscopy and 1–2 nm gold particles as a nanomarker.

Key words Gold nanoparticle, Delivery of nano-objects, Intracellular sub-domains, Electron microscopy

1 Introduction

Therapeutic delivery of genes and drugs to intracellular sub-domains with nanocarriers requires the creation of reliable delivery systems (1, 2). The methods available for monitoring the delivery of nano-objects could be divided into retrieval of the products of the nano-object delivery and visualization of nano-objects. The former is especially important in the case of gene delivery when the level of corresponding proteins clearly confirms the delivery. The latter, visualization, has to show the location of nano-objects directly.

The direct visualization of the nano-object requires use of microscope and corresponding probe which has to be conjugated with the nano-object. There are two most popular methods of the visualization: confocal and electron microscopy. They require correspondingly fluorescent or electron-dense probe to be tagged to nano-objects. Unfortunately, confocal microscopy could be used

Volkmar Weissig et al. (eds.), *Cellular and Subcellular Nanotechnology: Methods and Protocols*, Methods in Molecular Biology, vol. 991, DOI 10.1007/978-1-62703-336-7_4, © Springer Science+Business Media New York 2013

mostly for in vitro experiments. Also, the precise localization of nano-objects with confocal microscopy is significantly complicated due to diffraction limited resolution of confocal systems and three-dimensional convolution of fluorescence (3). This along with folding of cellular membranes and clamping of nanoparticles makes practically impossible precise localization of nano-objects within structures smaller than 0.5 μm. For example, diameter of isolated mitochondria is about 1 μm, and mitochondrial inner membrane makes multiple invaginations. It is clear that confocal microscopy cannot resolve between fluorescent dot located inside mitochondrial matrix and another dot located in the mitochondrial intermembrane space.

Hence, for purposes of localization of nanoparticles more accurate confocal microscopy methods, such as Förster resonance energy transfer (FRET) or more accurate systems, as superresolution structured illumination microscopy (SSIM) or electron microscopy (EM), should be used. However, both FRET and SSIM has some limitations of confocal microscopy and cannot be used for localization of nanoparticles in tissue. The EM allows studying of tissue fragments after in vivo experiments and precise localization of electron-dense objects >10 nm in diameter within ultrathin sections of the tissue (50–90 nm).

To make this localization of nanogold particles more precise, we slightly modified the usual procedure of EM preparation (4, 5). Namely, we employed water-soluble resin for cell polymerization and silver enhancement within ultrathin sections (4, 5). Figure 1 shows the major steps of the method. The silver enhancement significantly increases size of gold nanoparticles and makes their distribution obvious. In addition, the silver enhancement is more effective closer to the surface of the slice; therefore, the size of silver grains allows deciding about location of the gold sol within the ultrathin section. Our approach allows precise localization of 1–2 nm gold particles that could be tagged to any nano-object.

Here we describe the method for precise localization of nano-object within the cell using the EM. The method could be useful for electronopaque nano-objects and nanoparticles tagged to electron-dense marker, such as gold sols.

2 Materials

1. LR White resin.
2. Silver Enhancing Kit (Ted Pella, Inc., Redding, CA).
3. Gold nanoparticles. To prevent aggregation of the gold particles in experimental solution and their binding to cell proteins, the nanoparticles should be pretreated (coated) with polyvinylpyrrolidone (PVP). For that we incubated gold sols in 1% PVP (MW 10,000) for 10 min with gentle agitation (see Note 1).

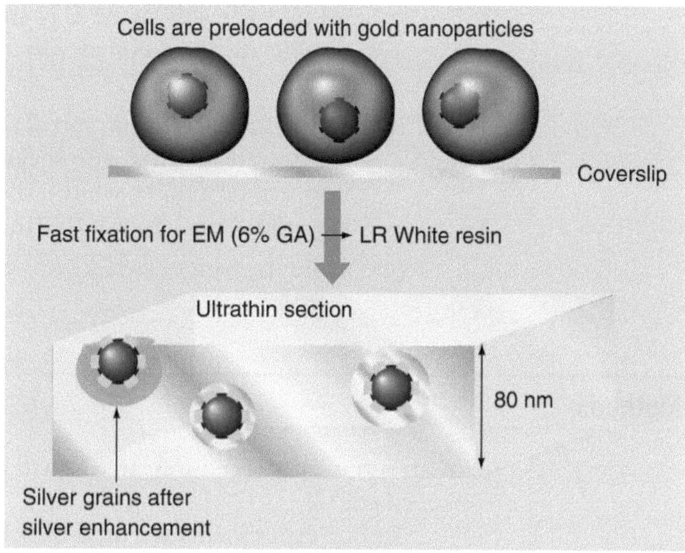

Fig. 1 Simplified schematic representation of a method for localization of PVP-coated gold nanoparticles within cells (reproduced from ref. 1 with permission from Future Medicine)

4. Ultramicrotome.
5. Transmission electron microscope.
6. Cell culture. Freshly isolated cells should be plated on coverslips coated with laminin.
7. 0.1 M Na^+-cacodylate buffer:
   (a) Prepare 0.2 M stock solution of sodium cacodylate in double distilled water (21.4 g/500 ml).
   (b) Add 27 ml of 0.2 M HCl per 500 ml cacodylate stock solution.
   (c) Add double distilled water to a final volume of 1 L.
8. Fixation buffer: 6 ml of 25% glutaraldehyde in 19 ml of 0.1 M Na^+-cacodylate buffer.
9. Rinse buffer: Na^+-cacodylate buffer supplemented with 0.2 M RNAase-free sucrose in 500 ml of 0.1 M Na^+-cacodylate buffer.
10. Postfix buffer: 1% osmium tetroxide in the 0.1 M Na^+-cacodylate buffer.
11. 2% aqueous uranyl acetate solution.
12. Lead citrate solution (Reynold's lead citrate stain):
    (a) 50 ml lead solution: 0.19 M $Pb(NO_3)_2$ in double distilled boiled (30 min, CO_2-free) and filtered H_2O.
    (b) 50 ml of 0.28 M tribasic sodium citrate solution in double distilled boiled (30 min, CO_2-free) and filtered H_2O. Add one drop of the lead solution.

(c) 50 ml of freshly made 1 N sodium hydroxide solution in double distilled boiled (30 min, CO_2-free) and filtered H_2O.

(d) Lead citrate solution: mix 21 ml lead solution and 21 ml lead citrate solution and shake vigorously for 2 min (solution will be a milky white); then in 30 min of gentle shaking, add 8 ml 1 N NaOH (the solution should be clear).

(e) Store in syringes (needle down into the rubber cork, without air) at 4°C.

3 Methods

Carry out all procedures at room temperature unless otherwise specified.

In our experiments we used primary culture of rat ventricular myocytes (single freshly isolated cells); however, any monolayer cell culture of any confluency could be used.

1. Fix cells or small pieces of tissue (~1 mm^3) in 2 ml (per sample unit) 6% glutaraldehyde in 0.1 M Na^+-cacodylate buffer (pH = 7.4), for 20 min. Rinse two times with the Na^+-cacodylate buffer supplemented with 0.1 M sucrose. Postfix the cells with 1% osmium in Na^+-cacodylate buffer for 1 h.
2. Stain samples en bloc with 1% uranyl acetate in 25% ethanol for 1 h. Dehydrate cells in ethanol and acetone step by step as shown:
   (a) Increase ethanol concentration by moving the cells from one solution to another. Amounts of ethanol in water solution: 30, 40, 50, 60, 70, 80, 90% (every step is 10 min, 1 time), and 100% (10 min, 3 times).
   (b) Acetone: 100%—10 min, 3 times.
3. Embed the cells in increasing concentrations of LR White resin. Proportions of LR White to acetone (use 50 mm glass Petri dishes): (1) 1 to 3, (2) 2 to 3, (3) 3 to 1, and (4) fresh resin. Every step is 12 h.

   To embed cultured cells (on coverslips) in LR White resin for the final step, use a 1.5-ml tube with cap cut off. Fill the tube with the resin; cover (seal) the tube with the cover slip, so that cells face the resin; and tightly bind the construction with parafilm, scotch tape, and foil to prevent the resin from exposure to air.
4. For resin polymerization, put the tubes upside down into thermostat (+60°C) for 24 h.
5. Remove the bandage from the tube. To separate embedded cells from coverslip, dip the coverslip into liquid nitrogen for

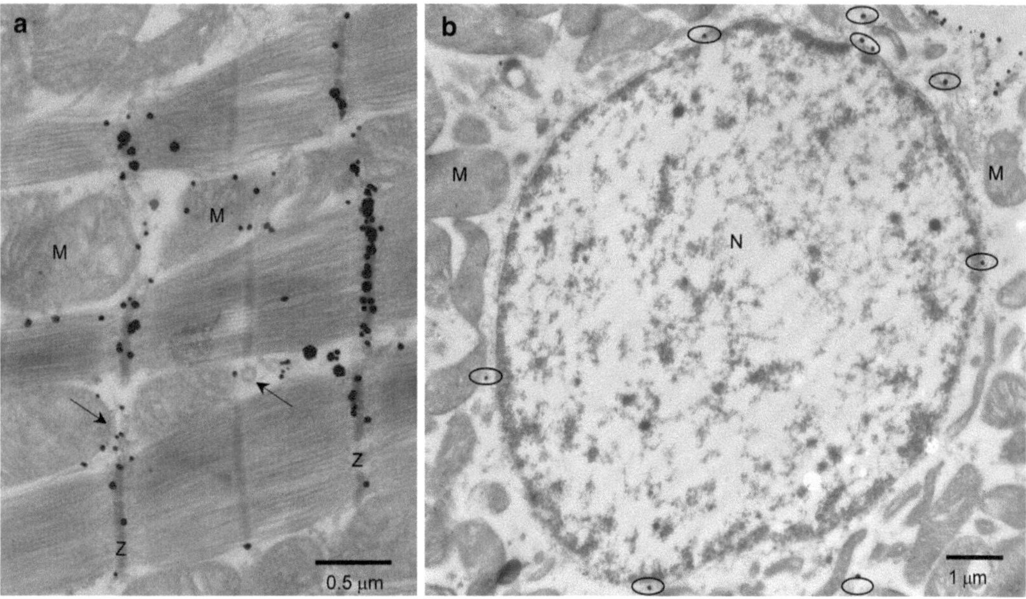

Fig. 2 Distribution of gold nanoparticles in permeabilized cardiomyocytes. Representative electron micrographs show the distribution of the nanoparticles in the cytoplasm along Z lines (**a**), but not in the nucleus or mitochondria (**b**). *M* mitochondrion; *Z* line; *N* nucleus; *arrows*—T-tubules; *black ovals* show nanoparticles located deeper inside the ultrathin section and therefore having smaller diameters after silver enhancement (reproduced from ref. 4 with permission from Cell press)

about 5 s. Now your cells are on the top of polymerized resin, and you can see them with binocular microscope.

6. Sharpen the block for ultramicrotome cutting.
7. Obtain ultrathin sections with ultramicrotome. Most valuable slices are 85–90 nm thick (they have a light gold color). Collect the slices on formvar-coated nickel grids.
8. Perform silver enhancement with Silver Enhancing Kit (follow instructions from Ted Pella, Inc.) (see Note 2).
9. After drying, stain slices for 15 min with 2% aqueous uranyl acetate and then for 2 min with lead citrate.
10. Dry them. Now the slices are ready for electron microscopy.
11. Store images in tiff format (see Notes 3 and 4). Representative micrographs (Fig. 2) show the typical distribution of nanoparticles within ventricular cardiomyocytes.

4 Notes

1. To prevent aggregation of gold nanoparticles and their binding to proteins, after conjugation, gold sols have to be coated with polyvinylpyrrolidone (PVP) or polyethylene glycol PEG (Fig. 3).

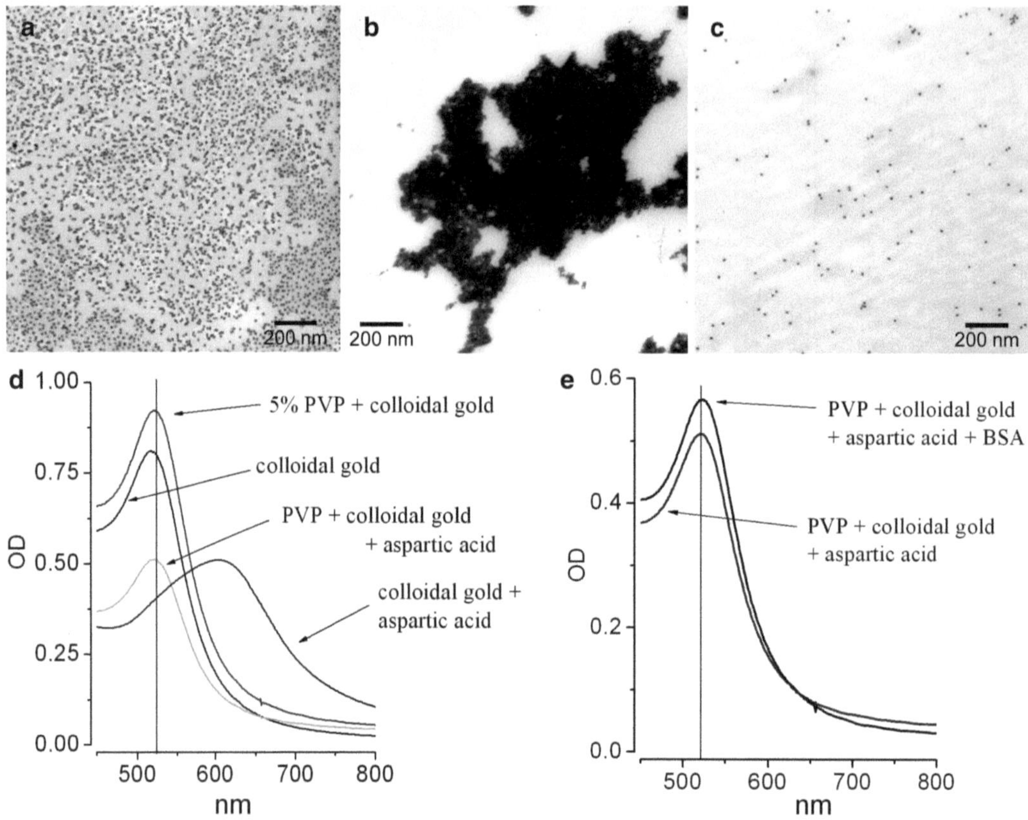

Fig. 3 Stabilizing the effect of polymer polyvinylpyrrolidone (PVP) on gold nanoparticles. Electron micrographs (without silver enhancement) show 10 nm gold sols (**a**) in water, (**b**) in 150 mM potassium aspartate solution (pH 7.2), and (**c**) in particles pretreated with 1% PVP and in aspartic acid solution. The suspensions were air-dried on formvar-coated grids. No staining. (**d**) The absorption spectra for colloidal gold (peak at 520 nm) and mixtures of gold and 150 mM potassium aspartate and/or 5% PVP. *OD* optical density (depends on gold particles concentration); *dashed lines* show the absorbance maximum for colloidal gold (523 nm). (**e**) The absorption spectra for 10 nm colloidal gold stabilized with 5% PVP before and after adding 1% bovine serum albumin (BSA) to the cuvette (reproduced from ref. 4 with permission from Cell press)

For that, coating the sols has to be pretreated for 10 min in 1% PVP10 (*neutral*, MW = 10,000). To measure the size of the PVP-coated nanoparticles, we employed dynamic light scattering (Protein Solutions Ltd., England). These measurements showed that 1% PVP increased the diameter of the gold particles, adding about 2 nm to their original size (4).

2. The size of the silver grains depends on the time of exposure and the accessibility (i.e., depth of position within the section) of the gold. We applied the solution for 8 min.
3. To find real silver grains on the EM micrograph, reduce contrast by 70–80% (until less dense cell structures vanish) with Adobe

Photoshop (Adobe Systems Incorporated, San Jose, CA). Then mark them and retune the image contrast to normal.

4. To calculate the number of particles, we recommend using Image J 1.31v (National Institutes of Health, Bethesda, USA).

References

1. Lukyanenko V (2007) Delivery of nano-objects to functional sub-domains of healthy and failing ventricular myocytes. Nanomedicine 2: 831–846
2. Lukyanenko V (2010) Therapeutic nano-object delivery to sub-domains of cardiac myocytes. In: Weissig V, D'Souza GGM (eds) Organelle-specific pharmaceutical nanotechnology. Artech House, Norwood MA, pp 433–448
3. Pratusevich VR, Balke CW (1996) Factors shaping the confocal image of the calcium spark in cardiac muscle cells. Biophys J 71: 2942–2957
4. Parfenov AS, Salnikov V, Lederer WJ, Lukyanenko V (2006) Aqueous diffusion pathways as a part of the ventricular cell ultrastructure. Biophys J 90:1107–1119
5. Salnikov VV, Lukyanenko YO, Frederick CA, Lederer WJ, Lukyanenko V (2007) Probing the outer mitochondrial membrane in cardiac mitochondria with nanoparticles. Biophys J 92: 1058–1071

Chapter 5

Immunoisolation of Nanoparticles Containing Endocytic Vesicles for Drug Quantitation

Ari Nowacek, Irena Kadiu, JoEllyn McMillan, and Howard E. Gendelman

Abstract

Cell-mediated nanoparticle delivery has recently emerged as an efficacious method of delivering therapeutic agents across physiological barriers. Use of cells as nanodelivery vehicles requires accurate assessment of their loading capacity and identification of intracellular compartments where nanoparticles are sequestered. This is of great interest since specific endocytic trafficking routes can ultimately influence the mode of nanoparticle release and their efficacy and function. Here, we describe a technique that allows for the isolation of individual populations of nanoparticle-containing endosomes for subsequent quantitative analysis and more accurate description of where nanoparticles are stored on a subcellular level.

Key words Nanoparticles, Immunoisolation, Endosomes, Subcellular trafficking, Macrophage

1 Introduction

For over 20 years, nanoparticles (NP) have been researched for their use in drug delivery (1, 2). Drug-loaded NP have the potential to increase efficacy, reduce toxicity, and improve clinical outcomes of diseases. These nanoparticles tend to be designed to deliver drugs or other therapeutic compounds, such as protein or DNA, to specific cell populations. One of the important questions when researching drug-carrying nanoparticles is determining precisely how much drug the target cells are able to take up. Generally, this question is not difficult to answer. However, of even greater importance than how much drug the target cells are able to pick up is where within the cells are the nanoparticles being trafficked and stored.

Ari Nowacek and Irena Kadiu have contributed equally.

Volkmar Weissig et al. (eds.), *Cellular and Subcellular Nanotechnology: Methods and Protocols*, Methods in Molecular Biology, vol. 991, DOI 10.1007/978-1-62703-336-7_5, © Springer Science+Business Media New York 2013

2 Materials

Prepare all solutions using ultrapure water (prepared by purifying deionized water to attain a sensitivity of 18 MΩ cm at 25°C) and analytical grade reagents. Prepare and store all reagents at room temperature (unless indicated otherwise). Diligently follow all waste disposal regulations when disposing waste materials.

2.1 Conjugation of Magnetic Beads to Antibodies

1. PureProteome Protein A and Protein G Paramagnetic Beads (Millipore).
2. Antibodies to endosome surface markers of interest (see Note 1).
3. Bovine serum albumin fraction V (10%).
4. Sterile 1× phosphate-buffered saline (PBS).
5. Microcentrifuge tubes (1.7 mL).
6. Microcentrifuge tube tumbler rotator.
7. Magnetic separator rack.
8. Refrigerated tabletop centrifuge.

2.2 Cellular Treatment Components

1. Cells in culture (see Note 2).
2. Cell incubator.
3. Serum-free DMEM (or other appropriate serum-free culture medium).
4. Nanoparticles (see Note 3).
5. Sterile PBS.

2.3 Homogenization of Nanoparticle-Loaded Cells

1. Homogenization buffer: 10 mM HEPES–KOH, pH 7.2, 250 mM sucrose, 1 mM EDTA, and 1 mM $Mg(OAc)_2$.
2. Cell scrapers.
3. Dounce homogenizer (7 mL).
4. 15 mL centrifuge tubes.
5. Refrigerated centrifuge.

2.4 Isolation of Nanoparticle-Containing Endosomes

1. Homogenate from nanoparticle-treated cells (from Subheading 2.3).
2. Magnetic beads with attached antibodies (from Subheading 2.1).
3. Magnetic separator rack.
4. Sterile PBS.
5. Refrigerated tabletop centrifuge.

2.5 Quantification of Drug Content by HPLC

1. HPLC-grade methanol.
2. Sonicator disruptor with probe tip.
3. Refrigerated tabletop centrifuge.
4. 0.5 mL microcentrifuge tubes.
5. HPLC autoinjector vials with low-volume inserts.

3 Methods

3.1 Conjugate Antibodies to Magnetic Beads

1. In a 1.7-mL microcentrifuge tube, combine 1 mL of 10% bovine serum albumin in PBS with 20 μL of magnetic bead slurry and 20 μg of antibody of interest (see Note 1).
2. Place tubes on a microcentrifuge tube tumbler rotator and rotate at 15 rpm for 12 h at 4°C.
3. Place tubes in the magnetic separator rack for up to 1 h at 4°C. Remove supernatant and resuspend antibody-bead conjugates in sterile PBS. Repeat wash cycle two more times. Finally, resuspend antibody-bead conjugates in 500 μL of sterile PBS and use within 24 h.

3.2 Treat Cells with Nanoparticles

1. Wash cells three times for 10 min with 37°C serum-free medium to remove residual serum protein.
2. Add nanoparticles in sterile serum-free DMEM (or other appropriate cell medium) to cells at desired concentration (Fig. 1a).
3. Incubate at 37°C to allow cells to take up nanoparticles (see Note 4).
4. Once maximum nanoparticle uptake has been reached, wash the cells three times with 37°C sterile PBS to remove any nanoparticles that have not been taken up. Keep cells in PBS and immediately begin homogenization.

3.3 Homogenization of Nanoparticle-Loaded Cells

1. Remove PBS and add enough homogenization buffer to cover the bottom of each well or flask being used.
2. Detach cells from bottom of well or flask using cell lifter.
3. Add scraped cells to Dounce homogenizer and grind cells with 15 strokes (Fig. 1b) (see Note 5).
4. Add entire volume to a 15 mL centrifuge tube and remove nuclei and unbroken cells by centrifuging at 400 × *g* for 10 min at 4°C (Fig. 1c).
5. Remove supernatant, place in 1.7 mL microcentrifuge tubes, and use for immune isolation of endocytic compartments.

3.4 Isolate Nanoparticle-Containing Endosomes

1. In a 1.7-mL microcentrifuge tube, combine 1 mL of the cellular homogenate from Subheading 3.3, step 5, and the entire volume (500 μL) of one antibody-bead combination

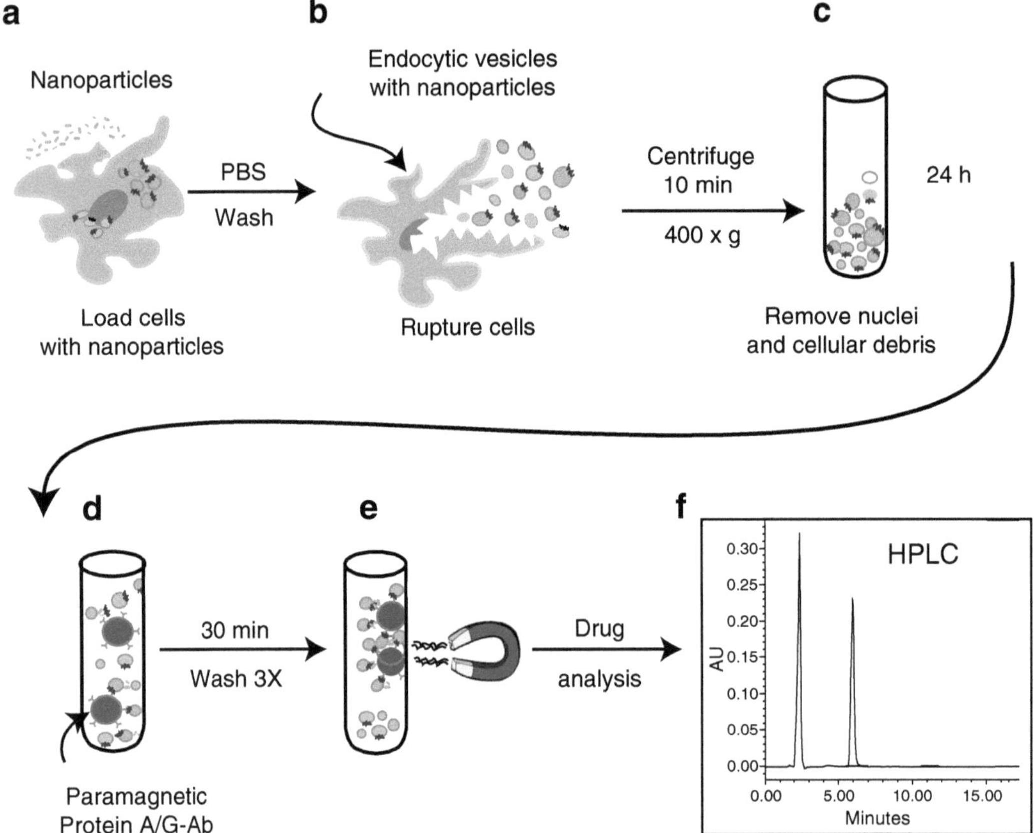

Fig. 1 Schematic diagram of immunoisolation of nanoparticle-containing endosomes. Cells are first treated with nanoparticles (**a**). After maximum nanoparticle endocytosis, cells are ruptured in homogenization buffer using a Dounce tissue homogenizer (**b**). Cell homogenate is then slowly centrifuged in order to remove nuclei, organelles, and cellular debris (**c**). Following enrichment, the endosome fraction is exposed to protein A/G paramagnetic beads conjugated to antibodies and allowed to bind with the endosomes carrying the surface markers of interest. (**d**) The endosome population bound to the beads is isolated by magnetic separation (**e**). The isolated fraction undergoes several washes with sterile cold PBS prior to drug quantitation (**f**)

from Subheading 3.1, step 3 (Fig. 1d). Be sure to include one tube with blank beads (i.e., fresh beads with no antibody conjugate) as a control.

2. Place tubes on a microcentrifuge tube tumbler rotator and rotate at 15 rpm for 18–24 h at 4°C.
3. Place tubes on microcentrifuge tube magnetic separator and allow for magnetic separation for 1 h at 4°C (Fig. 1e). The solution should become clear as the beads form a layer in the tube wall facing the magnet.
4. Carefully remove the solution without disturbing the beads. Add 1 mL of cold sterile PBS and resuspend the beads (see Note 6). Repeat this wash cycle two more times. Endosomes are now ready for quantitative analysis (store at 4°C until ready for analysis).

3.5 Quantification of Drug Content of Nanoparticle-Containing Endosomes by HPLC

1. Take samples from Subheading 3.4, step 4, and centrifuge at 10,000 × *g* for 10 min at 4°C. Remove the supernatant and add 400 μL of 100% methanol (see Note 7).
2. Sonicate solution with sonicator probe for 3 s at 20% amplitude.
3. Centrifuge solutions at 20,000 × *g* for 10 min at 4°C. Remove supernatant and add to a clean 0.5 mL microcentrifuge tube.
4. Transfer 70 μL to an HPLC autosampler vial and inject three 20 μL aliquots onto the HPLC for drug quantitation (see Notes 8 and 9).
5. Determine drug content by comparing peak area of drug in sample to peak areas of known concentrations of drug standards.

4 Notes

1. There are hundreds of potential endosomal markers to choose from, which can be very overwhelming. When first performing this experiment, it may be wise to select an antibody to isolate each of the major endosomal compartments, for example, early endosomes (early endosome antigen-1), recycling endosomes (Rab-11), late endosomes (Rab-7), and lysosomes (lysosome-associated membrane protein-1). In this way one can find out which major endosomal compartment the nanoparticles are trafficked to. Afterwards, a more thorough search of that specific endosomal compartment can be performed.
2. This protocol has been used primarily with primary human monocytes and human monocyte-derived macrophages. However, any cells that will take up the nanoparticle being tested could be used. Adjust the protocol appropriately to accommodate for cell type being used. It is suggested to use at least 10×10^6 cells in order to isolate enough nanoparticle-containing endosomes for drug analysis.
3. This protocol has been used only for crystalline antiretroviral nanoparticles coated in lipophilic surfactants (3). For this method drug-containing nanoparticles are being used because the final step of the protocol involves quantitation of drug levels by HPLC. However, this protocol could easily be used to isolate nanoparticle-containing endosomes for any type of nanoparticle. Although, if the nanoparticle contains a nondrug therapeutic compound (such as protein or DNA), a method other than HPLC will be necessary to quantify the amount of therapeutic material contained within each endosomal compartment.
4. Both the amount of nanoparticles to be added to serum-free DMEM (or other appropriate cell medium) and the duration of treatment must be determined ahead of time by previous experiments. It is suggested to allow the cells enough time for maximum nanoparticle uptake before harvesting for endosomal

isolation in order to ensure that sufficient material will be available.

5. When using the Dounce homogenizer, press and pull the piston hard enough to keep the solution flowing past the glass rod at a steady rate. Only light force is necessary to effectively homogenize the cells and excess force could break the homogenizer.
6. Be very careful not to disturb the beads when removing the solution otherwise endosomes will be lost during the wash process. Place the pipette tip on the wall of the tube opposite from the side with the beads just beneath the surface and remove the solution from the top down. If the beads are disturbed, place the magnetic rack back in the refrigerator for about 10 min to allow the beads to gather at the magnet again.
7. An internal standard of known quantity may be added to the methanol (or other extraction solvent) prior to addition to the samples. Acetonitrile extraction of drug may also be used. To concentrate samples prior to HPLC, sample extracts can be evaporated to dryness using a SpeedVac concentrator and resuspended in methanol or mobile phase.
8. Low-volume inserts for autosampler vials are available for many 1–4 mL vial types. Vial types appropriate for your system will be identified in the autoinjector user manual. Glass inserts are preferred because of their inertness to solvents used for drug extraction. Add enough volume to the autosampler vials to provide for at least two injections of sample onto the HPLC system, although three injections are preferred.
9. Depending on the drug of interest, HPLC with UV/Vis or mass spectrometry detection may be used.

Acknowledgments

The work was supported by the National Institutes of Health grants 1P01 DA028555, 2R01 NS034239, 2R37 NS36126, P01 NS31492, P20RR 15635, P01MH64570, and P01 NS43985 (to H.E.G.) and a research grant from Baxter Healthcare. The authors thank Ms. Robin Taylor for critical reading of the manuscript and outstanding graphic and literary support.

References

1. Speiser PP (1991) Nanoparticles and liposomes: a state of the art. Methods Find Exp Clin Pharmacol 13:337–342
2. Douglas SJ, Davis SS, Illum L (1987) Nanoparticles in drug delivery. Crit Rev Ther Drug Carrier Syst 3:233–261
3. Kadiu I, Nowacek A, McMillan J, Gendelman HE (2011) Macrophage endocytic trafficking of antiretroviral nanoparticles. Nanomedicine (Lond) 6:975–994

Chapter 6

Methods for Isolation and Identification of Nanoparticle-Containing Subcellular Compartments

Ari Nowacek, Irena Kadiu, JoEllyn McMillan, and Howard E. Gendelman

Abstract

Nanoparticle-based drug delivery systems have considerable potential for improvement of drug stability, bioavailability, and reduced dosing frequency. Important technological advantages of nanoparticles include high carrier capacity across biological membranes and controlled drug release. Ultimately, success of nano-delivery systems depends on toxicologic issues associated with the understanding of the fate of nanocarriers and their polymeric constituents within the targeted cells. Here we describe a method for determining subcellular distribution of nanoparticles by isolation and identification of organelles that come in direct contact with these structures.

Key words Macrophage, Nanoparticle, Drug delivery, Endosome, Subcellular trafficking

1 Introduction

A wide variety of nanoparticles have been developed for the cellular delivery of various therapeutic compounds and the potential clinical benefits of these particles are great (1, 2). However, very little is known about the subcellular distribution of nanoparticles in the targeted cells. This information is necessary if we are to explain how nanoparticles function on a subcellular level and to identify any potential sources of cellular toxicity. In order to accomplish this, a method must be used that can simultaneously allow for the isolation and subsequent identification of proteins that interact with a nanoparticle while it is in a cell. Here, we demonstrate that the proteins that come into contact with a nanoparticle can be individually labeled, isolated, and then identified by liquid chromatography–mass spectrometry (LC-MS/MS). This relatively simple method involves four basic steps: (1) labeling of the nanoparticles

Equal contributions were made by Ari Nowacek and Irena Kadiu.

Volkmar Weissig et al. (eds.), *Cellular and Subcellular Nanotechnology: Methods and Protocols*, Methods in Molecular Biology, vol. 991, DOI 10.1007/978-1-62703-336-7_6, © Springer Science+Business Media New York 2013

with a visible dye, (2) treatment of cells with the nanoparticles, (3) isolation of nanoparticle-laden subcellular compartments on a sucrose gradient, and (4) identification of the proteomes of subcellular compartments by LC/MS-MS. This method provides the user with a broad view of the subcellular distribution of nanoparticles within the same experiment. It is appropriate for use by researchers who do not know the fate of their nanoformulations within the targeted cells or their mechanisms of release. It can also be used successfully to identify the subcellular trafficking pathways of crystalline antiretroviral nanoparticles in human monocyte-derived macrophages (3). Alternative approaches such as immunostaining and confocal imaging of every cellular organelle and internalized nanoparticles as well as measurement of their fluorescence overlap are time consuming and costly.

2 Materials

Prepare all solutions using ultrapure water (prepared by purifying deionized water to attain a sensitivity of 18 MΩ cm at 25°C) and analytical grade reagents. Prepare and store all reagents at room temperature (unless indicated otherwise). Diligently follow all waste disposal regulations when disposing waste materials.

2.1 Components to Label Nanoparticles

1. Crystalline nanoparticles (see Note 1).
2. Coomassie Brilliant Blue R250 (CBB) (see Note 2).
3. Sterile 1× phosphate buffered saline (PBS).
4. 0.5 or 1.7 mL microcentrifuge tubes.
5. Microcentrifuge tube tumbler rotator.
6. Table-top refrigerated centrifuge that can reach 20,000 ×*g*.
7. Sonicator with probe.
8. Method to measure nanoparticle size and charge (see Note 3).

2.2 Cellular Treatment Components

1. Cells in vitro (see Note 4).
2. Cell incubator.
3. Serum-free DMEM (or other appropriate serum-free cell culture medium).
4. Labeled nanoparticles.
5. Sterile PBS.

2.3 Homogenization of Nanoparticle-Loaded Cells

1. Homogenization buffer: 100 mM sucrose, 10 mM imidazole, pH 7.4.
2. Dounce homogenizer (7 mL).
3. 15 mL centrifuge tubes.
4. Refrigerated centrifuge.

2.4 Enrichment of Nanoparticle-Laden Compartments

1. Homogenization buffer: 100 mM sucrose, 10 mM imidazole, pH 7.4.
2. Sucrose solutions: 10, 20, 35, and 60% weight/volume, 10 mM imidazole, pH 7.4 in sterile water.
3. Transparent ultracentrifuge tubes (12 mL; Beckman-Coulter).
4. Refrigerated ultracentrifuge that can reach 100,000×g and swinging bucket.
5. 3 mL syringe with an 18-gauge needle.
6. Sterile PBS.

2.5 Sample Processing for 1D Electrophoresis and Mass Spectrometry Analysis

1. Lysis buffer: 30 mM Tris-Cl, 7 M urea, 2 M thiourea, 4% (w/v) 3-((3-cholamidopropyl)dimethylammonio)-1-propanesulfonate, 20 mM dithiothreitol, 1× protease inhibitor cocktail, pH 8.5 (Sigma-Aldrich) (see Note 11).
2. ReadyPrep™ 2D Cleanup Kit (Bio-Rad Laboratories, Inc.).
3. 2D Quant kit (GE Healthcare).
4. Bis–Tris 4–12% and 7% Tris-Glycine gels (Invitrogen).
5. Fixation buffer: 10% methanol and 7% acetic acid in distilled–deionized water.
6. Colloidal coomassie (GE healthcare).
7. Destaining buffer: 20% methanol and 10% acetic acid in distilled–deionized water.
8. Single edge razor blades.
9. Sterile glass autosampler vials (Thermo-Fisher Scientific).
10. Vacuum concentrator centrifuge (SpeedVac) with cooling trap.
11. In-Gel Tryptic Digestion Kit (Thermo-Fisher Scientific).
12. μC18 ZipTip® pipette tips (Millipore; see Note 12).
13. Resuspension buffer: 0.5% trifluoroacetic acid (TFA; Sigma-Aldrich).
14. Wetting solution: 100% acetonitrile (ACN; Thermo-Fisher Scientific).
15. Equilibration/wash solution: 0.1% TFA.
16. Elution solution: 50% ACN, 0.1% TFA.

3 Methods

3.1 Label Nanoparticles with Coomassie Brilliant Blue R250

1. Combine nanoparticles (see Note 1) with 0.01% (weight/volume) of CBB (see Note 2) in sterile PBS. Mix on a microcentrifuge tube tumbler rotator at 15 rpm for 12 h at room temperature (see Note 5).

2. Centrifuge the mixture at 20,000 × *g* for 5 min to pellet the particles and then remove the supernatant with a pipette. Add sterile PBS to the tube and briefly resuspend the particles by sonicating them at 20% amplitude for 1–3 s with a sonicator probe (see Note 6). This will remove the excess dye. Repeat the PBS wash and sonication cycle five times or until no more dye is visible in the supernatant (Fig. 1a).
3. Store labeled particles in sterile PBS at 4°C until ready for use.

3.2 Treat Cells with CBB-Labeled Nanoparticles

1. Wash cells three times for 10 min with 37°C sterile PBS to remove residual serum protein.
2. Add nanoparticles in sterile PBS to cells at desired concentration.
3. Incubate at 37°C to allow cells to take up nanoparticles. The cells will visibly become blue as they take up the labeled nanoparticles (Fig. 1b) (see Note 7).
4. Once the cells have taken up the nanoparticles, wash the cells three times with 37°C sterile PBS to remove any residual non-internalized nanoparticles. Keep cells in PBS and immediately begin homogenization.

3.3 Homogenization of Nanoparticle-Loaded Cells

1. Remove PBS and add 6 mL of homogenization buffer to each flask if working in a T75 culture flask. Adjust the buffer volume to the minimum necessary for covering the dish surface if working with other culture systems.
2. Detach cells from bottom of flask using a cell lifter.
3. Add entire volume to Dounce homogenizer and grind cells with 15 strokes (see Note 8).
4. Add entire volume to a 15 mL centrifuge tube and remove cellular debris and nuclei by centrifuging at 500 × *g* for 10 min at 4°C.
5. Remove supernatant and centrifuge for 1 h at 100,000 × *g* at 4°C. Resuspend the pellet in 3 mL of 10% sucrose, 10 mM imidazole, pH 7.4, solution and store on ice.

3.4 Enrichment of Nanoparticle-Laden Compartments

1. Take sucrose solutions and set up sucrose gradient in 12 mL thin-walled ultracentrifuge tube. Take 3 mL of 60% sucrose solution and place it in the bottom of the tube followed by layering 3 mL each of 35 and 20% sucrose solutions one on top of the next in the order given (see Note 9).
2. Carefully add the supernatant from the last step of Subheading 3.3 to the top of the sucrose gradient and centrifuge at 100,000 × *g* at 4°C for 1 h.
3. Using a 3 mL syringe with an 18-gauge needle, carefully perforate the tube at the end of the blue sucrose band and aspirate

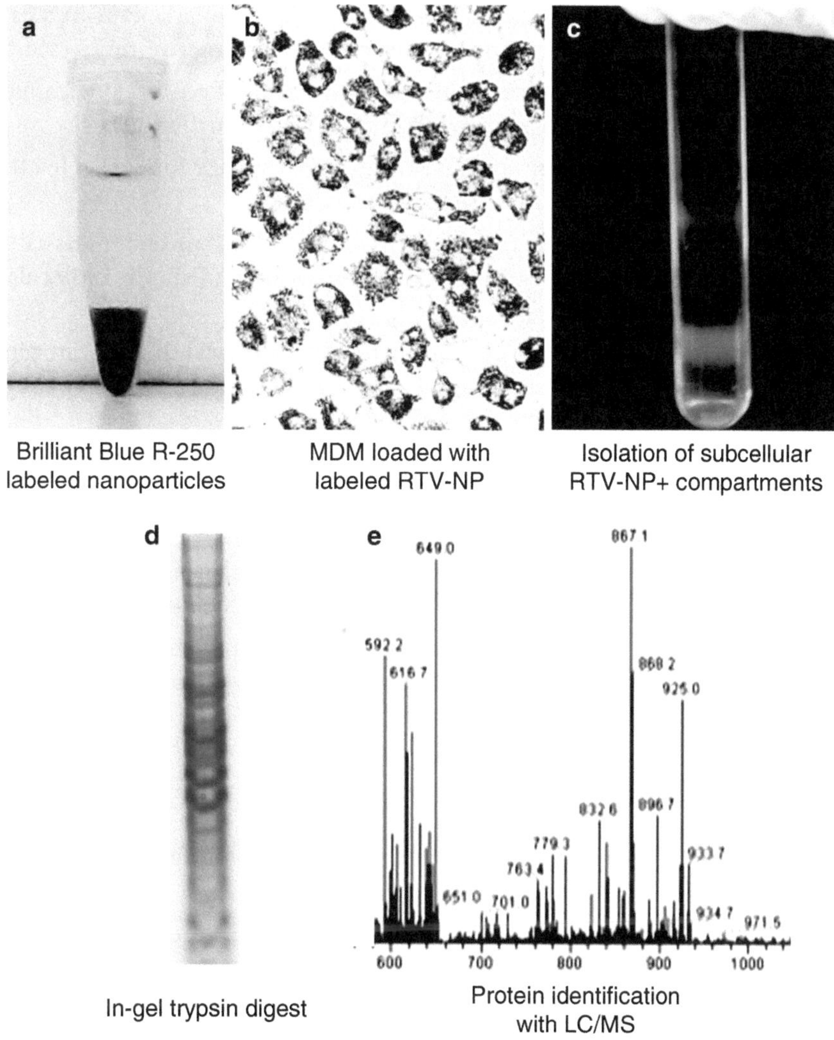

Fig. 1 Representative images of processes for nanoparticle staining, cell treatment, endosome enrichment and protein processing and identification. Crystalline nanoparticles that have been labeled with CBB with all unbound dye washed away (**a**). Human monocyte-derived macrophages after being treated with CBB-labeled nanoparticles. Note the cells have developed a *purple color* after ingesting the labeled nanoparticles (**b**). Enriched endosomes containing proteins stained by CBB-labeled nanoparticles are seen as bands on a sucrose gradient after being centrifuged at 100,000 × *g* for 1 h at 4°C (**c**). One lane of a gel showing labeled proteins separated by molecular weight (**d**). Chromatogram of protein fractionation followed by identification using LC/MS-MS (**e**)

the solution until the color disappears (Fig. 1c). Transfer each band to a separate ultracentrifuge tube (see Note 10).

4. Pellet the nanoparticle-enriched subcellular compartments by centrifuging at 100,000 × *g* at 4°C for 1 h. Remove supernatant and wash pellet with PBS and subsequent centrifugation at 100,000 × *g* at 4°C for 1 h to remove residual sucrose.

3.5 Sample Processing for 1D Electrophoresis and Mass Spectrometry Analysis

1. Solubilize enriched subcellular compartments in lysis buffer by resuspending the pellet and pipetting five times (see Note 11).
2. Precipitate proteins using a ReadyPrep™ 2D Cleanup Kit (GE Healthcare) per manufacturer's instructions.
3. Quantify protein using a 2D Quant kit (GE Healthcare) per the manufacturer's instructions.
4. Run samples on Bis–Tris 4–12% and 7% Tris-Glycine gels (Invitrogen) to separate low and high molecular weight proteins.
5. Incubate gels in fixation buffer for 1 h at room temperature followed by staining with colloidal coomassie for 24 h at room temperature (Fig. 1d).
6. Destain gels by rinsing with destaining solution until solution is light blue or clear.
7. Manually excise the bands using a razor blade and place each band in a separate glass autosampler vial. Excise each gel band in several pieces to increase surface contact (see Note 12).
8. Perform in-gel tryptic digestion using the In-Gel Tryptic Digestion Kit per the manufacturer's instructions.

3.6 Peptide Purification and Concentration for Mass Spectrometry Analysis and Protein Identification

1. Resuspend the extracted peptides from the in-gel tryptic digestion procedure in 30 μL of resuspension buffer and vortex vigorously for 5 min.
2. Wet Zip-Tip pipette tip by aspirating and releasing wetting solution three times.
3. Equilibrate tip by pipetting and discarding equilibration/wash solution three times.
4. Bind peptides to the zip tip by pipetting 15 times inside the sample tube then discard sample.
5. Wash tip three times (and discard fluid) in equilibration/wash solution.
6. Elute the peptides into a vial insert by pipetting 10 μL at a time of the elution solution (100 μL). Keep vials on ice until all samples are zip tipped.
7. Freeze samples briefly at −80°C then SpeedVac to dryness.
8. Resuspend peptides with the appropriate volume of 0.1% formic acid and analyze by LC-MS/MS (see Note 13).

4 Notes

1. This protocol has been used only for crystalline antiretroviral nanoparticles coated in lipophilic surfactants such as poloxamer-188 (P188), 1,2-distearoyl-phosphatidyl ethanolamine-methyl-polyethyleneglycol-2000 ($mPEG_{2000}$-DSPE), and

1,2-dioleoyl-3-trimethylammonium-propane (DOTAP) (3). We suggest using rigid nanoparticles that have a lipophilic coating because the dye will easily label the particle without disrupting its structure. If other types of nanocarriers are used, it is highly encouraged that the nanoparticles are re-characterized after labeling to ensure that they have not been altered.

2. This protocol has been used only with Brilliant Blue R250. However, the goal is to coat the particles with a visible dye that will label the proteins that the nanoparticles come into contact with. Thus, other dyes that label proteins and are readily seen in the visible light spectrum, such as bromophenol blue, could also be used. It is unknown how using Brilliant Blue R250 (or other such dyes) to label different types of nanoparticles will affect their physical properties. Thus, it is highly encouraged that the nanoparticles be re-characterized after labeling to ensure that they have not been altered.
3. There are a number of methods for measuring nanoparticle size and charge. Companies such as Horiba, Malvern, TSI, and many others offer equipment that will simultaneously measure both the size and charge of nanoparticles. No one method or machine is preferred for this protocol.
4. This protocol has been used primarily with primary human monocytes and human monocyte-derived macrophages. However, any cells that will take up the nanoparticle being tested could be used. Adjust the protocol appropriately to accommodate for the cell type being used. It is suggested to use at least 100×10^6 cells in order to purify enough protein for proteomic analysis.
5. Typically the labeling procedure can be carried out in a 0.5 mL microcentrifuge tube. When using this small of a volume, using just a few grains (about 1–3 grains) of Brilliant Blue R250 dye will be enough to sufficiently label the particles without altering their physical characteristics. If too much dye is used, there is a risk of nanoparticle aggregation. It does not take much dye to label the particles so use less dye rather than more.
6. The purpose of sonication is to resuspend the particles so that they can be efficiently washed. However, it is possible to overheat or dissolve the particles with too much sonication. Therefore, a brief sonication (10 s) to resuspend the particles is all that is necessary.
7. The investigators will need to adjust the cell-nanoparticle exposure time based on endocytic activity of the targeted cells and the size and coating of their nanoformulations. It is suggested to allow for maximal nanoparticle uptake since this will allow for a better identification of nanoparticle-laden compartments.
8. When using the Dounce homogenizer, press and pull the piston hard enough to keep the solution flowing past at a

steady rate. Excess force is not necessary and could break the homogenizer.

9. When forming the sucrose gradient, carefully add each subsequent layer of sucrose by gently pouring it down the side of the tube. This will help to prevent mixing of the layers and will increase the intensity of the protein bands that form during centrifugation.
10. Do not disturb the contents of the ultracentrifuge tube when removing it from the centrifuge. The fractions need to be collected within 15 min after centrifugation; otherwise the bands will diffuse. When collecting the fractions (2–4 blue bands representing nanoparticle-laden compartments), use separate needles and syringes for each band. Start from the top band. Insert the needle-syringe at the bottom of band line facing up and aspirate until very little blue is left. Do not remove the syringe after band aspiration. Insert a new needle-syringe at the level of the next lower band and repeat.
11. Use nitrile gloves when handling the gel and the in-gel tryptic digestion solutions. Latex can interfere with downstream mass spectrometry analysis. Use a clean surface and equipment and avoid contact of gloves with skin or hair during the processing of gel bands. Dust and shedding epithelial cells can be a major contaminant of the samples, and it compromises the mass spectrometry analysis.
12. HPLC-grade reagents and HPLC-grade water should be used to make solutions used for in-gel tryptic digestion and peptide extraction (zip-tipping).
13. The volume necessary for resuspension of samples depends on the method used for mass spectrometry analysis and the instrument configuration (nanospray or electrospray). Typically, peptides from in-gel tryptic digestion are resuspended in 4–8 μL of 0.1% formic acid and are run in a nanospray configuration.

Acknowledgments

The work was supported by the National Institutes of Health grants 1P01 DA028555, 2R01 NS034239, 2R37 NS36126, P01 NS31492, P20RR 15635, P01MH64570, and P01 NS43985 (to H.E.G.) and a research grant from Baxter Healthcare. The authors thank Ms. Robin Taylor for critical reading of the manuscript and outstanding graphic and literary support.

References

1. Nowacek A, Gendelman HE (2009) NanoART, neuroAIDS and CNS drug delivery. Nanomedicine (Lond) 4:557–574
2. Nowacek A, Kosloski LM, Gendelman HE (2009) Neurodegenerative disorders and nano-formulated drug development. Nanomedicine (Lond) 4:541–555
3. Kadiu I, Nowacek A, McMillan J, Gendelman HE (2011) Macrophage endocytic trafficking of anti-retroviral nanoparticles. Nanomedicine (Lond) 6:975–994

Chapter 7

Permeabilization of Cell Membrane for Delivery of Nano-objects to Cellular Sub-domains

Valeriy Lukyanenko

Abstract

Delivery of nano-objects to specific cellular sub-domains is a challenging but intriguing task. There are two major barriers on the way of a nano-object to its intracellular target: (1) the cell membrane and (2) the intracellular barriers. The former is a common issue for all nanomedicine and a matter of very intense research. The latter is the primary problem for targeted delivery of nano-objects to specific cellular sub-domains and can be studied more easily using permeabilized cells. Membrane permeabilization for nano-medical research requires (1) perforation of the outer membrane, (2) development of a solution that will keep cellular sub-domains in the functional state, and (3) modification of the perimembrane cytoskeleton. We developed a very successful model of saponin membrane permeabilization of cardiomyocytes. This allowed us to deliver particles up to 20 nm in size to perinuclear and perimitochondrial space. Here we describe the method.

Key words Saponin permeabilization, Perimembrane cytoskeleton, Gold nanoparticle, Delivery of nano-objects, Intracellular sub-domains, Electron microscopy

1 Introduction

Successful delivery of genes and drugs to intracellular sub-domains depends on solving two major problems: transport of nano-objects through both cellular and intracellular membranes (1, 2). While the former is a common problem for nanomedicine (and a matter of intensive research with multiple very promising results), the intracellular pathways for delivery of nano-objects to certain cellular organelles remain to be elucidated (1). This research could be significantly facilitated with careful (without damage of cellular sub-domains) permeabilization of cell membrane.

Excluding the development of a specific protocol, membrane permeabilization for nanomedical research could be divided into three separate tasks:

1. Cell membrane permeabilization without permeabilization or damage of intracellular membranes.

Volkmar Weissig et al. (eds.), *Cellular and Subcellular Nanotechnology: Methods and Protocols*, Methods in Molecular Biology, vol. 991, DOI 10.1007/978-1-62703-336-7_7, © Springer Science+Business Media New York 2013

2. Development of "intracellular" solution that will keep cellular organelles functional and maintain the intracellular distances unchanged (for instance, will prevent mitochondrial swelling).
3. Modification of the cytoskeleton along the cell membrane to allow particles up to 20 nm to enter the cell.

In our experiments, we used rat ventricular myocytes obtained by enzymatic dissociation (3). Spatio-temporal characteristics and the frequency of elementary Ca^{2+} release events (Ca^{2+} sparks) are very sensitive to any mistake in permeabilization. For example, high $[Ca^{2+}]$ in the intracellular solution results in an almost immediate increase in amplitude and frequency of Ca^{2+} sparks (4), while permeabilization of the endoplasmic reticulum (SR in muscle cells) membrane will abolish the sparks. In our experiments, we monitored the Ca^{2+} sparks, the amount of Ca^{2+} in the SR and contraction of the myocytes with confocal microscopy (4, 5). Intracellular distances in intact and permeabilized cardiomyocytes were measured with calibrated gold nanoparticles using electron microscopy (6, 7).

There are many methods of cell membrane permeabilization, from mechanical skinning to making membrane holes only for special ions with corresponding ionophores. Gentle saponin permeabilization allows the removal of the outer cell membrane without damaging intracellular membranes.

Figure 1 shows saponin permeabilization in cardiac myocytes. The upper panel (**a**) presents the same myocyte before and after permeabilization. Note that the myocyte after permeabilization has the same shape and size. The middle panel (**b**) confirms the permeabilization. The myocyte was preloaded with membrane-permeable form of Ca^{2+}-sensing fluorescent dye. Therefore, before permeabilization (**b** *a*), Ca^{2+} sparks (local event) and waves (global Ca^{2+} release events) could be seen very well. Within 1 min after permeabilization (**b** *b*), the fluorescent dye left the cell and the sparks disappeared. The addition to the bathing solution of the same but membrane-impermeable Ca^{2+}-sensing fluorescent dye (**b** *c, d*) again allows seeing the Ca^{2+} release events, which have the same velocity, frequency, and spatio-temporal properties (**c**). This demonstrates that intracellular membranes were not damaged with our method of saponin permeabilization.

This method of membrane permeabilization is suitable for nano-objects as well. Figure 2 shows saponin permeabilization of the cardiac cell transfected with green fluorescent protein (GFP), which is a 4.2 nm long cylinder with a cylindrical diameter of 2.4 nm (8). The cell was exposed to 0.01% saponin for 30 s. Multiple GFP-filled membrane blebs indicate gentle permeabilization. Therefore, GFP takes much longer to leave the cell after permeabilization (7), and GFP stays within cellular sub-domains (including nuclei; N) even after 5 min. This reminds us that membrane is not the only barrier for free diffusion of nano-objects inside the cell.

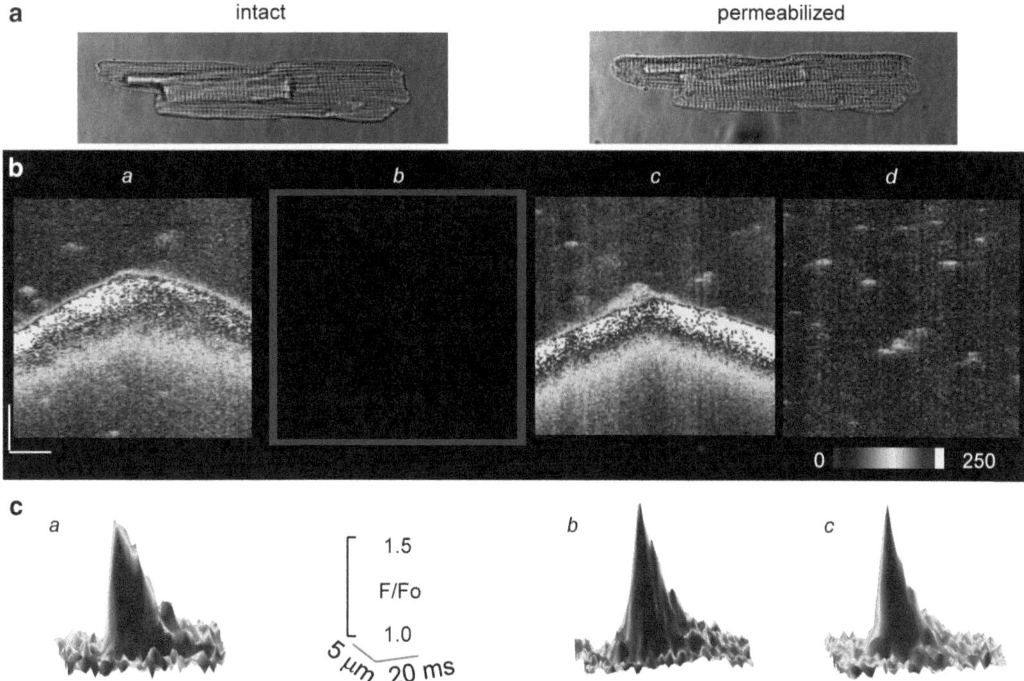

Fig. 1 Saponin permeabilization of cardiomyocytes has no effect on Ca^{2+} release from sarcoplasmic reticulum. (**a**) Images of a cardiac myocyte obtained in transmitted light before and after permeabilization with saponin. (**b**) Line scan images of fluorescence in a portion of the same cell preloaded with fluo-3 AM measured before permeabilization (a), after permeabilization in an internal solution with no dye (b), and after addition to the internal solution 30 μM fluo-3 potassium salt in the presence of 0.1 (c) or 0.5 mM EGTA (d) (pCa 7). *Calibration bars*: horizontal 10 μm, vertical 0.4 s, the pseudo scale bar represents changes in units of absolute fluorescence. (**c**) Surface plots of averaged Ca^{2+} sparks measured before permeabilization (a) and after permeabilization (reproduced from ref. 5 with permission from Wiley-Blackwell)

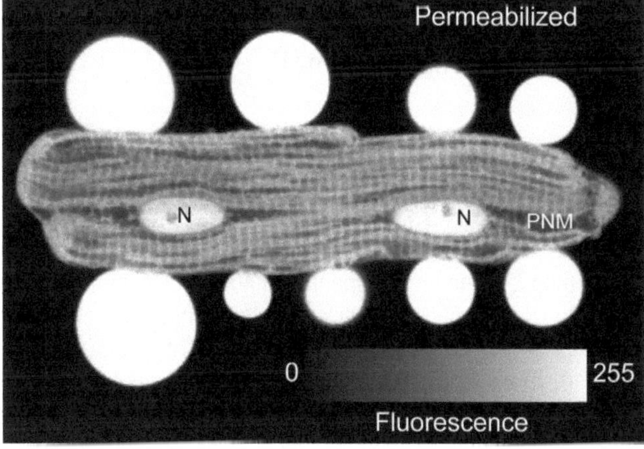

Fig. 2 Light permeabilization. Before the permeabilization, the cardiomyocytes were transfected with GFP. *N* nucleus, *PNM* perinuclear mitochondria (reproduced from ref. 7 with permission from Cell press)

Our experiments showed that the cytoskeleton of permeabilized myocytes does not allow nano-objects ≥11 nm in diameter to enter the cell. We should note that this problem pertains only to permeabilized cells because of a lack of transmembrane transporting mechanisms. In the case of saponin-permeabilized cells, the mesh of actin filaments located along the cell membrane creates an additional barrier for nano-objects. However, short pretreatment of intact cells with cytochalasin D significantly reduces the mesh integrity and allows particles up to 20 nm to diffuse inside of the myocytes (6). Electron micrographs in Fig. 3 show the effect of 40 μM cytochalasin D on the distribution of nanoparticles in permeabilized ventricular myocytes. Note that the silver grains are only markers for the location of calibrated gold nanoparticles (the deeper in the slice the particle was located, the smaller the silver grain produced with the silver enhancement procedure).

Here we describe the method of saponin permeabilization of the cell membrane. The method is shown to be useful for the delivery of nano-objects to perinuclear and perimitochondrial space.

2 Materials

1. Tyrode solution: 140 mM NaCl, 5.4 mM KCl, 0.5 mM $MgCl_2$, 1 mM $CaCl_2$, 10 mM Hepes, 0.25 mM NaH_2PO_4, 5.6 mM glucose, pH 7.3 (6).
2. The permeabilization solution: 100 mM K^+ aspartate (see Note 1), 20 mM KCl, 3 mM MgATP, 0.81 mM $MgCl_2$ ($[Mg^{2+}]_{free}$ = ~1 mM), 0.5 mM EGTA, 0.114 mM $CaCl_2$ ($[Ca^{2+}]_{free}$ = ~100 nM), 20 mM Hepes, 3 mM glutamic acid, and 3 mM malic acid, pH 7.2 (see Note 2) (6).
3. 1% saponin in permeabilization solution.
4. The solution for permeabilized cells: 100 mM K^+ aspartate (see Note 1), 20 mM KCl, 3 mM Mg ATP, 0.81 mM $MgCl_2$ ($[Mg^{2+}]_{free}$ = ~1 mM), 0.1 mM EGTA, 0.03 mM $CaCl_2$

Fig. 3 Remodeling of cytoskeleton with cytochalasin D allows particles up to 20 nm to diffuse inside of cells. Representative micrographs show the distribution of the nanoparticles before (**a**) and after (**b**) partial ablation of the cytoskeleton (20 min pretreatment with 40 μM cytochalasin D). (**c**) Graphs representing the density of nanoparticles in intact ventricular cells before (*gray*) and after (*dark gray*, only for particles ≥11 nm) 20 min of pretreatment with 40 μM cytochalasin D. *Asterisks* indicate data that are statistically different from the corresponding control. *M* mitochondrion, *Z* Z line. *Arrows on* (**b**) indicate T-tubules and *black ovals* show nanoparticles located deeper inside the ultrathin section, which therefore have smaller diameters after silver enhancement (reproduced from ref. 6 with permission from Cell press)

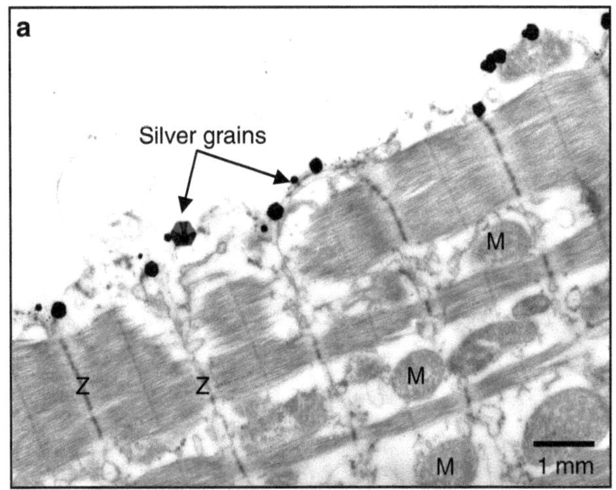
a
Silver grains
M
M
Z
Z
M
1 mm

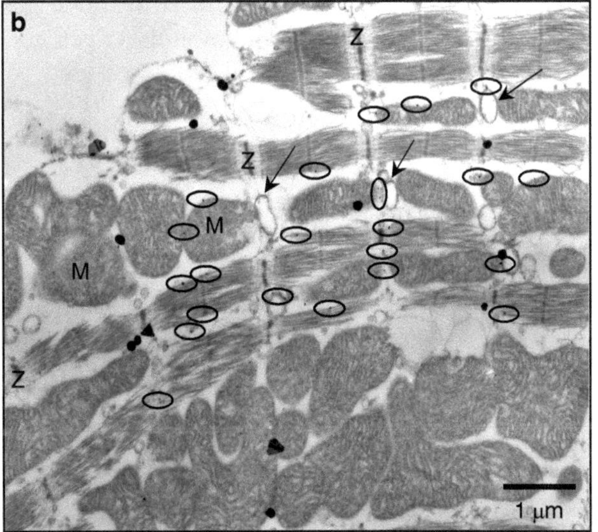
b
Z
Z
M
M
Z
1 μm

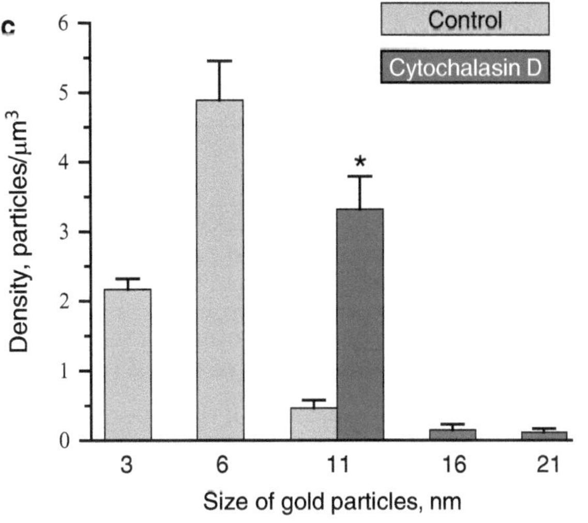
c
Control
Cytochalasin D
Density, particles/μm³
Size of gold particles, nm
0
1
2
3
4
5
6
3
6
11
16
21
*

($[Ca^{2+}]_{free}$ = ~60 nM), 20 mM Hepes, 3 mM glutamic acid, 3 mM malic acid, 10 mM phosphocreatine, 5 U/ml creatine phosphokinase, and 1% polyvinylpyrrolidone (PVP10; MW 10,000), pH 7.2 (see Note 2) (7).

5. Cells have to be attached to the bottom. To do that, we coated cover slips with laminin as recommended by the manufacturer (Molecular Probes; Invitrogen) and allowed cells 30 min to attach.

3 Methods

Carry out all procedures at room temperature.

In our experiments, we used Tyrode solution and primary culture of rat ventricular myocytes (single freshly isolated cells). However, any monolayer cell culture (any confluency) and corresponding media could be used.

1. Pretreat cells 20 min with 40 μM cytochalasin D in 1 ml of Tyrode solution.
2. Replace the Tyrode solution with 1 ml of permeabilization solution for 1 min (see Notes 3 and 4).
3. Replace the permeabilization solution with 1 ml of the same (permeabilization) solution containing 0.01% saponin for 30–60 s.
4. Replace the permeabilization solution with 1 ml of the solution for permeabilized cells containing nanoparticles.

To visualize cells, we used C-Apochromat 63×/1.2 W corr objective. Under the mentioned conditions, intracellular organelles (such as mitochondria and sarcoplasmic reticulum) of cardiac myocytes remain functional at room temperature for at least 2 h (see Note 5).

4 Notes

1. For both permeabilization solution and the solution for permeabilized cells, we used DL-aspartic acid potassium salt.
2. During the preparation of the solution for permeabilized cells, pay attention to the pH. The pH is very important for Ca^{2+} buffering power of EGTA.
3. All steps of the saponin permeabilization have to be performed under visual control. We used an inverted microscope equipped with at least ×40 objective.
4. The moment of permeabilization is seen (within a minute) as a sharp reduction in cell shining (cells become gray; Fig. 1a).

After that, the permeabilization solution should be immediately replaced by the solution for permeabilized cells.

5. We found that entry of 3-nm particles into VDAC pore (located in the outer mitochondrial membrane) is significantly restricted in permeabilized cardiomyocytes in comparison to isolated mitochondria (7).

References

1. Lukyanenko V (2007) Delivery of nano-objects to functional sub-domains of healthy and failing ventricular myocytes. Nanomedicine 2:831–846
2. Lukyanenko V (2010) Therapeutic nano-object delivery to sub-domains of cardiac myocytes. In: Weissig V, D'Souza GGM (eds) Organelle-specific pharmaceutical nanotechnology. Artech House, Norwood, MA, pp 433–448
3. Györke S, Lukyanenko V, Györke I (1997) Dual effects of tetracaine on spontaneous calcium release in rat ventricular myocytes. J Physiol 500:297–309
4. Lukyanenko V, Viatchenko-Karpinski S, Smirnov A, Wiesner TF, Györke S (2001) Dynamic regulation of the SR Ca^{2+} content by lumenal Ca^{2+}-sensitive leak through RyRs in rat ventricular myocytes. Biophys J 81:785–798
5. Lukyanenko V, Györke S (1999) Ca^{2+} sparks and Ca^{2+} waves in saponin-permeabilized cardiac myocytes. J Physiol 521:575–585
6. Parfenov AS, Salnikov V, Lederer WJ, Lukyanenko V (2006) Aqueous diffusion pathways as a part of the ventricular cell ultrastructure. Biophys J 90:1107–1119
7. Salnikov VV, Lukyanenko YO, Frederick CA, Lederer WJ, Lukyanenko V (2007) Probing the outer mitochondrial membrane in cardiac mitochondria with nanoparticles. Biophys J 92:1058–1071
8. Yang F, Moss LG, Phillips GN Jr (1996) The molecular structure of green fluorescent protein. Nat Biotechnol 14:1246–1251

Chapter 8

A Method to Encapsulate Molecular Cargo Within DNA Icosahedra

Dhiraj Bhatia, Saikat Chakraborty, Shabana Mehtab, and Yamuna Krishnan

Abstract

DNA self-assembly has yielded various polyhedra based on platonic solids. DNA polyhedra can act as nanocapsules by entrapping various molecular entities from solution and could possibly find use in targeted delivery within living systems. A key requirement for encapsulation is that the polyhedron should have maximal encapsulation volume while maintaining minimum pore size. It is well known that platonic solids possess maximal encapsulation volumes. We therefore constructed an icosahedron from DNA using a modular self-assembly strategy. We describe a method to determine the functionality of DNA polyhedra as nanocapsules by encapsulating different cargo such as gold nanoparticles and functional biomolecules like FITC dextran from solution within DNA icosahedra.

Key words DNA icosahedron, Polyhedra, Nanocapsules, Encapsulation, Gold nanoparticles, FITC dextran

1 Introduction

Encapsulation of a range of molecular cargo inside a given biomolecular host is highly challenging. This is due to the large size and sensitive nature of the cargo, and retention of functionality of cargo post encapsulation (1). DNA has been shown to be capable of assembling into various polyhedral structures using one pot assembly (2), modular self-assembly (3), and origami based approaches (4). Encapsulation is an attractive property of DNA polyhedra and we describe methods to characterize DNA encapsulated cargo using single molecule methods as well as bulk biophysical methods. Here we describe a detailed method to study encapsulation characteristics of DNA icosahedra. These methods can be generalized for encapsulation of various entities like biomacromolecules and functional nanoparticles, etc.

We outline first a strategy to assemble the molecular host, the DNA icosahedron in high yields using a modular approach (5).

Volkmar Weissig et al. (eds.), *Cellular and Subcellular Nanotechnology: Methods and Protocols*, Methods in Molecular Biology, vol. 991, DOI 10.1007/978-1-62703-336-7_8, © Springer Science+Business Media New York 2013

This assembly involves a step-wise association of different modules (5-way junctions) having programmable overhangs into two half or hemi-icosahedra. These two hemi-icosahedra then self-assemble into icosahedral DNA nanocapsules that enclose a hollow cavity. We describe the use of complementary hemi-icosahedra to encapsulate two types of (1) inorganic cargo such as gold nanoparticles (GNPs) and (2) biomacromolecules such as fluorescently labeled dextran **FD10** inside DNA icosahedra (5, 6).

This method of cargo encapsulation inside DNA polyhedra is advantageous because (a) It is not limited to molecules that need to undergo molecular recognition with the host scaffold. This affords the following advantages: (1) Larger varieties of molecules may be encapsulated provided they have a size compatibility with the polyhedron. (2) The size of the polyhedron can also be easily altered to encapsulate differently sized molecules. (3) Guest molecules do not need to undergo a chemical reaction for encapsulation. (b) The DNA scaffold is amenable to site specific chemical modifications using multiple orthogonal chemistries. This affords the following advantages: (1) The ability to uniformly functionalize DNA polyhedra in a precisely tunable manner, with multiple tags in bulk. (2) Greater homogeneity of functionalized DNA polyhedra carrying cargo internally, and carrying surface displayed tags for targeting. We describe in detail how one may characterize such loaded DNA polyhedra both at the single molecule level and using bulk biophysics.

2 Materials

2.1 Oligonucleotide Sample Preparation

1. Prepare 1 mM stocks in Milli-Q (MQ) water (Millipore, USA) of all oligonucleotides shown in Table 1. Oligonucleotides are obtained from Sigma, HPLC purified and lyophilized.
2. Ethanol, absolute from Merck.
3. 3.0 M Potassium chloride solution: 2.23 g KCl dissolved in 10 mL MQ water.
4. Phosphate buffers: 100 mM (10×) NaH_2PO_4: 1.2 g NaH_2PO_4 dissolved in 100 mL MQ water. 100 mM (10×) Na_2HPO_4: 1.42 g Na_2HPO_4 dissolved in 100 mL MQ water. The phosphate buffer (10 mM, pH 6) is prepared by mixing appropriate amounts of Na_2HPO_4 and NaH_2PO_4.
5. Magnesium chloride (10 mM): 0.203 g $MgCl_2$ dissolved in 100 mL MQ water. Sodium chloride (1 M): 5.84 g NaCl dissolved in 100 mL MQ water.
6. Heat Block.
7. Adenosine triphosphate (ATP) stock, 100 mM.

Table 1
Oligonucleotide sequences for icosahedron

Name	Sequence
V1	5′-GCCTGGTGCCACCGGTGACGTTCCGC-3′
V2	5′-GCCTGGTGCCCCGCGTCCTCACCGGT-3′
V3	5′-GCCTGGTGCCGCCACGCTTTGGACGCGG-3′
V4	5′-GCCTGGTGCCGCGAGTGCAAAGCGTGGC-3′
V5	5′-GCCTGGTGCCGCGGAACGAAGCACTCGC-3′
U1	5′-CATCAGTCGCACCGGTGACGTTCCGC-3′
U2	5′-TTATAGGACTCCGCGTCCTCACCGGT-3′
U3	5′-TTATAGGACTGCCACGCTTTGGACGCGG-3′
U4	5′-GCGACTGATGGCGAGTGCAAAGCGTGGC-3′
U5	5′-GGCACCAGGCGCGGAACGAAGCACTCGC-3′
L1	5′-CATCAGTCGCACCGGTGACGTTCCGC-3′
L2	5′-AGTCCTATAACCGCGTCCTCACCGGT-3′
L3	5′-AGTCCTATAAGCCACGCTTTGGACGCGG-3′
L4	5′-GCGACTGATGGCGAGTGCAAAGCGTGGC-3′
L5	5′-GGCACCAGGCGCGGAACGAAGCACTCGC-3′

8. T4 Polynucleotide kinase (10 U/μL) and associated buffer—500 mM Tris–HCl, pH 7.6 at 25°C, 100 mM $MgCl_2$, 50 mM DTT, 1 mM EDTA, 1 mM Spermidine.

2.2 Gel Electrophoresis

1. 50× Tris-Acetate-EDTA (TAE) buffer: 24.2 g Tris buffer, 5.71 mL glacial acetic acid and 10 mL of 0.5 M EDTA (pH 8.0) dissolved in 100 mL of MQ water.
2. 10× Tris-Boric Acid–EDTA (TBE) buffer: 5.4 g Tris base, 2.75 g Boric acid and 20 mL of 0.5 M EDTA (pH 8.0) dissolved in 100 mL of MQ water.
3. 40% polyacrylamide stock: 29.9 g Acrylamide, 0.8 g *N,N′*-Bis-methylene acrylamide dissolved in 100 mL of MQ water.
4. 0.8% Agarose gel: 0.8 g agarose powder (Bangalore Genei, India) dissolved in 98 mL MQ water and 2 mL 50× TAE buffer.
5. 10 and 15% Polyacrylamide gel: Different percentages of polyacrylamide gels can be made from 40% stock in 0.1% Ammonium persulfate and 20 μL Tetramethylethylenediamine (TEMED).

6. Ammonium persulfate, TEMED, and Ethidium bromide (EtBr) (Sigma, USA).
7. All gels are run in a cold room at 4°C and visualized by EtBr staining under UV illuminator.

2.3 Ligation: Preparation of *N*-Cyano Imidazole

1. 5.5 g Cyanogen bromide (BrCN) and 3.2 g Imidazole are dissolved in 25 mL and 50 mL of dry benzene respectively.
2. A solution of 5.5 g BrCN is added drop wise with stirring to a solution of 3.2 g imidazole in 50 mL Benzene.
3. The reaction mixture is warmed to 50°C during the addition and for 5 min after the addition is done.
4. The reaction mixture is cooled at 4°C for 8 h. This may be preferably left overnight.
5. The resultant yellow solid is filtered through Whatman filter paper and the supernatant solution is collected.
6. The filtrate is concentrated to dryness under reduced pressure.
7. A white crystalline solid remains which is collected and purified by sublimation. The sublimate is pure *N*-Cyano imidazole that is aliquoted in eppendorf tubes and stored at −20°C (7).

2.4 Gold Nanoparticles

1. Auric chloride, Tri-sodium citrate, Tannic Acid, Potassium bicarbonate.
2. 2 mg Auric chloride is taken in 16 mL MQ water in a round bottom flask. Heat up to 60°C in an oil bath with constant stirring. This is solution A.
3. In three different eppendorfs the following solutions are taken: 20 mg Trisodium citrate in 2 mL MQ water, 20 mg tannic acid in 2 mL MQ water, and 6.9 mg Potassium carbonate in 2 mL MQ water.
4. Now 0.8 mL citrate, 1 mL tannic acid and 1 mL Potassium carbonate and 1.2 mL water are mixed together to form 4 mL of solution B.
5. Solution B is also heated to 60°C.
6. Solution B is added to solution A with constant stirring. The color changes from yellow to wine red.
7. The temperature is increased to 100°C and the solution is refluxed for 30 min.
8. This protocol gives homogeneous gold nanoparticles of 5 nm size. The size of GNPs can be changed by changing the amounts of citrate and tannic acids keeping the amount of Auric chloride constant.
9. The sizes of gold nanoparticles may be checked by transmission electron microscopy (TEM) or dynamic light scattering (DLS).
10. Using this protocol, GNPs from 2 to 15 nm can be made (8).

2.5 Transmission Electron Microscopy

1. 400 mesh carbon coated and glow discharged grids (Ted Pella, USA).
2. 1% Uranyl acetate (Ted Pella, USA).
3. Transmission electron microscope—JEOL 100 CX II operating at acceleration voltage 80 kV; Tecnai 12 Biotwin, FEI, Netherlands operating at acceleration voltage 120 kV.
4. Images are acquired using side-mount 1,024×768 pixel resolution CCD camera.

2.6 Size Exclusion Chromatography

1. Biosep-SEC-S3000 (Phenomenex) of dimensions 300×4.6 mm, 5 μm bead size and 29 nm pore size.
2. Shimadzu HPLC system equipped with a temperature controller, a photodiode array detector, fraction collector, and auto-injector (Shimadzu, Japan).
3. Acetonitrile (HPLC grade), Degassed MQ water.

2.7 Fluorometer and Anisotropy Setup

1. Fluorolog 3 L instrument (Horiba Jobin Yvon, Japan) having the polarizing angle fixed (90°).
2. g factor calibrated using Fluorescein (pH 7, 50 nM) as standard (r=0.018).
3. The sample is excited at 488 nm and emission is collected at 515 nm with the slit widths adjusted accordingly.

2.8 Dynamic Light Scattering

1. DynaPro-99 unit (Protein Solutions, USA) operating at 25°C.
2. Buffers and samples are first filtered through 0.02 μm filters and 0.22 μM filters, respectively and spun at 9300 rcf for 10 min prior to use.
3. Experimental settings used an acquisition time of 3 s; S/N threshold of 2.5 and sensitivity of 70%.
4. Samples were illuminated with 829.4 nm laser and scattering intensity at 90° was measured.
5. Fluctuations greater than 15% in the scattering intensity are excluded from the analysis.
6. The DynaLS software (Protein Solutions, USA) is used to resolve acquisitions into well-defined Gaussian distributions of hydrodynamic radii.

2.9 Quenchers

1. Quenchers used: Iodide (0.5 nm), Amino TEMPO (1 nm), Nanogold (1.5 nm), Gold nanoparticles (GNPs) (2, 3, 4, and 5 nm), TEMPO-Dextran 1 kDa (2.5 nm).
2. TEMPO Dextran (1 kDa) is obtained by coupling Dextran (10 mg) to carboxy TEMPO (50 mg) using dicyclohexylcarbodiimide (20 mg) in dry DMSO at 20°C for 8 h and purified by SEC-HPLC.
3. All GNPs are synthesized by the procedure given in Subheading 2.4 and characterized by TEM.

2.10 Lifetime Measurements

1. Frequency domain Fluorolog Tau 3 (Horiba Jobin Yvon, Japan) operating at 25°C and at 10 MHz frequency.
2. The S and T channels are calibrated using glycogen as a standard.
3. For each sample, the frequency and modulation are spanned from 10 to 150 MHz using 7–10 intermediate frequency readings.
4. The data is fitted using the associated software and only readings showing χ^2 value less than 1.2 are selected for analysis.

3 Methods

DNA icosahedra are constructed from three distinct five way junction (5WJ) components **V**, **U** and **L**, with programmable overhangs (Fig. 1a). Each 5WJ module **V**, **U** and **L** are constructed from equimolar ratios of the respective five phosphorylated single strands. **V** forms a 1:5 complex with **L** to give $\mathbf{VL_5}$ (Fig. 1b). The complementary module $\mathbf{VU_5}$ is similarly synthesized from components **V** and **U** in a 1:5 ratio. At this stage, contiguously hybridized strands in $\mathbf{VU_5}$ and/or $\mathbf{VL_5}$ are chemically ligated with *N*-Cyano imidazole (NCI), to enhance stability. The two different hemi-icosahedra, $\mathbf{VU_5}$ and $\mathbf{VL_5}$, each have ten identical overhangs where the overhangs in $\mathbf{VL_5}$ are complementary to the ones in $\mathbf{VU_5}$. $\mathbf{VU_5}$ and $\mathbf{VL_5}$ complex with each other in a 1:1 ratio to yield the DNA icosahedron and the contiguous termini are ligated again with NCI (Fig. 1c).

In order to encapsulate cargo inside these DNA icosahedra, the two halves $\mathbf{VU_5}$ and $\mathbf{VL_5}$ are annealed together in a 1:1 ratio in

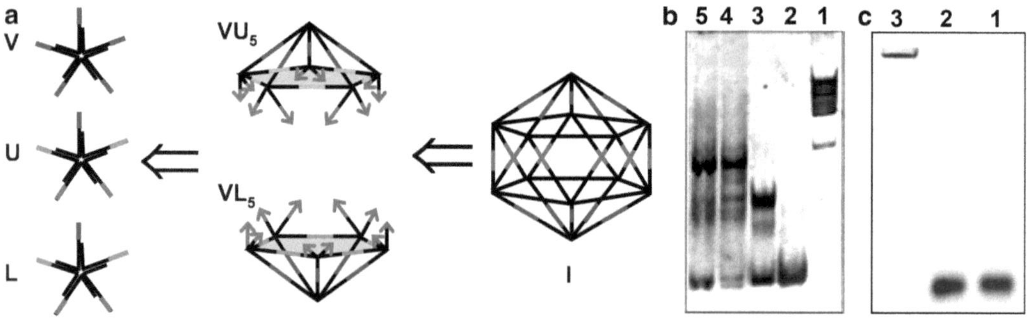

Fig. 1 Construction and characterization of DNA icosahedron. (**a**) The icosahedron **I** (*right*) is constructed from two half icosahedra $\mathbf{VU_5}$ and $\mathbf{VL_5}$. Each half icosahedron (*middle*) is made from two types of 5WJs: **V** and **U** for $\mathbf{VU_5}$ and **V** and **L** for $\mathbf{VL_5}$ (*left*). Complementary overhangs are color coded. (**b**) 10% PAGE showing formation of 5WJ and half icosahedra. *Lane 1*: $\mathbf{VL_5}$; *lane 2*: $\mathbf{VU_5}$; *lane 3*: 5WJ **V**; *lane 4*: **V1** oligonucleotide; *lane 5*: DNA marker; (**c**) 0.8% Agarose gel showing the formation of the icosahedron from $\mathbf{VU_5}$ and $\mathbf{VL_5}$. *Lane 1*: ligated icosahedron; *lane 2*: ligated $\mathbf{VL_5}$; *lane 3*: ligated $\mathbf{VU_5}$

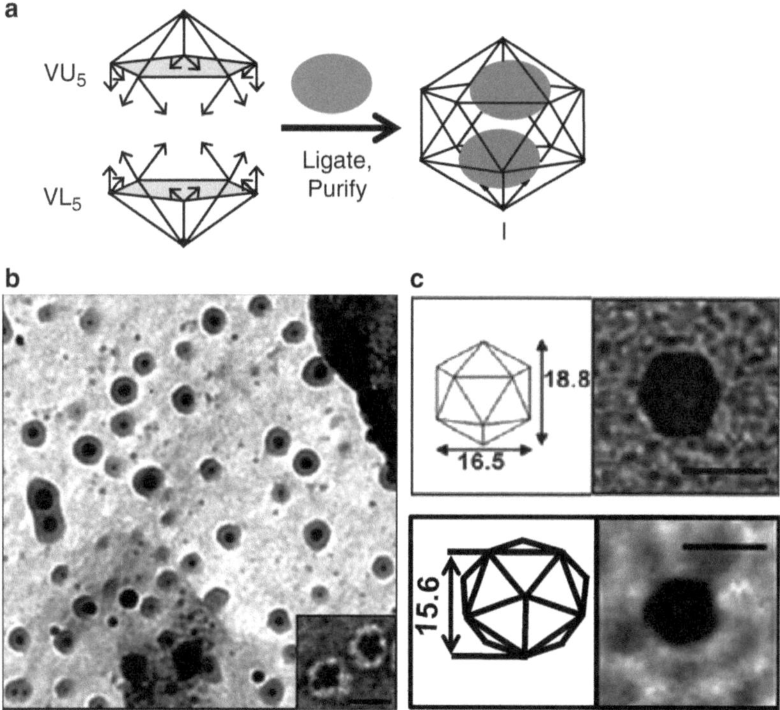

Fig. 2 Encapsulation of molecular cargo like gold nanoparticles (GNPs) within DNA icosahedron. (**a**) General schematic showing encapsulation of various molecular cargo inside DNA icosahedra. The two half icosahedra (**VU**$_5$ & **VL**$_5$) are mixed in 1:1 ratio in presence of excess of desired cargo so that few molecules of cargo are encapsulated within the icosahedron. These loaded icosahedra are purified from bulk of free molecules. (**b**) A representative low-resolution TEM image shows the dense core of gold nanoparticles encapsulated within DNA icosahedra. The *inset* shows representative high-resolution image in which the individual gold nanoparticles can be seen to be present within the icosahedral cages. Scale bar: 50 nm. (**c**) Representative TEM micrographs of platinum shadowed icosahedra showing hexagonal (*top*) and pentagonal (*bottom*) symmetries. Corresponding theoretically calculated distances (in nm) are shown in *left*. Scale bar: 20 nm

presence of an aqueous solution of cargo like GNPs or **FD10** (Fig. 2a). The complex, i.e., cargo loaded DNA icosahedron, is separated from bulk, unencapsulated cargo using size separating techniques like dialysis and/or gel electrophoresis. Electron dense cargo such as GNPs inside DNA icosahedra may be characterized by single molecule methods such as electron microscopy (TEM) (Fig. 2b, c), while biomolecular cargo such as **FD10** inside icosahedra may be characterized by bulk biophysical methods such as fluorescence spectroscopy (Fig. 3c, d).

3.1 Sample Preparation

3.1.1 Phosphorylation of Oligonucleotides

1. To an eppendorf tube is added 2 μL oligonucleotide (from 1 mM stock), 10 μL MQ water, 2 μL 10× T4 PNK (Polynucleotide kinase) buffer (500 mM Tris–HCl, pH 7.6 at 25°C, 100 mM $MgCl_2$, 50 mM DTT, 1 mM EDTA, 1 mM Spermidine).

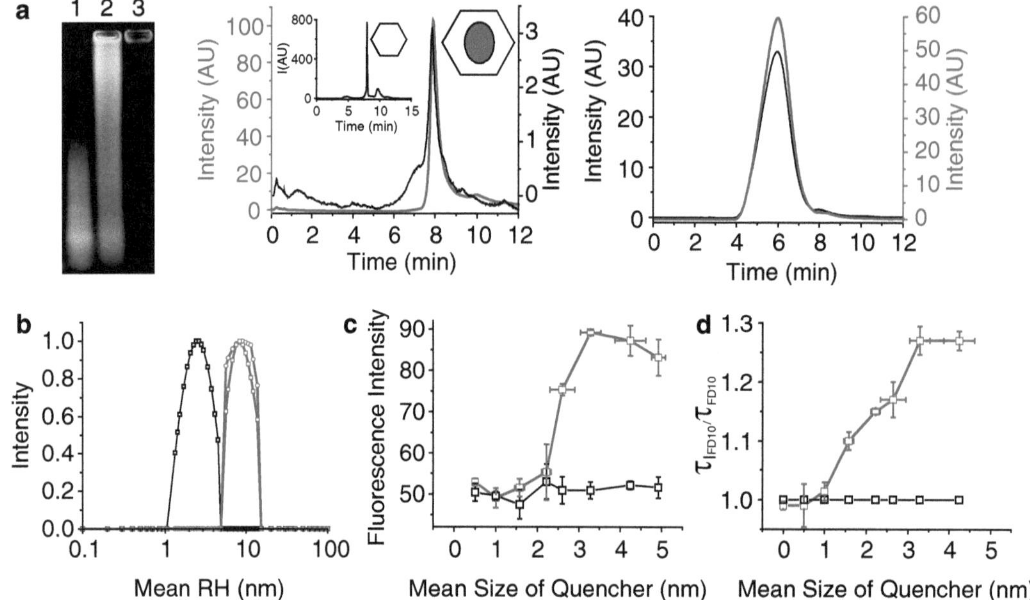

Fig. 3 Encapsulation of **FD10** inside DNA icosahedron. (**a**) (*Left*) Gel electrophoretic mobility shift assay for the formation of I_{FD10}. 0.8% agarose gel (1× TAE) showing association of **FD10** with icosahedron: *lane 1*. **FD10**, *lane 2*. 1:1 (VU_5: VL_5) + 2 mM **FD10** post ligation, *lane 3*. Purified I_{FD10}. Gel was visualized using 488 nm excitation. (*Middle*) Size exclusion chromatogram (SEC-HPLC) of I_{FD10} complex post gel excision. SEC traces were followed at 254 nm (*grey*) and 488 nm (*black*). *Inset*: SEC of standard, reference sample of unlabeled, unloaded icosahedron **I** (retention time 8 min). (*Right*) SEC trace of free **FD10** was followed at 254 nm (*black*) and 488 nm (*grey*). (**b**) Dynamic light scattering (DLS) traces of free **FD10** (*black squares*), the standard sample of DNA icosahedra, **I** (*open grey circles*) and purified I_{FD10} complex (*grey squares*). (**c**) Fluorescence intensity-based quenching assay for free **FD10** (*black squares*) and I_{FD10} complex (*grey squares*). (**d**) Fluorescence lifetime measurements of free **FD10** (*black squares*) and I_{FD10} complex (*grey squares*) with the same quenchers. Mean values of two independent experiments are presented, along with their s.d

2. 2 μL T4 PNK enzyme is added to this mixture. The concentration of enzyme is 10 U/μL.
3. 4 μL of 1 mM ATP is added to the above solution, vortexed for 1 min to mix the solution well and the tube is incubated at 37°C in a heat block for 1 h.
4. Post-incubation, the enzyme is deactivated by heating the mixture to 75°C for 10 min.
5. The DNA is precipitated by addition of 80 μL absolute ethanol and 20 μL 3 M potassium acetate, and incubated at −20°C for 8 h.
6. Post incubation at −20°C, the tubes are spun at 20800 rcf for 40 min at 4°C.
7. The DNA pellet is visible at the bottom of the tube. All supernatant solution is carefully removed and discarded.
8. The pellet is washed with 100 μL of 70% ethanol and spun at 14 k for 2 min.

9. All ethanol is removed carefully and eppendorfs are air dried for 15 min to remove any remaining ethanol.
10. The pellet is dissolved in 20 μL Nuclease free water and the phosphorylated oligos are quantified by their UV absorbance at 260 nm.

3.1.2 5-Way Junctions

1. The buffer used to make the 5WJ is 10 mM phosphate buffer (pH 6), 1 mM $MgCl_2$ and 100 mM NaCl.
2. From the stocks of phosphorylated oligos for **V** 5WJ, the oligos **V1**:**V2**:**V3**:**V4**:**V5** are mixed together in a 1:1:1:1:1 ratio in 50 μL of buffer containing 20 μM concentration of each oligo. In a similar manner, 5WJ of **U** and **L** are made from oligos **U1**–**U5** and **L1**–**L5**, respectively.
3. Once all solutions are added, the eppendorfs are heated to 90°C for 15 min. After 15 min, the sample is annealed from 90°C at the rate of 1°C/3 min till room temperature, incubated at room temperature for 2 h and then stored at 4°C for 48 h.
4. The individual 5WJ are characterized by 15% PAGE stained with EtBr (see Subheading 2.2) (Fig. 1b).

3.1.3 Half Icosahedra (VU_5 and VL_5)

1. In an eppendorf tube, 50 μL of **U** 5WJ (20 μM) and 10 μL of **V** 5WJ (20 μM) are mixed to form half icosahedron $\mathbf{VU_5}$.
2. The tube is heated in a heat block at 45°C for 4 h, and then the temperature is decreased at the rate of 1°C/3 min till room temperature (20°C), where the samples are incubated for 2 h and then stored at 4°C for 72 h.
3. $\mathbf{VL_5}$ can be similarly made by mixing **V** and **L** 5WJs.
4. The half icosahedra are characterized by 15% PAGE stained with EtBr as described in Subheading 2.2 (Fig. 1b).

3.1.4 Ligation

1. Half icosahedra $\mathbf{VU_5}$ or $\mathbf{VL_5}$ are ligated using NCI as described below.
2. 50 μL of sample ($\mathbf{VU_5}$ or $\mathbf{VL_5}$, 3.33 μM) is taken in an eppendorf tube.
3. To this 0.3 mg solid NCI and 2 μL $NiCl_2$ (from a 1 M solution of $NiCl_2$) is added and incubated for 48 h.
4. After 48 h, step 3 is repeated and the tubes containing ligated $\mathbf{VU_5}$ or $\mathbf{VL_5}$ are stored at 4°C for 72 h (see Note 1).

3.1.5 Icosahedron

1. 10 μL of $\mathbf{VU_5}$ and $\mathbf{VL_5}$ (3.33 μM) each are mixed in an eppendorf to form icosahedra.
2. The tube is heated in a heating block at 45°C for 4 h, and the temperature is decreased at a rate of 1°C/3 min till room temperature (20°C) followed by incubation at 20°C for 2 h. Then the sample is transferred to 4°C to equilibrate for 72 h.

3. The half icosahedra and full icosahedra are characterized on 0.8% agarose gel (see Subheading 2.2) (Fig. 1c).
4. The icosahedron is then ligated as described in Subheading 3.1.4.

3.2 Encapsulation of GNPs

1. Citrate capped GNPs of diameters 2, 3.5, 8 nm are prepared according to procedure given in Subheading 2.4.
2. 15 μL $\mathbf{VU_5}$ and $\mathbf{VL_5}$ (200 nM) were mixed with 30 μL solution of GNPs (at 400 nM GNP concentration) of desired size in 10 mM phosphate buffer.
3. The tube is heated in a heating block at 45°C for 4 h. The temperature is then decreased at the rate of 1°C/3 min till room temperature (20°C), where it is incubated for 2 h and then equilibrated for 72 h at 4°C.
4. Finally, the solution is ligated using NCI following the procedure described earlier (see Subheading 3.1.4).

3.3 Purification

1. The DNA icosahedra loaded with GNPs (I_{GNP}) are separated from free GNPs in solution using dialysis.
2. 50 μL of the solution of GNPs and post-ligated icosahedron is loaded in a dialysis membrane size: 3.4×15 cm, 50 kDa MWCO, sealed from bottom using a plastic clip as provided by the supplier.
3. This sample was further diluted to 1 mL with a buffer containing 10 mM phosphate, pH 6 and 100 mM NaCl (see Note 2).
4. The resultant solution is dialyzed against buffer containing 10 mM phosphate buffer, pH 6 and 100 mM NaCl for 24 h at 20°C, where the external buffer is changed every 6 h.
5. Post-dialysis, the sample is vacuum concentrated (Labconco Centrivac Console) at 20°C and this solution containing loaded, purified icosahedra is taken for further characterization.

3.4 Transmission Electron Microscopy

1. TEM is a good method to characterize icosahedra whether they contain GNP cargo or whether investigating icosahedra that have not been subjected to any encapsulation process.
2. For visualizing the icosahedral shell, 10 μL of icosahedron solution (10 nM) is adsorbed on the carbon-coated, glow discharged copper grid of 400 mesh size for 20 min.
3. The excess solution is wicked off using a Whatman filter paper.
4. The grid is stained by placing a drop of 1% uranyl acetate solution for 2 s and immediately wicking off with a Whatman filter paper.
5. The grid is then loaded onto a holder and visualized by TEM using low beam current (4–8 nA).
6. Alternatively, the samples can also be visualized by platinum shadowing instead of uranyl acetate staining.

7. For platinum shadowing, the samples are loaded on the grids as described, and these grids are rotary platinum shadowed for 10 s in a vacuum evaporator. These icosahedra are visualized as pentagonal and hexagonal particles (icosahedral symmetry) in bright field EM (Fig. 2c).
8. For TEM characterization of icosahedra loaded with gold nanoparticles, only uranyl acetate staining is used. The samples are prepared as described in point 2 of this section.
9. At low magnification (50k×), particles with highly electron dense core are seen. At high magnification (160k×), individual gold nanoparticles can be seen present inside the DNA icosahedra (Fig. 2b) (see Note 3).

3.5 Encapsulation of FITC Dextran, 10 kDa (FD10)

1. 5 mM stock of **FD10** is prepared by dissolving 5 mg of **FD10** in 100 μL phosphate buffer, 10 mM, pH 7.
2. In an eppendorf tube, 15 μL of $\mathbf{VU_5}$ and $\mathbf{VL_5}$ (3.33 μM) and 20 μL of 5 mM **FD10** are mixed. This will result in a final concentration of **FD10** of 2 mM. This concentration of **FD10** was selected because at 2 mM we have one **FD10** molecule per 1,000 nm^3. This is the internal volume of a single DNA icosahedron.
3. The tube is heated in a heating block at 45°C for 4 h and the temperature is decreased at the rate of 1°C/3 min till 20°C where it is incubated for 2 h and equilibrated for 72 h at 4°C.
4. Finally, the solution is ligated using NCI (see Subheading 3.1.4).

3.6 Purification

The DNA icosahedron loaded with **FD10** ($\mathbf{I_{FD10}}$) is separated from free **FD10** using a two-step purification—(a) gel electrophoresis followed by (b) Size exclusion chromatography (SEC-HPLC).

3.6.1 Gel Electrophoresis

1. The ligated mixture of $\mathbf{I_{FD10}}$ (1 μM, DNA) is loaded on 0.8% agarose gel (Fig. 3a left).
2. When resolved on gel, free **FD10**, being an unstructured polymer, migrates as a smear along the lane.
3. The band corresponding to $\mathbf{I_{FD10}}$ (Fig. 3a) is excised and eluted in 100 mM NaCl, 1 mM $MgCl_2$ solution for 24 h at room temperature (Fig. 3a, left).
4. This sample is vacuum concentrated and then subjected to a second round of purification by size exclusion chromatography (SEC).

3.6.2 Size Exclusion Chromatography

1. The gel purified and vacuum concentrated $\mathbf{I_{FD10}}$ sample is subjected to SEC.
2. 50 μL solution of $\mathbf{I_{FD10}}$ (1 μM, DNA) is injected onto the HPLC column which is pre-rinsed with degassed Acetonitrile/water mixture.

3. The elution is carried out at a flow rate of 0.5 mL/min and absorbance at 488 nm (FITC absorbance) and 254 nm (FITC + DNA absorbance) are monitored.
4. The plain icosahedron sample without any **FD10** (i.e., **I**) is used as reference for $\mathbf{I}_{\mathbf{FD10}}$. The empty icosahedron elutes close to 8 min in the chromatogram (Fig. 3a middle inset).
5. Free **FD10**, being a polydisperse, unstructured polymer, elutes as a broad peak from 5 to 8 min (Fig. 3a, right).
6. $\mathbf{I}_{\mathbf{FD10}}$ elutes at 8 min as a sharp peak showing absorbance both at 488 nm as well as 254 nm indicating the formation of $\mathbf{I}_{\mathbf{FD10}}$ complex.
7. The ratio of 254 and 488 nm can also be used to calculate the number of **FD10** molecules encapsulated per DNA icosahedron on average (Fig. 3a, middle).
8. The eluted fractions are collected, vacuum-concentrated, and adjusted to buffer conditions (see Subheading 3.1.2).

3.7 Dynamic Light Scattering

1. DLS has been used extensively to study particle sizes and also to study the interactions between various biomolecules. DLS can be used to study the sizes of DNA polyhedra and also it can be used as a tool to predict the association of various cargo molecules with DNA icosahedra.
2. Filter water and buffer (10 mM phosphate buffer with 1 mM $MgCl_2$ and 100 mM NaCl) through 0.02 μm Whatman syringe filter paper. The DNA samples and FD10 should be filtered through 0.22 μm Millipore filter.
3. 100 μL of 1 μM sample of DNA icosahedron, $\mathbf{I}_{\mathbf{FD10}}$ complex and 1 mM sample of **FD10** in buffer above are used for investigation.
4. All samples including water and buffers are spun at 9300 rcf for 10 min using a centrifuge at room temperature.
5. The DLS cuvettes are washed rigorously with water and then with buffer.
6. 50 μL of buffer in a cuvette is taken and the counts in DLS are measured using 100% beam intensity. The average counts should be less than 5 and stay constant for at least 2 min. This indicates that the solution and cuvette are clean and free of dust particles.
7. 50 μL of I is taken and the laser intensity is fixed at 70%, the S/N ratio at 2.5, and the acquisition time at 3 s. The sample readings are initiated by monitoring autocorrelation functions and 40 continuous readings are taken.
8. Individual readings are checked by the shape of autocorrelation function and only those readings that show sharp, straight ends of the autocorrelation function are chosen for further analysis.

9. Then, the distribution is selected. In the distribution, all readings below 1 nm are discarded since these are due to solvent scattering.
10. All other readings are used to plot the distribution of intensity as a function of hydrodynamic radius (R_H) (Fig. 3b).

3.8 Quencher Characterization

1. Each quencher has an intrinsically different ability to collisionally quench fluorescence.
2. This is corrected for by using that concentration of the quencher which results in a 50% decrease in fluorescence intensity of the sample. This is obtained from the reciprocal of their measured Stern–Volmer constants (9).
3. Quenchers of different sizes are selected based on literature reports. These include Iodide (0.5 nm), Amino TEMPO (1 nm), and Nanogold (1.5 nm). Quenchers in the regime 2–5 nm are all gold nanoparticles and are synthesized as described in Subheading 2.4.
4. 400 μL of 50 nM solution of free **FD10** is taken in a quartz cuvette and its emission is scanned from 500 to 600 nm when it is excited at 488 nm in a Fluorolog 3 L instrument. The fluorescence intensity at 515 nm is taken as F_0.
5. Then the quenchers are added in small aliquots. After each addition of quencher, the solution is equilibrated for 2 min.
6. Then, the fluorescence value, F, at 515 nm is recorded. The value of F obtained is further corrected for dilution by multiplying with the dilution factor.
7. After approximately ten such readings, the plot of F_0/F is plotted against the quencher concentration. This yields a straight line whose y-intercept is 1.
8. The slope of the line gives the Stern–Volmer Constant, K_{sv}.
9. Thus, the concentration that is used in further experiments, to observe 50% quenching will be $(1/K_{sv})$ M of the relevant quencher.

3.9 Intensity Based Quenching

1. The $\mathbf{I_{FD10}}$ complex could have the **FD10** externally associated, or internally entrapped within the DNA Icosahedron.
2. In order to confirm the encapsulation of **FD10** within the DNA icosahedral cavity of **I**, free **FD10** and $\mathbf{I_{FD10}}$ are subjected to external quenchers of various sizes ranging from 0.5 to 5 nm and their extents of quenching is studied (Fig. 3c).
3. 400 μL of 50 nM solution of free **FD10** is taken in a quartz cuvette and its emission is recorded from 500 to 600 nm when it is excited at 488 nm in a Fluorolog 3 L instrument. The fluorescence intensity at 515 nm is taken as F_0. This value is chosen for normalization for all other readings and is taken as 100%.

4. $1/K_{sv}$ amount of each quencher is added, equilibrated for 2 min and the emission at 515 nm is recorded. This value should be half of the original F_0 value (Fig. 3c).
5. The cuvette is cleaned after each reading rigorously with water and ethanol to remove trace amounts of quenchers present on the walls of the cuvette.
6. Then, 400 μL of 50 nM, pH 7, solution of $\mathbf{I}_{\mathbf{FD10}}$ is taken in a quartz cuvette and its emission is recorded from 500 to 600 nm upon excitation at 488 nm. The fluorescence intensity at 515 nm is taken as F_0. This value is chosen for normalization.
7. $1/K_{sv}$ amount of each quencher is added to this cuvette, equilibrated for 2 min and the emission at 515 nm is recorded.
8. In case of $\mathbf{I}_{\mathbf{FD10}}$, only quenchers smaller than 2 nm, which can diffuse through the pores of the DNA icosahedron, access the fluorophore and quench it to 50%. Quenchers larger than 3 nm cannot access the internal fluorophore and fluorescence from **FD10** is resistant to quenching.
9. From the plot of percentage fluorescence against size of the quencher, the pore size of the icosahedron is evident.

3.10 Lifetime Based Quenching

1. Intensity quenching can be supported by similar studies of fluorescence lifetimes.
2. 400 μL of 5 μM **FD10**, pH 7, is taken in a quartz cuvette. For lifetime measurements, we need two cuvettes, one containing a standard sample and other containing the sample of interest (see Note 4).
3. Glycogen is used as the standard for instrument response factor (IRF) measurements.
4. For the standard sample the settings are maintained as, excitation and emission at 488 nm, while for the desired sample excitation and emission were 488 nm and 515 nm, respectively. The instrument measures values for both standard and sample at all operating frequencies.
5. A frequency range from 10 to 150 MHz is selected and ten intermediate values are recorded with each value repeated in quadruplicate.
6. The instrument measures the lifetime and modulation at each frequency and gives the raw data.
7. The associated data analysis program allows one to change component lifetimes.
8. FITC exhibits a two component lifetime and the average life time is calculated from two lifetimes using formula:

$$<\tau_{avg}> = (\tau_1 f_1^2 + \tau_2 f_2^2)/(\tau_1 f_1 + \tau_2 f_2)$$

where τ_1 and τ_2 are the lifetimes of two components and f_1 and f_2 are the respective fractions of the component (10).

9. For each average lifetime, the instrument gives a χ^2 value which is called the goodness of fit parameter. The average lifetime for which the χ^2 value is less than 1.2 is selected. Two such readings are taken for each sample and the lifetime is presented as mean lifetime with associated standard deviation.
10. The lifetimes of **FD10** and $\mathbf{I}_{\mathbf{FD10}}$ are first measured without any quencher to investigate the effect of the DNA polyhedron on the lifetime of the encapsulated **FD10**.
11. Free **FD10** exhibits an average lifetime of 3.8 ns at pH 7. To measure the lifetime quenching, the quencher is added in small amounts and the life time is measured till it decreases to 2 ns (this will correspond to the $1/K_{sv}$ amount of quencher).
12. This amount of quencher is added to the samples of $\mathbf{I}_{\mathbf{FD10}}$ and the lifetimes are determined.
13. The results from lifetime measurements give the same results as those obtained from intensity based quenching and both should be consistent with the pore size of the DNA capsule (Fig. 3d).

4 Notes

1. Sometimes, post-addition to the sample to be ligated, NCI forms a white precipitate. The precipitate formed is nickel phosphate which solubilizes by the addition of dilute acid. So in case of precipitate, 5 μL of acetic acid is added along with 10 μL water, the eppendorf is vortexed and then the sample can be used further.
2. In samples containing GNPs, Mg^{2+} is avoided since it causes aggregation of GNPs. Hence dialysis is performed against only phosphate buffer and NaCl.
3. GNPs below 3 nm diameter cannot be encapsulated inside DNA icosahedron as the effective pore size of the DNA icosahedron is in the range of 2.5–3 nm.
4. Even though the **FD10** is encapsulated within DNA icosahedra in phosphate buffer of pH 6, for all the fluorescence studies the pH should be adjusted to 7 since FITC is a pH sensitive fluorophore and its fluorescence and lifetime are maximal at pH 7.

Acknowledgments

We thank Dr. S.S. Indi and Dr. Atanu Basu at Department of Microbiology and Cell Biology, IISc and NIV, Pune, respectively, for providing electron microscopy facilities, Prof. Dipanker Chatterji, MBU, IISc for use of the lifetime instrument. D.B.,

S.M., and S.C. thank CSIR, Government of India (GoI) for research fellowships. This work was funded by the Nano Science and Technology Initiative, DST, GoI, and the Innovative Young Biotechnologist Award, DBT (GoI) to Y.K.

References

1. Holliday BJ, Mirkin CA (2001) Strategies for the construction of supramolecular compounds through coordination chemistry. Angew Chem Int Ed Engl 40:2022–2043
2. Goodman RP, Schaap IA, Tardin CF, Erben CM, Berry RM, Schmidt CF, Turberfield AJ (2005) Rapid chiral assembly of rigid DNA building blocks for molecular nanofabrication. Science 310:1661–1665
3. He Y, Ye T, Su M, Zhang C, Ribbe AE, Jiang W, Mao C (2008) Hierarchical self-assembly of DNA into symmetric supramolecular polyhedra. Nature 452: 198–202
4. Douglas SM, Dietz H, Liedl T, Högberg B, Graf F, Shih WM (2009) Self-assembly of DNA into nanoscale three-dimensional shapes. Nature 459:414–418
5. Bhatia D, Mehtab S, Krishnan R, Indi SS, Basu A, Krishnan Y (2009) Icosahedral DNA nanocapsules by modular assembly. Angew Chem Int Ed Engl 48:4134–4137
6. Bhatia D, Surana S, Chakraborty S, Koushika SP, Krishnan Y (2011) A synthetic, icosahedral DNA-based host-cargo complex for functional in vivo imaging. Nat Commun 2:339. doi:10.1038/ncomms1337
7. Ghodke HB, Krishnan R, Vignesh K, Kumar GV, Narayana C, Krishnan Y (2007) The I-tetraplex building block: rational design and controlled fabrication of robust 1D DNA scaffolds via non-Watson Crick self-assembly. Angew Chem Int Ed Engl 46:2646–2649
8. Mirkin CA (2000) Programming the assembly of two- and three-dimensional architectures with DNA and nanoscale inorganic building blocks. Inorg Chem 39:2258–2272
9. Lakowicz JR, Weber G (1973) Quenching of fluorescence by oxygen: probe for structural fluctuations in macromolecules. Biochemistry 12:4161–4170
10. Lakowicz JR (2006) Principles of fluorescence spectroscopy, 3rd edn. Springer, New York, pp 353–380

Chapter 9

Delivery of Plasmid DNA to Mammalian Cells Using Polymer–Gold Nanorod Assemblies

James Ramos, Huang-Chiao Huang, and Kaushal Rege

Abstract

Functionalized and surface-modified gold nanorods (GNRs) have emerged as promising vehicles for the delivery of several therapeutic agents. Ease of functionalization, increased stability, biocompatibility, and size-dependent plasmonic properties make gold nanorods attractive in sensing, imaging, and delivery to different cellular types. Here, we demonstrate the use of polyelectrolyte-coated GNRs (PE-GNRs) for delivering plasmid DNA to mammalian cells for transgene expression.

Key words Gold nanoparticle, Nonviral gene delivery, Cationic polymers, Polyelectrolytes, Stability

1 Introduction

Many novel nanomaterials, including gold nanoparticles, are showing strong potential as nanocarriers for delivery of therapeutic agents (1–5). In particular, gold nanorods (GNRs) are being increasingly investigated for biological sensing (6, 7), imaging (8, 9), photothermal therapy (10, 11), and drug (12, 13) and gene delivery (14–16), due to their unique optical properties and facile methods of surface modification and functionalization (17). Functionalized gold nanorods provide a high surface area ratio which allows for high payload (e.g., plasmid DNA)-to-carrier ratios. Furthermore, hydrophobicity and charge can be tuned through polymer monolayer coverage of the gold nanorods, which can result in improved cellular uptake as well as decreased cytotoxicity (5). Specifically, functionalization of gold nanorods with polymers further increases their potential for use in noninvasive therapeutic applications as it results in increased stability, lower cytotoxicity, controllable surface properties, and the possibility for further surface functionalization (e.g., with targeting

Volkmar Weissig et al. (eds.), *Cellular and Subcellular Nanotechnology: Methods and Protocols*, Methods in Molecular Biology, vol. 991, DOI 10.1007/978-1-62703-336-7_9, © Springer Science+Business Media New York 2013

biomolecules) (18). Thus, several strategies have been pursued in which gold nanoparticles have been functionalized with polymers for use as a delivery vehicle (19–21).

Here, we demonstrate the generation of polyelectrolyte-coated GNRs (PE-GNRs) based on candidates from a library of cationic polymers recently synthesized in our laboratory (22–25). These cationic polyelectrolytes engender high colloidal stabilities of gold nanorods and also facilitate high loadings of plasmid DNA on the nanorods by means of electrostatic interactions. In vitro studies with human prostate cancer cell lines demonstrated that subtoxic concentrations of PE-GNRs, loaded with exogenous plasmid DNA, are capable of mediating transgene delivery and expression.

2 Materials

Prepare all solutions using nanopure water (18.2 MΩ-cm, resistivity) unless otherwise specified. Prepare and store all reagents at room temperature unless otherwise specified. Diligently follow all waste disposal regulations when disposing waste materials.

2.1 Polymer Synthesis

1. 0.01× PBS: Prepare 500 mL of 10× phosphate-buffered saline (PBS) by mixing 40.9 g NaCl, 7.098 g Na_2HPO_4, and 1.006 g KCl in 500 mL of water. Dilute desired amount to 0.01× PBS and adjust the pH to 7.4 using 6N HCl and 3 M NaOH.
2. 20 mL glass scintillation vials.
3. 1,4-cyclohexanedimethanol Diglycidyl Ether.
4. 1,4-bis(3-aminopropyl)piperazine.

2.2 Gold Nanorod Synthesis

1. Gold nanorod (GNR) synthesis technique was adapted from the seed-growth method described by El-Sayed et al. (26).
2. Gold(III) chloride trihydrate ($HAuCl_4 \cdot 3H_2O$).
3. Cetyltrimethyl ammonium bromide (CTAB).
4. L-ascorbic acid.
5. Sodium borohydride.
6. Silver nitrate.

2.3 Generation of Polyelectrolyte-Gold Nanorod Assemblies

1. Poly(styrene sulfonic acid) (PSS): Dissolve PSS in 0.01× PBS to a concentration of 10 mg/mL (see Note 1).
2. Serum-free media: RPMI-1640 media supplemented with 1% penicillin–streptomycin antibiotic.
3. Water bath sonicator.

2.4 Determination of Polyelectrolyte–Gold Nanorod Assembly Cytotoxicity

For the purpose of this discussion, we describe the determined cytotoxicity of the PE-GNRs towards PC3 and PC3-PSMA prostate cancer cells.

1. PC3 cells (ATCC, Manassas, VA).
2. PC3-PSMA cells (a generous gift from Dr. Michel Sadelain of the Memorial Sloan Cancer Center (27)).
3. Tissue culture treated 24 well plates (Costar).
4. MTT Cell Proliferation Assay Kit (ATCC, Manassas, VA): Includes MTT reagent and detergent reagent.
5. Serum-free media as described above.
6. Aluminum foil.

2.5 PE-GNR-Mediated Plasmid DNA Delivery and Transgene Expression

For the purpose of this discussion, we describe the transfection and determination of transgene expression in PC3 and PC3-PSMA cells using the firefly luciferase encoding pGL3 plasmid DNA and appropriate assays.

1. PC3 cells.
2. PC3-PSMA cells.
3. 96 well white flat bottom polystyrene assay plates.
4. Clear 96 well plates.
5. One 1.5 mL microcentrifuge tube for each treated well of the 24 well plate.
6. 1× PBS: Dilute previously described 10× PBS to 1× PBS by adding nanopure water.
7. 1× lysis solution: Dilute 5× Cell Culture Lysis Reagent (Promega, Madison, WI) to 1× Cell Culture Lysis Reagent with water.
8. Luciferase Assay System (Promega, Madison, WI): Contains luciferase assay substrate and luciferase assay buffer. One of each will be needed for a single luciferase assay.
9. Pierce® BCA Protein Assay Kit (Thermo-Fisher Scientific, Rockford, IL): Contains Pierce® BCA Protein Assay Reagent A, Pierce® BCA Protein Assay Reagent B, and Albumin standards.
10. Serum-free media as described above.
11. pGL3 plasmid DNA control vector (Promega, Madison, WI).

3 Methods

Carry out all procedures at room temperature unless otherwise specified.

3.1 Polymer Synthesis

Cationic polymers can be synthesized using the following described methods of ring opening of diglycidyl ethers by amines (22, 25). For the purpose of this discussion, we will use the cationic polymer synthesized via the polymerization of 1,4-cyclohexanedimethanol Diglycidyl Ether (1,4C) and 1,4-bis(3-aminopropyl)piperazine (1,4Bis) as an example.

1. In 20 mL glass scintillation vials, mix equimolar amounts of 1,4C (238.2 μL) and 1,4Bis (269.5 μL) and mix well.
2. Set scintillation vials aside for 12–16 h (see Note 2).
3. Weigh out and dissolve polymerized mixture in 0.01× PBS at a concentration of 10 mg/mL (see Note 3). Adjust the pH to 7.4 and place solutions on a rotator overnight (see Note 4).
4. Check and adjust pH of polymer solutions next day to a value of 7.4 (see Note 5).

3.2 Gold Nanorod Synthesis

1. Prepare a gold "seed" solution by adding 5 mL of 0.5 mM $HAuCl_4 \cdot 3H_2O$ to 5 mL of 200 mM CTAB in a 15 mL polypropylene conical tube with gentle mixing by inversion for 2 min.
2. Prepare 0.6 mL of 0.01 M sodium borohydride and allow to chill to 4°C, following which, immediately add this solution to the gold seed dispersion (see Note 6).
3. Prepare a growth solution by mixing 5 mL of 1 mM $HAuCl_4 \cdot 3H_2O$ with 5 mL of 200 mM CTAB solution containing 280 μL of 0.004 M silver nitrate in a 15 mL polypropylene conical tubes.
4. Add 70 μL of 0.0788 M L-ascorbic acid to the growth solution (see Note 7).
5. Add 12 μL seed solution to the growth solution and allow to mix for 4 h under continuous stirring at 28°C, 150 rpm (see Note 8).
6. Pellet the as-prepared GNRs, by centrifugation at 6,000 rcf for 10 min.
7. After sedimentation, remove the clear supernatant and re-disperse GNR pellet to the initial volume with nanopure water.

3.3 Synthesis of Polyelectrolyte–Gold Nanorod Assemblies

1. Adjust GNR optical density to 0.5 absorbance units (a.u.) (Fig. 1) in nanopure water and centrifuge 0.5 mL of GNR dispersion in 1.5 mL microcentrifuge tubes at 6,000 rcf for 10 min (Microfuge 18 centrifuge, Beckman Coulter). Remove excess CTAB surfactant.
2. Re-disperse GNR pellet in 100 μL of a poly(styrene sulfonate) (PSS) solution (10 mg/mL in 0.01× PBS).

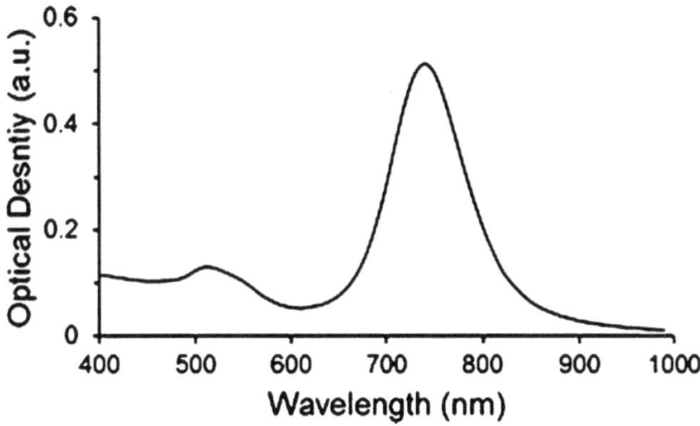

Fig. 1 Absorption spectrum of gold nanorods possessing a transverse band at ~520 nm and a longitudinal band at ~750 nm

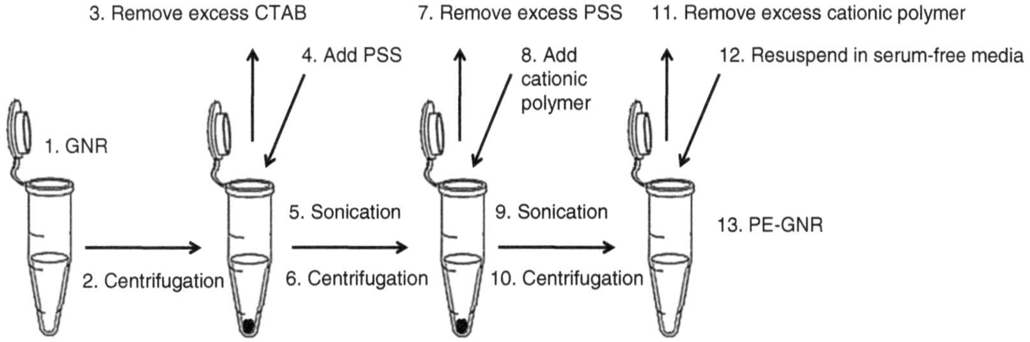

Fig. 2 Schematic of PE-GNR generation process

3. Immediately sonicate in a water bath sonicator for 30 min at room temperature to allow for the formation of PSS-coated GNRs (PSS-CTAB-GNRs).
4. Centrifuge PSS-CTAB-GNRs at 6,000 rcf for 10 min. Remove excess PSS by removing the supernatant.
5. Re-disperse PSS-CTAB-GNRs pellet in 300 μL of nanopure water.
6. Add 200 μL of cationic polymer to dispersion (see Note 9).
7. Immediately sonicate for 30 min to allow for the formation of the cationic polymer (polyelectrolyte)-coated PSS-CTAB-GNRs or PE-GNRs.
8. Centrifuge PE-GNRs at 6,000 rcf for 10 min. Remove excess cationic polymer by removing the supernatant (Fig. 2).
9. Disperse PE GNR in serum-free media and adjust the optical density to 0.25 a.u. (see Note 10).

3.4 Determination of Cytotoxicity of Polyelectrolyte–Gold Nanorod Assemblies

1. Plate 50,000 cells per well in a tissue culture treated 24 well plate and incubate overnight at 37°C and 5% CO_2 to allow cells to attach.
2. After overnight incubation, remove cell growth media and replace with 500 μL of PE-GNR dispersions in serum-free media set to different optical densities (see Note 11).
3. Allow to incubate for 6 h. Use untreated cells (i.e., those not treated with PE-GNRs) as a live control and prepare a dead control (i.e., treated with 5 μL of 30% hydrogen peroxide).
4. After 6 h, remove serum-free media from wells and replace with fresh growth media.
5. Add 20 μL of MTT reagent to each well, wrap each plate in aluminum foil, and incubate at 37°C for 4 h (see Note 12).
6. Following 4 h, add 150 μL detergent reagent to each well. Wrap plates in aluminum foil and incubate for 4 h at room temperature.
7. Following 4 h, mix solution in wells and transfer 150 μL to a clear 96 well plate (see Note 13). Read absorbance and 570 nm.
8. Determine cell viability by reporting values as a percentage of live and dead controls.
9. Determine a cutoff value for subtoxic treatment conditions of PE-GNRs (see Note 14).

3.5 Determination of PE-GNR-Mediated Transgene Expression

1. Plate 50,000 cells per well in a tissue culture treated 24 well plate and incubate overnight at appropriate conditions to allow cells to attach.
2. Synthesize PE-GNRs (see Subheading 3.3, step 9) and aliquot previously determined subtoxic doses of PE-GNRs into a well of a 96 well plate. Add desired amounts of plasmid DNA (see Note 15) and allow to co-incubate for 30 min (Fig. 3).
3. Remove cell growth media and replace with 500 μL minus the PE-GNR treatment volume of serum-free media.
4. Add PE-GNR/DNA assemblies to wells and allow to incubate for 6 h.
5. Following 6 h, remove serum-free media and replace with fresh growth media. Incubate at 37°C and 5% CO_2 conditions for 48 h.
6. Following 48 h, remove growth media and place in a microcentrifuge tube (see Note 16).
7. Wash each well with 150 μL of 1× PBS. Remove PBS and place into same microcentrifuge that growth media from that well was placed in.

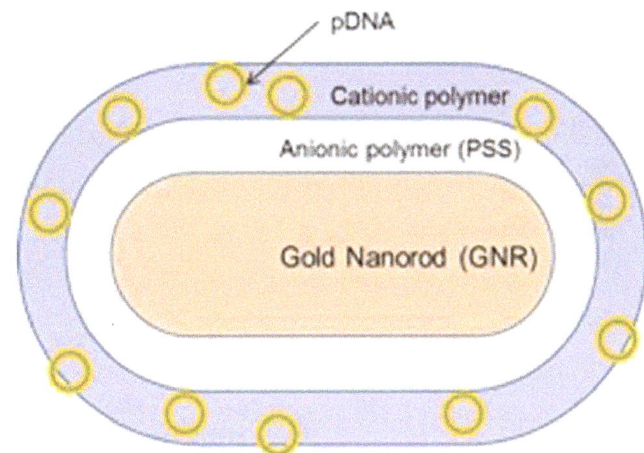

Fig. 3 Schematic of PE-GNR loaded with plasmid DNA

8. Add 150 μL of 1× Cell Culture Lysis Reagent. Incubate for approximately 3 min at room temperature and check cells under a microscope to ensure complete lysis.
9. Add growth media/PBS mixture in the microcentrifuge tubes back to the corresponding wells in the 24 well plate.
10. Aliquot 50 μL of the cell lysate into four wells of the white plate for luciferase assay. This will allow for four readings of the cell lysate.
11. Aliquot 100 μL of the cell lysate back into the corresponding microcentrifuge tube. Add 900 μL of water to each tube for use in BCA assay.
12. Luciferase Assay:
    (a) Add one full bottle (10 mL) Luciferase Assay Buffer to Luciferase Assay Substrate (see Note 17) and immediately add 100 μL to each well in the 96 well plate (see Note 18).
    (b) Immediately read the luminescence in plate reader and record relative light (or luminescence) units (RLU).
13. BCA Assay:
    (a) Aliquot 10 μL from albumin standards and the microcentrifuge tubes to a well in a clear 96 well plate.
    (b) Mix Pierce® BCA Protein Assay Reagent A and Pierce® BCA Protein Assay Reagent B at a 50:1 ratio to make working reagent (see Note 19).
    (c) Add 190 μL of working reagent to each well and incubate at 37°C for 30 min (see Note 20).
    (d) Following 30 min incubation, read absorbance at 562 nm. Using standards, back-calculate the concentration of each sample in mg/mL.

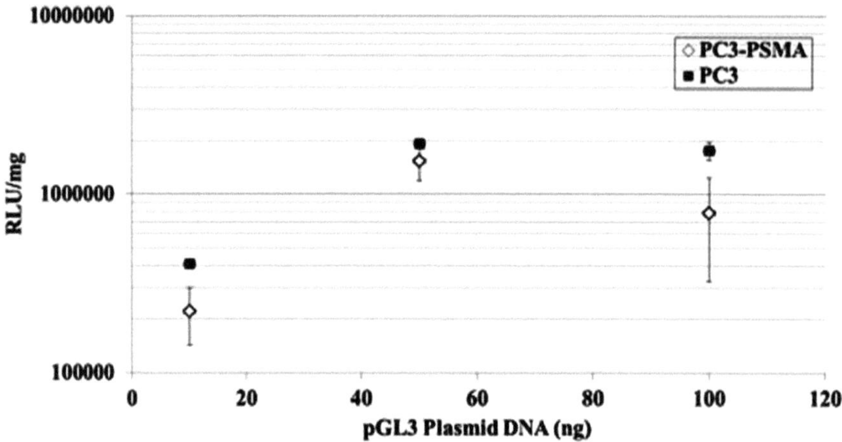

Fig. 4 Transgene (luciferase) expression in PC3 and PC3-PSMA cells with PE-GNR loaded with different amounts of pGL3 plasmid DNA (ng). Luciferase expression, in relative luminescence units (RLU), was analyzed 48 h post transfection and normalized to total protein content (RLU/mg) (28)

14. Using calculated mg/mL for each sample, determine amount of protein in each well used for luciferase assay.
15. Normalize the measured relative light units measured by dividing by the calculated amount of protein in each well and report values as RLU/mg.
16. We show an example of the transgene expression (Fig. 4). Higher DNA amounts lead to higher transgene expression up to a point at which a maximum is reached. Following that, there is a slight decrease as more DNA is loaded. This is likely due to shielding of the polymer's positive charge by the negatively charged plasmid DNA.

4 Notes

1. We find that preparing this fresh works best.
2. This mixture will be used following polymerization to make polymer solutions. Depending on the monomers used, polymerization may take varying amounts of times.
3. Polymers should be viscous; if they have "hardened" (cured) or remain "watery," then it will be necessary to repeat the process with a shorter or longer incubation time respectively.
4. Polymer solutions will be basic. When dissolving polymers in solution, they may appear milky or may not completely dissolve initially. Adjusting the pH may allow them to dissolve

further. Lowering polymer concentration may also help increase solubility. Discard any polymers that do not dissolve.

5. Polymer solutions may oftentime become slightly more basic following the initial pH adjustment.
6. Seed solution should become pale brown.
7. After introduction of the L-ascorbic acid, growth solution should become colorless.
8. During incubation, solution color should change from colorless to purple in between 20 min.
9. Check to make sure the dispersion is not milky pink in color. This is typically indicative of undesirable aggregation due to very high local concentrations of the cationic polymer, and the process must be restarted. To avoid this, addition of more than 300 μL of water in the previous step can prevent this or removal of more of the excess PSS.
10. It is best to disperse the PE-GNRs in half the volume of the initial GNR sample (250 μL in this case) and slowly add serum-free media until the desired optical density is reached.
11. Recommended PE-GNR dispersions in serum-free media optical densities are 0.0025, 0.005, 0.0075, 0.0125, 0.025, 0.0375, 0.05, and 0.125.
12. After 4 h incubation, metabolically active cells should become a purple color following addition of the MTT reagent.
13. It is recommended that this step is done in triplicate for each well.
14. Concentrations that result in >70% cell viability may be treated as subtoxic.
15. Recommended amounts of pGL3 plasmid DNA range from 10 to 200 ng.
16. A microcentrifuge tube should be "assigned" for each treated well. Do not dispose of any solution.
17. Luciferase assay reagents are stored at −20°C and need adequate time for thawing. It is recommended that they are taken out of storage at Subheading 3.5, step 6 and kept at room temperature to thaw.
18. Once mixed, the luciferase substrate decays rapidly. It is recommended that these steps are carried out as fast as possible (within 3–5 min if possible).
19. Working reagent should be green in color.
20. After 30 min incubation, the solutions should turn from green to a purple color.

Acknowledgments

This work was supported by the National Science Foundation (Grant CBET-0829128) and National Institutes of Health (Grant 5R21CA133618-02).

References

1. Jabr-Milane L et al (2008) Multi-functional nanocarriers for targeted delivery of drugs and genes. J Control Release 130:121–128
2. Heath J, Davis M (2008) Nanotechnology and cancer. Annu Rev Med 59:251–265
3. Cho K et al (2008) Therapeutic nanoparticles for drug delivery in cancer. Clin Cancer Res 14:1310–1316
4. Smith A et al (2008) Bioconjugated quantum dots for in vivo molecular and cellular imaging. Adv Drug Deliv Rev 60:1226–1240
5. Ghosh P et al (2008) Gold nanoparticles in delivery applications. Adv Drug Deliv Rev 60:1307–1315
6. Guo L, Zhou X, Kim D (2011) Facile fabrication of distance-tunable Au-nanorod chips for single nanoparticle plasmonic biosensors. Biosens Bioelectron 26:2246–2251
7. Castellana E, Gamez R, Russell D (2011) Label-free biosensing with lipid functionalized gold nanorods. J Am Chem Soc 133: 4182–4185
8. Pan D et al (2010) A facile synthesis of novel self-assembled gold nanorods designed for near-infrared imaging. J Nanosci Nanotechnol 10:8118–8123
9. Ha S et al (2011) Detection and monitoring of the multiple inflammatory responses by photoacoustic molecular imaging using selectively targeted gold nanorods. Biomed Opt Express 2:645–657
10. Huang XH et al (2006) Cancer cell imaging and photothermal therapy in the near-infrared region by using gold nanorods. J Am Chem Soc 128:2115–2120
11. Choi WI et al (2011) Tumor regression in vivo by photothermal therapy based on gold-nanorod-loaded, functional nanocarriers. ACS Nano 5:1995–2003
12. Alkilany AM et al (2012) Gold nanorods: Their potential for photothermal therapeutics and drug delivery tempered by the complexity of their biological interactions. Adv Drug Deliv Rev 64(2):190–199. doi:10.1016/j.addr.2011.03.005
13. Min Y et al (2010) Gold nanorods for platinum based prodrug delivery. Chem Commun (Camb) 46:8424–8426
14. Braun GB et al (2009) Laser-activated gene silencing via gold nanoshell-siRNA conjugates. ACS Nano 3:2007–2015
15. Chen C et al (2006) DNA-gold nanorod conjugates for remote control of localized gene expression by near infrared irradiation. J Am Chem Soc 128:3709–3715
16. Salem AK, Searson PC, Leong KW (2003) Multifunctional nanorods for gene delivery. Nat Mater 2:668–671
17. Huang X, Neretina S, El-Sayed M (2009) Gold nanorods: From synthesis and properties to biological and biomedical applications. Adv mater 27:1–31
18. Wei Q, Ji J, Shen J (2008) Synthesis of near-infrared responsive gold nanorod/PNIPAAm core/shell nanohybrids via surface initiated ATRP for smart drug delivery. Macromol Rapid Commun 29:645–650
19. Bonoiu AC et al (2009) Nanotechnology approach for drug addiction therapy: Gene silencing using delivery of gold nanorod-siRNA nanoplex in dopaminergic neurons. Proc Natl Acad Sci USA 106:5546–5550
20. Sandhu KK et al (2002) Gold nanoparticle-mediated transfection of mammalian cells. Bioconjug Chem 13:3–6
21. Thomas M, Klibanov AM (2003) Conjugation to gold nanoparticles enhances polyethylenimine's transfer of plasmid DNA into mammalian cells. Proc Natl Acad Sci USA 100: 9138–9143
22. Barua S et al (2009) Parallel synthesis and screening of polymers for nonviral gene delivery. Mol Pharm 6:86–97
23. Huang H-C et al (2009) Simultaneous enhancement of photothermal stability and gene delivery efficacy of gold nanorods using polyelectrolytes. ACS Nano 3: 2941–2952
24. Barua S, Rege K (2010) The influence of mediators of intracellular trafficking on transgene

expression efficacy of polymer-plasmid DNA complexes. Biomaterials 31:5894–8902
25. Kasman L et al (2009) Polymer-enhanced adenoviral transduction of CAR-negative bladder cancer cells. Mol Pharm 6:1612–1619
26. Nikoobakht B, El-Sayed MA (2003) Preparation and growth mechanism of gold nanorods (NRs) using seed-mediated growth method. Chem Mater 15:1957–1962
27. Gong MC et al (1999) Cancer patient T cells genetically targeted to prostate-specific membrane antigen specifically lyse prostate cancer cells and release cytokines in response to prostate-specific membrane antigen. Neoplasia 1:123–127
28. Ramos J, Rege K (2012) Transgene delivery using poly(aminoether)-gold nanorod assemblies. Biotechnol Bioeng 109:1336–1346.

Chapter 10

Lipophilic-Formulated Gold Porphyrin Nanoparticles for Chemotherapy

Puiyan Lee and Kenneth K.Y. Wong

Abstract

Lipophilic formulation is an invaluable technique for the delivery of cancer drugs. Incorporation of poorly soluble and toxic compounds into a lipophilic carrier vehicle improves both the stability and compatibility in blood and body fluids. Currently, although a large proportion of novel cancer drugs are poorly water soluble, most existing drug carriers are only able to encapsulate hydrophilic drugs. As the ultimate goal of drug delivery (in particular cancer drug delivery) is to achieve high therapeutic effect with minimal toxicity, it would thus be beneficial to invest substantial efforts in the development of lipophilic carrier systems. Here we describe our technique to synthesize a lipophilic carrier for hydrophobic and toxic potent cancer drugs, such as gold(III) porphyrin.

Key words Lipophilic formulation, Gold porphyrin, Hydrophobic drugs, Carrier, Nanoparticles

1 Introduction

Research in various methods of drug delivery has been a hot topic since early 1900s (1–3), as evident by the enormous number of published articles in the literature. The idea of using drug delivery system was first initiated with the aim of raising the drug concentration in blood (1, 4).

With the increase in the mean age of the population, diseases such as cancer, diabetes, and obesity have become the focus for pharmaceutical companies to meet the demand of market. Cancer is a common disease which kills 13% of the population worldwide every year, according to statistics from World Health Organization (WHO). Discovery of novel and effective cancer drugs represents significant advances in academic or pharmaceutical research (5–7). It is, however, a long way before a drug can be sold on the market. Indeed, although many novel and potent cancer drugs are undergoing the clinical trial, only a few become pharmaceutical products due to adverse side effect of the cancer drugs. Nonetheless, this has

Volkmar Weissig et al. (eds.), *Cellular and Subcellular Nanotechnology: Methods and Protocols*, Methods in Molecular Biology, vol. 991, DOI 10.1007/978-1-62703-336-7_10, © Springer Science+Business Media New York 2013

not hampered the enthusiasm in researching various models of cancer drug delivery systems. Considerable amounts of efforts and money have been put in designing novel drug delivery systems (8–11), with the ultimate aim to facilitate the effectiveness and reduce the side effects of cancer drugs.

1.1 Drug Delivery System and Drug Encapsulation

A drug delivery system is necessary to overcome the challenges in clinical and pharmaceutical areas, assuming that a greatly effective anticancer drug is available. The issues involving drug encapsulation release and targeting need to be tackled so that the effect of drug can be maximized in a site-specific manner. Drug encapsulation is the most important issue in improving drug delivery. Encapsulating a drug within a carrier stabilizes the drug and protects it against interaction with blood proteins, resulting in an increase in the circulation time. Toxic cancer drugs shielded within a carrier can further reduce the side effects to healthy tissues. A carrier vehicle also acts as a platform for anchoring tumor-targeting ligands to aid site-specific delivery. Since the therapeutic effect depends on the drug availability and concentration, degradation of the carrier has to be guaranteed for drug dissociation from the vehicle carrier at the tumor site.

To meet these criteria, an ideal drug carrier would be inert but biodegradable and biocompatible in the aqueous body environment. Among different carriers, which include polymers, lipid has been considered as a more physiological option and is expected to have higher biocompatibility. Lipid naturally becomes liposome vesicle during self-assembly in an aqueous environment. The use of liposome for drug delivery began in 1980 (12), with liposome-encapsulating doxorubicin, Doxil®, being the first FDA-approved liposomal chemotherapeutic agent in 1995 (13). Despite the success of Doxil®, approximately 40% of other potent and effective cancer drugs are hydrophobic (14). The encapsulation efficiency of hydrophobic drug within liposome is not satisfactory because, in most cases, the hydrophobicity of the drug during the self-assembly process renders the highly hydrophobic drug in the vesicle membrane, leaving it unencapsulated on the surface (15).

Here we describe the use of fatty acids to synthesize lipophilic formulation for the encapsulation of hydrophobic cancer drugs, such as the highly potent but hydrophobic nanometallic gold porphyrin. We were able to incorporate gold porphyrin into the lipophilic carrier, with >90% of the compound encapsulated in the lipophilic carrier (16). The size of the final particle was around 101.94 ± 27.9 nm, as measured by transmission electron microscopy (TEM) (Fig. 1). We used direct light scattering (DLS) method to measure the polydispersity index (PDI) of 0.32. This is important as nanoparticles <200 nm have been suggested to be the most effective in delivering drugs to tumor sites (17).

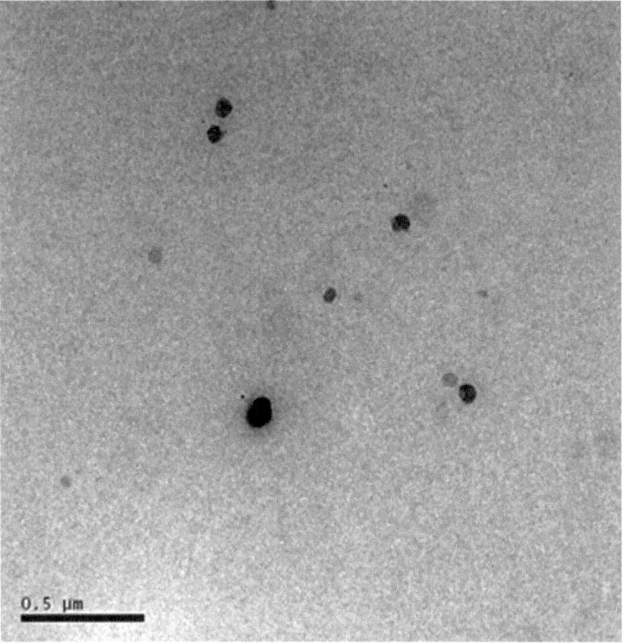

Fig. 1 Morphology and size of the gold porphyrin lipidic nanoparticle (GPNP) TEM shows the spherical morphology

2 Materials

Prepare the nanoparticle emulsion with clean glassware. Thoroughly clean the glassware and spatula with concentrated nitric acid followed by ethanol. Make sure no residue of white fatty acid and reddish gold porphyrin remain in the glassware. Furthermore, use double distilled water during the entire synthesis to guarantee the purified grade. Follow the organic disposal regulation for the disposal of emulsion residue if there is disposal residue with gold porphyrin.

1. Organic phase: fatty acid cetyl alcohol.
2. Polysorbate surfactant: Brij 78.
3. Double distilled (DDI) water.
4. 5 mL small beaker. Make sure the bottom is flat so the gold porphyrin (or any other drug to be used) can be evenly distributed in the organic phase when the nanoparticle template is formed.
5. Magnetic stir bar (<1 cm in length) so it can fit into the small 5 mL beaker. Rinse with ethanol before use.
6. Weigh balance (up to 0.001 mg in precision for accuracy).

7. Gold(III) porphyrin 1a was manufactured and kindly provided by the Department of Chemistry, HKU. The characterization identified with UV–vis (CH_2Cl_2) λ_{max}/nm (log): 409 (5.68), 521 (4.73). ^1H NMR ($CDCl_3$): δ 9.28 (s, 8H), 7.89 (m, 8H), 8.24 (d, 8H), and 7.89 (m, 4H). m/z=809 (see Note 1).

3 Methods

Perform all the procedure at 60°C unless specified.

1. Turn on the hot plate and adjust to 60°C. Prewarm the clean 5 mL beaker on the hot plate for 5 min (see Note 7).
2. Add the cetyl alcohol pellet into the prewarmed 5 mL beaker. Melt cetyl alcohol in the 5 mL beaker. When cetyl alcohol is melted, you will notice that the white cetyl alcohol pellet becomes clear solution.
3. Put 0.1 mg gold porphyrin into the melt organic phase cetyl alcohol (the maximum drug loading to organic phase does not exceed 1:1 ratio) (see Notes 1–3). Allow the gold porphyrin to dissolve in the organic phase for 5 min. No need to stir. When it is completely dissolved, you will notice mixture of the gold porphyrin fatty acid turning to clear reddish color.
4. Add the surfactant Brij 78 in reddish gold porphyrin-fatty acid platform. Stir the surfactant Brij 78 for 5 min with clean magnetic stir bar (see Notes 5–8). The mixture becomes clear red again when Brij 78 is melted into the wax platform.
5. Nanoemulsion—emulsifying the organic platform with aqueous solution to form nanosized particles: Use 1 mL pipette to add drop by drop 1 mL of DDI water into the fatty acid platform. This is to make sure the fatty acid platform will not cool down dramatically. Add a new drop promptly after the previous drop blends well into platform.
6. Keep stirring the platform gently in water for 20 min on hot plate (see Note 5).
7. Remove the beaker from the hot plate. Leave the beaker at room temperature for 10 min to cool down the nanoparticle solution (see Note 4).
8. Use 0.2 mm filter to sterilize the nanoparticles solution before in vivo injection (see Note 9).
9. Store the nanoparticle solution at 4°C for later use (see Note 10).
10. The morphology of the encapsulated nanoparticles can be observed using transmitting electron microscopy (TEM). The size of nanoparticles can be measured using dynamic light scattering (DLS).

4 Notes

1. Gold(III) porphyrin 1a was manufactured and kindly provided by the Department of Chemistry, HKU (18, 19).
2. Wear a mask when weighing and transferring the gold porphyrin (or any other toxic compounds) to avoid exposure by inhalation. Clean the balance and cover the weigh boat with plastic wrap to avoid gold porphyrin exposing to colleagues and labmates.
3. The addition of gold porphyrin in the beaker should be performed in the fume hood. The rest of synthesis is suggested to be in the fume hood to avoid the exposure of the gold porphyrin.
4. The surfactant has to be coated on the surface to be compatible with the aqueous environment.
5. Carefully keep track of the reaction time and temperature to control the size of the nanoparticles.
6. The stirring should be slow to avoid spillage from the beaker for safety and quality control.
7. The platform mixture sometimes accumulates at the edge at the bottom of the beaker and cannot be reached by the stir bar. This is due to the uneven thickness of bottom surface. Avoid loss of mixture with the help of spatula.
8. Make sure the bottom of the beaker is flat and even thickness. This also helps easier stirring with the stir bar and forming better emulsion.
9. Nanoparticles are best prepared fresh just before use.
10. The nanoparticle solution, if not in use, should be best stored at 4°C. Discard it if you notice increased turbidity.

References

1. Auer J, Gates FL (1916) The absorption of adrenalin after intratracheal injection. J Exp Med 23(6):757–772
2. Pearce L (1921) Studies on the treatment of human trypanosomiasis with tryparsamide (the sodium salt of N-phenylglycineamide-p-arsonic acid). J Exp Med 34(6 Suppl 1):1–104
3. Moore HF (1915) A further study of the bactericidal action of ethylhydrocuprein on Pneumococci. J Exp Med 22(5):551–567
4. Schemir FR, Gilson B, Hurst WW, Gilson JS, Camara AA, Layne JA (1955) Hypotonic intravenous solutions; their effect on the blood and their clinical usefulness. Rocky Mt Med J 52(9):791–800
5. Kupchan SM, Hemingway RJ, Coggon P, McPhail AT, Sim GA (1968) Crotepoxide a, novel cyclohexane diepoxide tumor inhibitor from Croton macrostachys. J Am Chem Soc 90(11):2982–2983
6. Kupchan SM, Hemingway RJ, Werner D, Karim A, McPhail AT, Sim GA (1968) Vernolepin, a novel elemanolide dilactone tumor inhibitor from Vernonia hymenolepis. J Am Chem Soc 90(13):3596–3597
7. Barry VC, Byrne J, Conalty ML, O'Sullivan JF (1967) Anticancer agents. II. Synthesis and antimycobacterial activity of some novel methylglyoxal derivatives. Proc R Ir Acad B 65(9):269–284

8. Cinti C, Taranta M, Naldi I, Grimaldi S (2011) Newly engineered magnetic erythrocytes for sustained and targeted delivery of anti-cancer therapeutic compounds. PLoS One 6(2): e17132
9. Greco F, de Palma L, Specchia N, Jacobelli S, Gaggini C (1992) Polymethylmethacrylate-antiblastic drug compounds: an in vitro study assessing the cytotoxic effect in cancer cell lines—a new method for local chemotherapy of bone metastasis. Orthopedics 15(2): 189–194
10. Yano T, Pinski J, Szepeshazi K, Milovanovic SR, Groot K, Schally AV (1992) Effect of microcapsules of luteinizing hormone-releasing hormone antagonist SB-75 and somatostatin analog RC-160 on endocrine status and tumor growth in the Dunning R-3327H rat prostate cancer model. Prostate 20(4):297–310
11. Miwa A, Ishibe A, Nakano M, Yamahira T, Itai S, Jinno S, Kawahara H (1998) Development of novel chitosan derivatives as micellar carriers of taxol. Pharm Res 15(12):1844–1850
12. Swenson CE, Popescu MC, Ginsberg RS (1988) Preparation and use of liposomes in the treatment of microbial infections. Crit Rev Microbiol 15(Suppl 1):S1–S31, Review
13. Petros RA, DeSimone JM (2010) Strategies in the design of nanoparticles for therapeutic applications. Nat Rev Drug Discov 9(8):615–627, Review
14. Tang R, Ji W, Panus D, Palumbo RN, Wang C (2010) Block copolymer micelles with acid-labile ortho ester side-chains: synthesis, characterization, and enhanced drug delivery to human glioma cells. J Control Release 151: 18–27
15. Li SD, Chono S, Huang L (2008) Efficient gene silencing in metastatic tumor by siRNA formulated in surface-modified nanoparticles. J Control Release 126(1):77–84
16. Lee PY, Zhang R, Li V, Liu XL, Sun RWY, Che CM, Wong KKY (2012) Enhancement of anti-cancer efficacy using modified lipophilic nanoparticle drug encapsulation. Int J Med 7:731–737
17. Lee PY, Zhu Y, Yan J et al (2010) The cytotoxic effects of lipidic formulated gold-porphyrin nanoparticles for the treatment of neuroblastoma. Nanotechnol Sci Appl 3:23–28
18. Che CM, Sun RW, Yu WY, Ko CB, Zhu N, Sun H (2003) Gold(III) porphyrins as a new class of anticancer drugs: cytotoxicity, DNA binding and induction of apoptosis in human cervix epitheloid cancer cells. Chem Commun (Camb) 14:1718–1719
19. To YF, Sun RW, Chen Y, Chan VS, Yu WY, Tam PK, Che CM, Lin CL (2009) Gold(III) porphyrin complex is more potent than cisplatin in inhibiting growth of nasopharyngeal carcinoma in vitro and in vivo. Int J Cancer 124(8):1971–1979

Chapter 11

Mitochondria-Specific Nano-Emulsified Therapy for Myocardial Protection Against Doxorubicin-Induced Cardiotoxicity

Amy Faulk, Volkmar Weissig, and Tamer Elbayoumi

Abstract

The quinonoid anthracycline, doxorubicin (Adriamycin), is a widely used potent antineoplastic agent, showing the broadest spectrum of antineoplastic activity against various types of solid carcinomas, hematological malignancies, and soft tissue sarcomas. Unfortunately, the clinical use of doxorubicin is associated with cumulative dose-limiting cardiac toxicity, manifested as cardiomyopathy and congestive heart failure, in which mitochondrial damage is primarily implicated. Free radical formation at and inside mitochondria, in particular the rise of reactive oxygen species (ROS), has long been hypothesized as the common mechanism by which doxorubicin causes this severe cardiotoxicity. Concomitant with newly gained insights into the central role of mitochondria in programmed cell death (apoptosis), irreversible destabilization of mitochondrial membrane permeability transition (mMPT), and disruption of mitochondrial Ca^{2+} homeostasis have been strongly implicated in triggering myocardial apoptosis, due to accumulated doxorubicin dosing.

Hence, our current protocols show the development of mitochondria-targeted nanoemulsions (NEs), based on previous work using nano-vesicle surface modification with mitochondriotropic triphenylphosphonium (TPP) ligands, which have successfully been demonstrated to target drug and DNA-loaded liposomes to mitochondria in living mammalian cells. Our mitochondria-specific TPP-coated therapeutic NEs are prepared using tocopherol oxygen scavengers and are highly loaded with mitochondria-stabilizing therapeutics, namely, cyclosporine A (CsA). Our targeted nano-formulation, proposed as injectable adjuvant therapy, is capable of reaching target affected mitochondria in sufficient therapeutic concentration, in order to revert or at least limit oxidative and non-oxidative doxorubicin induced mitochondrial damage, manifested in affected cardiac muscle tissues, Based on several encouraging studies using in vitro model rat cardiac muscle, H9C2 cardiomyocytes, and vascular media tunica media, A10, cell cultures, our proof-of principal mitochondriotropic nano-therapy demonstrates strong potential to improve not only the cardiac safety profile, through concurrent rescue administration of targeted nano-encapsulated FDA-approved cyclosporine A (CSA), but also dosing range of the currently available potent adriamycin/doxorubicin-based chemotherapy regimens.

Key words Anthracycline, Mitochondrial damage, Mitochondriotropic, Cyclosporine A, Cardiomyocytes, Chaotropic effect, Apoptosis

Volkmar Weissig et al. (eds.), *Cellular and Subcellular Nanotechnology: Methods and Protocols*, Methods in Molecular Biology, vol. 991, DOI 10.1007/978-1-62703-336-7_11, © Springer Science+Business Media New York 2013

1 Introduction

Doxorubicin (DOX) (Fig. 1) is an antitumor drug that is being widely used in the treatment of a wide variety of hematological malignancies and solid tumors (1). The clinical efficacy of this drug is often compromised due to the development of a cumulative dose-dependent cardiac toxicity, characterized by an irreversible dilated cardiomyopathy and congestive heart failure (2, 3). The risk of cardiotoxicity is proportional to the cumulative dose of doxorubicin received as follows (4): $<400\ mg/m^2 = 0.14\%$, $550\ mg/m^2 = 9\%$, $700\ mg/m^2 = 25\%$; for this reason, a total cumulative dose of $550\ mg/m^2$ is considered the therapeutic endpoint of doxorubicin therapy (5).

Although several mechanisms have been suggested to explain this cardiotoxicity, the exact mechanism and its metabolic consequences remain unclear. Biochemical and physiological data favor the hypothesis that DOX causes the formation of reactive free radicals (ROS) that stimulate lipid peroxidation and alter cellular membrane integrity (6).

The selective accumulation of DOX in mitochondria, coupled with increased ROS generation, renders cardiomyocytes more vulnerable to DOX toxicity.

Mitochondria are the most redox-active organelles and indispensable for cells to initiate or inhibit apoptosis, specifically, since more than 90% of the ATP utilized by cardiomyocytes is produced by mitochondrial respiration (7). Subsequently, any alterations in

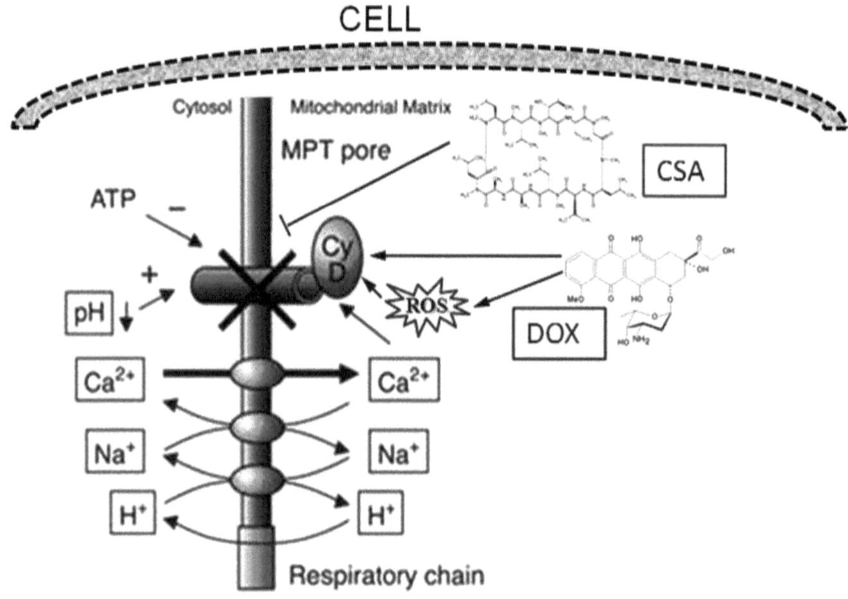

Fig. 1 Simplified diagram of different mitochondrial MPT activators and inhibitors

mitochondrial function, membrane structure, and the mitochondrial redox state are most important in determining cell survival and cell death (8, 9).

It was reported, at low concentrations (50 μM or less), doxorubicin can catalyze a site-specific oxidative damage to the NADH oxidation pathway. In contrast, tenfold higher doxorubicin concentrations (or more) are required for non-oxidative inactivation of the electron transport chain, probably via binding to cardiolipin and/or generalized membrane chaotropic effects. Mitochondria from DOX-treated rats under chronic doxorubicin treatment expressed a dose-dependent and irreversible decrease in calcium loading capacity that correlated with the accumulation of DNA adducts, histopathology, and clinically observed cardiomyopathy. These findings indicated that this cyclosporine-sensitive altered regulation of mitochondrial calcium homeostasis may be a critical factor involved in the pathogenic pathway of the cumulative and irreversible cardiomyopathy associated with long-term DOX cancer chemotherapy. At the same time, these mitochondria-perturbation events are shown to be insensitive to antioxidant treatments, which cannot reverse the precipitated cardiomyopathy signs at this late stage of cumulative doxorubicin cancer therapy.

Many reports proposed the direct correction of these MPT-dependent events in the cumulative doxorubicin toxicity via administration of MPT stabilizers, such as cyclosporine (CSA) (Fig. 1). CSA is potentially able to prevent or at least delay doxorubicin-induced cardiomyopathy first by directly inhibiting mMPT via cycophilin binding and second by restoring the mitochondrial calcium loading capacity. Restoring the intracellular calcium homeostasis in turn eliminates calcium as a facilitator for mMPT thereby also preventing or delaying cell death. The low-water solubility and membranotropic nature of CSA result in or diminished plasma concentration, as well as pronounced renal toxicity associated with high systemic doses of CSA, and generally stand against its sufficient intracellular localization. The main hurdle facing this approach is the need for relatively high local concentration of these lipophilic molecules into the mitochondria fraction of the cell, as well as the means for direct delivery of these drugs into the target mitochondria of affected cardiomyocytes.

We designed a nanocarrier system to mediate the selective delivery of cyclosporine A (CSA) to and into mitochondria within cardiomyocytes, which will prevent or at least delay doxorubicin-induced cardiomyopathy thus allowing to increase the total cumulative lifetime cap dose of DOX. We developed pharmaceutical formulation of CSA into long-circulating and mitochondria-targeted lipid-based nanocarrier system, namely, vegetable oil-based nanoemulsions (NEs). The advantages of nanoemulsion platform include an opportunity to solubilize highly hydrophobic compounds,

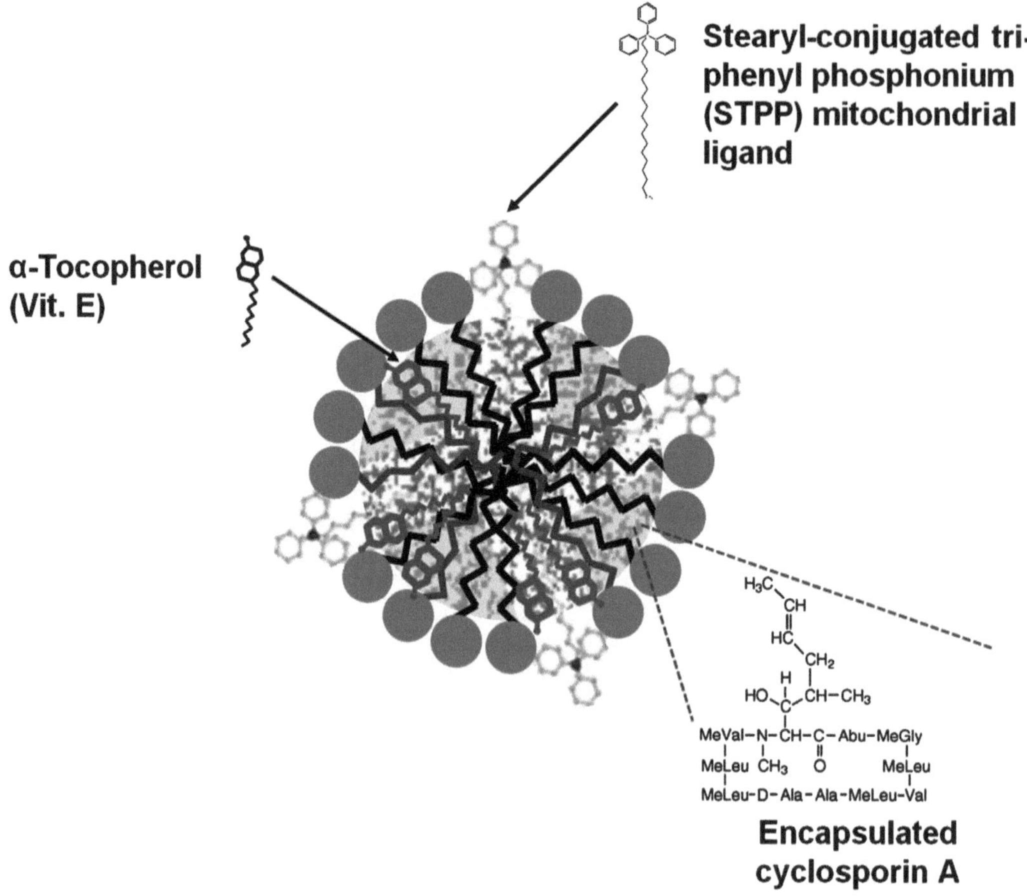

Fig. 2 Schematic representation of the incorporation of cyclosporine A and vitamin E into mitochondriotropic STTP-targeted lipidic nanocarriers

such as CSA (Fig. 2) and FK506 that serve as stabilizing ligands for mMPT. Triphenylphosphonium cations (TPPs) have been used previously to render long-circulating liposomes mitochondriotropic (10–13). TPPs were conjugated with stearyl residues producing stearyl-triphenylphosphonium cations (STPP) in order to facilitate the incorporation of TPPs into lipidic vesicles. The stearyl residue literally anchors the cation into the lipid bilayer and successfully onto the surface of nanoemulsion oil droplets. Based on previous in vitro and in vivo reports, STPP-coupled lipidic nanocarriers, such as liposomes and micelles, were capable of mitochondria-targeted delivery of proapoptotic agents, resulting in significantly increased anticancer activity of the corresponding agents. Analogously, our recent in vitro studies, our mitochondria-specific STPP-modified CSA-loaded nanoemulsions, as a targeted pharmaceutical lipid-based nanocarrier system (Fig. 2), were also capable of delivering a sufficient amount of CSA to mitochondria in cardiac myocytes in order to inhibit or delay doxorubicin-induced muscle damage and death.

Our targeted nano-therapy is capable of intracellular delivery of both therapeutic antioxidants and CSA, specifically and effectively preserving cardiac muscle mitochondria, acting as an adjuvant therapy to combat doxorubicin chemotherapy-induced cardiac failure; such mitochondriotropic nano-therapy has the potential to expand the eligible patient profile, effective dose range, and spectrum of malignancies, eligible for treatment with potent adriamycin-based chemotherapy regimens. Moreover, our targeted platform carries significant prospects for treating closely related oxidative damage molecular mechanisms involved in ischemia-reperfusion injury, congestive heart failure, cardiomyopathies, and acute cardiogenic shock.

2 Materials

2.1 Preparation and Characterization of Control and Targeted CSA-Loaded Nanoemulsions

1. High omega-3 fatty acid-containing argan oil (Jedwards International, Quincy, MA).
2. Phosphate buffered saline (PBS, 200 mM), pH 5.8: dissolve 137 mM NaCl, 2.7 mM KCl, 200 mM Na_2HPO_4, and 1.8 mM of KH_2PO_4 in 800 mL of H_2O, adjust pH to 5.8 using HCl, then add H_2O up to 1 L.
3. Doxorubicin HCl (DOX, LC Laboratories, Woburn, MA). Prepare stock solution of 1.7 M of DOX per 1 mL of 1× PBS, pH 5.8.
4. Chloroform (dry).
5. Ethanol (200% proof).
6. Cyclosporine A (CSA, LC Laboratories, Woburn, MA).
7. HEPES-buffered saline, pH 7.4.
8. Solutol HS-15 (Mutchler Inc.)
9. D-α-Tocopherol (Vitamin E, VE).
10. D-α-Tocopheryl polyethylene glycol 1000 succinate (TPGS, Antares Health Products Inc, St. Charles, Illinois).
11. Double distilled (DDI) water.
12. 10 and 25 mL pear-shaped glass flasks that fit rotary evaporator spout, for cosolvent evaporation.
13. Rotary evaporator (Labconco, Kansas City, MO).
14. Ultra-Turrax 10 homogenizer (IKA Works, Inc., Wilmington, NC).

15. Sonic probe dismembrator (Misonix XL-2000, Qsonica LLC, Newtown, CT).
16. Nitrogen gas-operated LIPEX™ extruder (Northern Lipids Inc., Burnaby, BC, CA).
17. LIPEX™—compatible polycarbonate filter disks size 100 and 200 nm (Northern Lipids Inc., Burnaby, BC, CA).
18. Weigh balance (up to 0.001 mg in precision for accuracy).
19. Malvern Zetasizer Nano ZS (Malvern Instruments, Westborough, MA).

2.2 Cell Viability Assay

1. One vial of 1 × 105 cells of model rat cardiomyocytes, H9C2 (American Type Culture Collection, ATCC catalogue# CRL-1446, Manassas, Virginia).
2. ATCC-formulated Dulbecco's Modified Eagle's Medium (DMEM).
3. Fetal bovine serum (FBS added to growth media as 10% vol/vol).
4. Complete serum-free medium (SFM).
5. Clinical centrifuge at 100–1,000 × *g*.
6. Cell culture plates 96-well opaque-walled plates compatible with fluorometer, with clear.
7. Multichannel pipettor.
8. CellTiter-Blue® Cell Viability Assay (Promega, Madison, WI).
9. Fluorescence plate reader with excitation 530–570 nm and emission 580–620 nm filter pair, Synergy 2 multi-mode microplate reader (BioTek Instruments, Winooski, VT).

2.3 MitoPT JC-1 Mitochondrial Polarization Assay

1. 15 mL polystyrene centrifuge tube (1 per sample).
2. Microfuge at 13,000 × *g*. Clinical centrifuge at 100–1,000 × *g*.
3. Pipette(s) capable of dispensing at 10 μL, 500 μL, and 1 mL.
4. Graduated cylinder.
5. Cell culture grade sterile dimethyl sulfoxide—DMSO.
6. Vortex mixer.
7. Hemocytometer.
8. Amber vials or polypropylene tubes for storage at −20°C.
9. Cell culture plates 96-well opaque-walled plates compatible with fluorometer, with clear.
10. Multichannel pipettor.
11. MitoPT™ JC-1 Assay Kit (Catalogue #: 924, ImmunoChemistry Technologies LLC, Bloomington, MN).

12. Fluorescence plate reader with excitation 530–570 nm and emission 580–620 nm filter pair, Synergy 2 multi-mode microplate reader (BioTek Instruments, Winooski, VT).

3 Methods

3.1 Preparations of Mitochondria-Specific CSA-Loaded Nanoemulsions

Prepare all nanoemulsions using only clean glassware. Thoroughly clean the glassware and spatulas with concentrated nitric acid followed by ethanol. Make sure no residue of white phospholipids or drug remains in the glassware. Furthermore, use double-distilled water (DDW) during the entire formulation processes to guarantee purified grade final product.

Perform all the procedure at 60°C unless specified as follows:

1. Turn on the hot plate and adjust to 60°C. Warm the clean 25 mL beaker on the hot plate for 5 min, filled with DDI water.
2. In a 25 mL pear-shaped glass flask, add 0.75 g of organic phase, composed of 65% wt/wt argan oil or flax seed oil, and 35% wt/wt a-tocopherol Vitamin E (VE). Mix oil component thoroughly, using vortex mixer and glass bead bath set at 60°C. In case of mitochondria-targeted formulation, the mitochondriotropic STTP ligand, dissolved in chloroform (1 mg/mL), will be mixed with the warm oil phase mixture, as 5% wt ratio.
3. To the organic phase, add 2 mL of CSA dissolved in 100% ethanol (10 mg/mL), and mix using vortexer, until clear solution is obtained.
4. Connect the pear-shaped glass to the rotary evaporator, and slowly evaporate solvent under 25 PSI vacuum, set at 30 rpm rotation, and 35°C water bath temperature, for approx. 20 30 min (see Note 1).
5. In a 15 mL glass tube, add 0.25 g total of surfactant mixture (composed of 0.075 g of vitamin E PGS and 0.175 g of Solutol HS-15, as weight ratio of 3:7, respectively). Using a heat gun, mix surfactant components using vortex mixture, while monitoring the mixture temperature not to exceed 65°C.
6. While warm, quickly transfer the CSA-oily mixture to the 15 mL glass tube containing the warm surfactant mixture, and vortex under heat gun for 15–20 s.
7. Using 1 mL pipette, gradually add drop wise 4 mL of warm DDI water onto the warm mixture inside the 15 mL glass tube, mixing thoroughly; this is to make sure the drug-oily platform will not cool down dramatically. Add a new drop promptly after the previous drop blends well into platform.

8. Keep stirring the platform gently using vortex mixer for 5 min, under distant air heat gun exposure, while monitoring the mixture temperature not to exceed 65°C.
9. The resulting milky macro-emulsion will them be homogenized for 10 min, at 20,000 rpm setting.
10. Following, the resulting microemulsion will be sonicated for using probe sonicator (8 W energy input, for three periods of 5 min each, resting for 1 min in between).
11. Finally, transfer the resulting nanoemulsion to the LIPEX™ extruder, and pass once, under 100 PSI nitrogen gas pressure using first 0.2 mm filter disk; then, run another pass using the 100 nm filter (see Note 2).
12. Store the NE formulations at 4°C for later use.

3.2 Physical Characterization of STPP-Coupled and Control Nanoemulsions

The nanoemulsion formulations will be characterized for particle size and size distribution using the dynamic light scattering technique with a Malvern nanosizer analyzer (Malvern instruments, Holtsville, NY) at 273° fixed angle and at 25°C temperature:

1. Dilute NE formulation, for particle-size analysis, using DDI water deionized at about 1,000-fold vol/vol, in plastic cuvette. The numbered average oil droplet hydrodynamic diameter and the polydispersity index will be determined (Fig. 3, *panel a*).
2. For the zeta potential, dilute NE samples in DDI water as 10,000-fold; then, employ a 1 ml syringe to inset solution inside the electrophoretic cell of the Malvern nanosizer to avoid inserting any air bubbles. The average surface charge will be measured (Fig. 3, *panel b*) (14).

3.3 Cell Viability Assay Protocol

Thaw CellTiter-Blue® reagent and bring to ambient temperature, and protect the CellTiter-Blue® reagent from direct light.

1. Set up 96-well assay flat bottom plates, containing approximately 2×10^4 cells in 100 μL of complete DMEM cell culture media per well (15).
2. Allow H9C2 cells to attach to the bottom of the plate for 24 h. Cell would be ready to the next step when they are about 60–70% confluent (see Notes 3 and 4).
3. Add doxorubicin HCl challenge (50 μM, diluted in serum-free medium, SFM) to both positive control wells, as well as all appropriate test well, so the final volume is 100 μL in each well. Then, co-incubate at 37°C, for 2 h, followed by removal of media from all wells (see Note 5).
4. Add the various NE vehicle controls, as well as test NEs containing the different CSA and VE treatment, all diluted 100× at minimum, in SFM, and applied to test wells in two-fold serial dilution pattern. Make sure the final volume is 100 μL in each well.

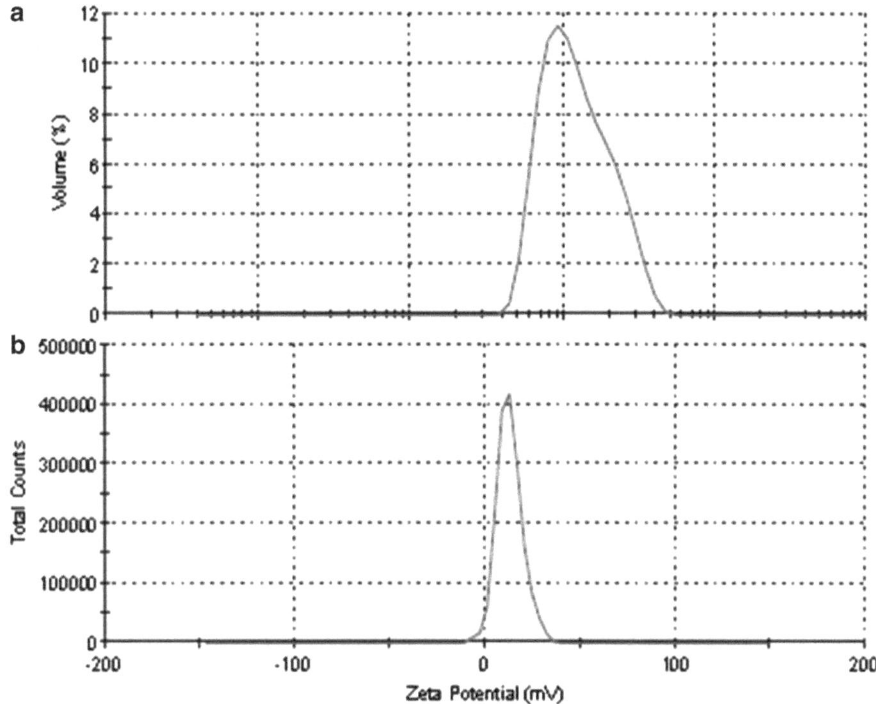

Fig. 3 Physicochemical characterization of STTP-coupled nano-emulsified lipid carriers, showing average particle-size distribution (**a**) and average zeta-potential measurement (**b**) of typical CSA-loaded nanoemulsion formulation

5. Culture cells for 30 h test exposure period.
6. Remove assay plates from 37°C incubator, and remove all NE media and SFM controls.
7. Apply cell wash to all wells containing cells, by carefully and slowly adding 100 μL of Hank's balanced buffer saline (HBS), leaving in well for 5 s and then removing HBS solution; repeat this step twice for the entire plate, taking extreme caution to minimize dislodging attached cells inside each well.
8. Finally, add 100 μL of SFM to each well, followed by the addition of 20 μL/well of CellTiter-Blue® reagent.
9. Shake plate for 10 s, to mix reagent well.
10. Incubate at 37°C, using the same standard cell culture conditions for 1–2 h (see Note 6).
11. Insert the developed plate in the plate reader, and set assay protocol/method to shake plate for 10 s; then, record end point fluorescence using ex.560/em.590 nm filter set.
12. Calculate results of fluorescence data; then, plot percent cell viability [(test well fluorescence—untreated cell control fluorescence)/100] vs. concentration of NE treatment (see Note 7) (16).

3.4 Modified MitoPT™ JC-1 Protocols

Mitochondrial membrane potential, Δψm, is an important parameter of mitochondrial function used as an indicator of cell health. The loss of mitochondrial membrane potential (ΔΨ) is a hallmark for apoptosis. The JC-1 Assay Kit measures the mitochondrial membrane potential in cells. JC-1(5,5′,6,6′-tetrachloro-1,1′,3,3′-tetraethylbenzimidazolylcarbocyanineiodide) is a lipophilic, cationic dye that can selectively enter into mitochondria and reversibly change color from green to red as the membrane potential increases.

In healthy cells with high mitochondrial Δψm, JC-1 spontaneously forms complexes known as J-aggregates with intense red fluorescence. On the other hand, in apoptotic or unhealthy cells with low Δψm, JC-1 remains in the monomeric form, which shows only green fluorescence.

3.4.1 Reconstitution of the 100× or 200× MitoPT™ JC-1 Stock

The MitoPT™ JC-1 dye reagent is supplied as a highly concentrated lyophilized powder. It must first be reconstituted, using cell culture grade (sterile) DMSO solvent:

1. For the 100 test kit, reconstitute the 100-test vial with 500 μL DMSO at room temperature (RT), forming a 100× stock.
2. Recap the vial and invert it several times to fully dissolve the MitoPT™ JC-1 dye reagent.
3. Immediately use the 100×, or aliquot and store it at −20°C (see Note 8).

3.4.2 Modified MitoPT™ JC-1 Protocol for 96-Well Fluorescence Plate Reader

1. Prepare the 1× working strength MitoPT™ JC-1 solution by diluting the stock 1:100 with cell culture media, warmed to 37°C (see Note 9).
2. Using clear 6-well flat bottom plates, seed 0.4×10^6 cells/well, in approx 2.0 mL volume of complete growth DMEM medium, and incubate at 37°C for 24–28 h, to allow A10 model cells to attach and become approx. 60–70% confluent (see Note 10).
3. When cells are ready, remove media, and then, assign triplicate wells as non-treated cell negative control that receive.
4. Add to cells in the rest of the wells doxorubicin HCl challenge (50 μM, diluted in serum-free medium, SFM) to both positive control wells, as well as all appropriate test wells; then, co-incubate plates for 1 h, at 37°C, using the same standard cell culture conditions.
5. Add the various NE vehicle controls, as well as test NEs containing the different CSA and VE treatment, all diluted equivocally, in SFM, and applied to test wells in twofold serial dilution pattern. Make sure the final volume does not exceed 2 mL in each well.
6. Culture cells for 20–24 h test exposure period (see Note 11).
7. Remove assay plates from 37°C incubator, and remove all NE media and SFM controls.

8. Apply cell wash to all wells containing cells, by carefully and slowly adding 100 μL of HBS, leaving in well for 5 s, and then removing HBS solution; repeat this step thrice, for all plates, taking extreme caution to minimize dislodging attached cells inside each well.
9. Remove cell wash solution, and then detach cells in each well using 2 mL of trypsin LE/well.
10. Pellet cells by centrifugation at approx. 200 × *g* for 5 min at RT (see Note 12).
11. Carefully remove and discard the supernatants.
12. Resuspend at between 0.5 and 1×10^6 cells in 1 mL of freshly prepared 1× working strength MitoPT™ JC-1 solution prepared fresh.
13. Gently vortex or pipette the cell pellets to disrupt any cell-to-cell clumping.
14. Incubate the H9C2 cells (staining with the MitoPT™ JC-1 dye reagent) at 37°C for 10–15 min in a CO_2 incubator.
15. Warm the working strength 1× assay buffer to 37°C .
16. Add 2 mL of 1× assay buffer to each tube, and mix well using vortex mixer.
17. Centrifuge the stained cells at <200 × *g* for 5 min at RT; then, carefully remove and discard the supernatants; and then, gently vortex the pellets to disrupt any cell-to-cell clumping.
18. Resuspend the cells in 1 mL of 1× assay buffer.
19. Take out a small (50 μL) aliquot of each cell population aliquot to determine the concentration of both the induced and non-induced cell populations. Then, add to 250 μL PBS (forming a 1:5 dilution of each), to count the cells using a hemocytometer (see Note 13).
20. Centrifuge the remaining stained cells at 400 × *g* for 5 min at RT; then, carefully remove and discard supernatants.
21. Gently vortex the pellets to disrupt any cell-to-cell clumping.
22. Adjust the volume of the induced cell suspension to match that of the non-induced suspension. A minimum of 1×10^5 cells/well is recommended to generate an adequate fluorescence signal using most 96-well plate readers (see Note 14).
23. For each sample to be tested, dispense 100 μL into each of 2–4 wells in a black round or flat bottom 96-well microtiter plate.
24. After inserting the plate, set the plate reader to perform an endpoint read, where the excitation wavelength is at 488–490 nm and then the emission wavelengths to 527 nm for green fluorescence and 590–600 nm for red fluorescence.

3.4.3 Modified MitoPT™ Fluorescence Microscopy JC-1 Staining Protocol for Adherent Cells

Following the fluorescence microscope protocol, each sample to be stained requires only 0.5 mL of 1× MitoPT™ JC-1 solution (equal to 5 μL of 100× MitoPT™ JC-1 stocks):

1. Using 6-well plates, culture cells at about seed 0.2×10^6 cells/well, in approx 2.0 mL volume of complete growth DMEM medium, and incubate at 37°C for 24–28 h, to allow H9C2 model cells to attach and become approx. 60–70% confluent (see Note 10). Cell density should not exceed the threshold where cell sloughing occurs.
2. Follow steps 3–8 from the previous method of MitoPT™ JC-1 protocol for 96-well fluorescence plate reader (see 3.4.2) to induce apoptosis.
3. Remove the cell wash solutions from both induced and non-induced monolayer cultures.
4. Add sufficient fresh 1× MitoPT™ JC-1 solution (approx 0.1 mL) to cover the cells on the cover slip, inside each well.
5. Incubate the cells, stained with the MitoPT™ JC-1 dye reagent, at 37°C for 15 min in a CO_2 incubator.
6. Warm the 1× assay buffer to 37°C; then, carefully remove and discard staining media.
7. Wash the monolayer culture on the cover slips with 1 mL of 1× assay buffer, twice, and then discard wash solution.
8. Add a drop of 1× assay buffer to cell culture slides; then, invert each cover slip on a glass slide, cell surface down.
9. Examine using fluorescence microscope.

4 Notes

1. Complete removal of ethanolic solvent is confirmed when a clear translucent running yellow color gel-like residue remains in the flask that gets somewhat thicker as the flask temperature cools down. The residue must be clear from any suspending white CSA drug precipitates.
2. Pre-warm the thermobarrel LIPEX™ extruder to about 40°C (measured externally), before running the NE sample, to guarantee smooth flow-pass through the filter. Make sure to run sample through the larger pore-size filter disk first before the smaller pore-size one, to avoid clogging of the filter disk.
3. Set up triplicate wells without cells to serve as the negative control to determine background fluorescence that may be present.
4. Set up quadruplicate wells with untreated cells to serve as a vehicle control. Add the same solvent used to deliver the test

compounds to the vehicle control wells (17). Since, test cells are subjected to both doxorubicin HCl and NE treatments diluted in serum-free cell culture medium, hence, use the same serum-free cell culture medium for untreated cell control wells.

5. DOX only treated cell wells will serve as the positive control for cytotoxicity, using quadruplicate wells containing cells treated with a DOX (50 μM, diluted in SFM) known to be toxic to the A10 cell model system. Only SFM media (no NE treatments) will be added to these wells, in the steps to follow in the assay.
6. Fluorescence generated can be stopped and stabilized by the addition of 3% SDS. Typically, add 40 μL per 100 μL in each well. The plate can then be stored at ambient temperature for up to 24 h before recording data, provided that the contents are protected from light and covered to prevent evaporation.
7. Optional: Subtract the average of fluorescence values of the culture medium background from all fluorescence values of experimental wells (17).
8. The 1X working solution must be prepared immediately prior to use; however, the reconstituted 100× or 200× stock can be stored at −20°C for 6 months and used twice during that time. As the 1× MitoPT™ JC-1 solution must be used immediately, prepare the MitoPT™ reagents at the end of your metabolic stress or apoptosis induction process.
9. Each sample to be stained requires only 0.5 mL of 1× MitoPT™ JC-1 solution (i.e., equivalent to 5 μL of 100× MitoPT™ JC-1 stock).
10. Cell density in the cell culture flasks should not exceed 10^6 cells per mL (i.e., about $1–1.5 \times 10^6$ cells/well of the 6-well plate). Cells cultivated in excess of this concentration may begin to naturally enter apoptosis. Optimal cell concentration will vary depending on the cell line used. Density can be determined by counting cell populations on a hemocytometer.
11. Generation of experimental apoptosis or $\Delta\psi$m-disrupted cells oxidative stress may take few hours up to 48 h, depending on model cell line, cell culture conditions, and the inducer test concentration.
12. When concentrated, cells should have been grown to yield a 0.5 mL concentrated pool between 1 and 2×10^6 cells/mL.
13. After counting, compare the density of each. The non-induced population may have more cells than the induced population, as some induced cells may be lost during the apoptotic process. If necessary, adjust the volume of the induced cell suspension to match that of the non-induced suspension.
14. Resuspend the non-induced cell pellets in 500 mL to 1 mL of 1× assay buffer to produce a cell suspension about 1 × 106 cells/mL,

which is generally sufficient to generate enough signal. The volume may vary depending upon cell density. It is possible to get by using a lower cell number/well, depending on the particular cell model system and fluorescence detection plate reader, but this should be predetermined for each individual case.

References

1. Lefrak EA, Pitha J, Rosenheim S, Gottlieb JA (1973) A clinicopathologic analysis of adriamycin cardiotoxicity. Cancer 32:302–314
2. Allen A (1992) The cardiotoxicity of chemotherapeutic drugs. Semin Oncol 19:529–542
3. Doroshow JH (1983) Effect of anthracycline antibiotics on oxygen radical formation in rat heart. Cancer Res 43:460–472
4. DiPiro JT, Talbert RL (2002) Pharmacotherapy: a pathophysiologic approach, 5th edn. McGraw-Hill, London
5. Gabizon AA, Lyass O, Berry GJ, Wildgust M (2004) Cardiac safety of pegylated liposomal doxorubicin (Doxil/Caelyx) demonstrated by endomyocardial biopsy in patients with advanced malignancies. Cancer Invest 22:663–669
6. Doroshow JH (1983) Anthracycline antibiotic-stimulated superoxide, hydrogen peroxide, and hydroxyl radical production by NADH dehydrogenase. Cancer Res 43:4543–4551
7. Tokarska-Schlattner M, Zaugg M, Zuppinger C, Wallimann T, Schlattner U (2006) New insights into doxorubicin-induced cardiotoxicity: the critical role of cellular energetics. J Mol Cell Cardiol 41:389–405
8. Wallace KB (2007) Adriamycin-induced interference with cardiac mitochondrial calcium homeostasis. Cardiovasc Toxicol 7:101–107
9. Zhou S, Starkov A, Froberg MK, Leino RL, Wallace KB (2001) Cumulative and irreversible cardiac mitochondrial dysfunction induced by doxorubicin. Cancer Res 61:771–777
10. Patel NR, Hatziantoniou S, Georgopoulos A, Demetzos C, Torchilin VP, Weissig V, D'Souza GG (2010) Mitochondria-targeted liposomes improve the apoptotic and cytotoxic action of sclareol. J Liposome Res 20:244–249
11. Boddapati SV, D'Souza GG, Erdogan S, Torchilin VP, Weissig V (2008) Organelle-targeted nanocarriers: specific delivery of liposomal ceramide to mitochondria enhances its cytotoxicity in vitro and in vivo. Nano Lett 8:2559–2563
12. Weissig V, Boddapati SV, Cheng SM, D'Souza GG (2006) Liposomes and liposome-like vesicles for drug and DNA delivery to mitochondria. J Liposome Res 16:249–264
13. Boddapati SV, Tongcharoensirikul P, Hanson RN, D'Souza GG, Torchilin VP, Weissig V (2005) Mitochondriotropic liposomes. J Liposome Res 15:49–58
14. Ganta S, Amiji M (2009) Coadministration of Paclitaxel and curcumin in nanoemulsion formulations to overcome multidrug resistance in tumor cells. Mol Pharm 6:928–939
15. Mu L, Elbayoumi TA, Torchilin VP (2005) Mixed micelles made of poly(ethylene glycol)-phosphatidylethanolamine conjugate and d-alpha-tocopheryl polyethylene glycol 1000 succinate as pharmaceutical nanocarriers for camptothecin. Int J Pharm 306:142–149
16. Elbayoumi TA, Pabba S, Roby A, Torchilin VP (2007) Antinucleosome antibody-modified liposomes and lipid-core micelles for tumor-targeted delivery of therapeutic and diagnostic agents. J Liposome Res 17:1–14
17. Lukyanov AN, Elbayoumi TA, Chakilam AR, Torchilin VP (2004) Tumor-targeted liposomes: doxorubicin-loaded long-circulating liposomes modified with anti-cancer antibody. J Control Release 100:135–144

Chapter 12

Formation of Pit-Spanning Phospholipid Bilayers on Nanostructured Silicon Dioxide Surfaces for Studying Biological Membrane Events

Indriati Pfeiffer and Michael Zäch

Abstract

Zwitterionic phospholipid vesicles are known to adsorb and ultimately rupture on flat silicon dioxide (SiO_2) surfaces to form supported lipid bilayers. Surface topography, however, alters the kinetics and mechanistic details of vesicles adsorption, which under certain conditions may be exploited to form a suspended bilayer. Here we describe the use of nanostructured SiO_2 surfaces prepared by the colloidal lithography technique to scrutinize the formation of suspended 1-palmitoyl-2-oleoyl-*sn*-glycero-3-phosphocholine (POPC) lipid bilayers from a solution of small unilamellar lipid vesicles (SUV_s). Atomic force microscopy (AFM) and quartz crystal microbalance with dissipation monitoring (QCM-D) were employed to characterize nanostructure fabrication and lipid bilayer assembly on the surface.

Key words Nanostructured surfaces, Phospholipid vesicles, POPC, BLM, QCM-D, AFM, Colloidal lithography, Biosensor

1 Introduction

Phospholipid bilayers have chemical and physical properties that closely resemble the biological cell membrane (1–3) and have therefore been frequently used in biosensor applications to tether membrane proteins to the sensor surface (4–6). The bilayer provides a natural environment for the protein and ideally functions as a cushion, which prevents direct interaction of the protein with the surface, two important prerequisites to preserve the protein's biorecognition sites and function (7, 8). On flat SiO_2 surfaces, small unilamellar phospholipid vesicles (SUV_s) are known to adsorb and rupture to form a supported phospholipid bilayer (9, 10). The kinetics and mechanistic details of this process on flat surfaces have been investigated using a number of surface analytical techniques, e.g., quartz crystal microbalance with dissipation monitoring (QCM-D)

Volkmar Weissig et al. (eds.), *Cellular and Subcellular Nanotechnology: Methods and Protocols*, Methods in Molecular Biology, vol. 991, DOI 10.1007/978-1-62703-336-7_12, © Springer Science+Business Media New York 2013

(11, 12) combined with atomic force microscopy (AFM) (13, 14), which together with computer simulations (15) have provided a thorough understanding of the process and thus ensure reproducible bilayer formation (14, 16). There has recently been a growing interest for preparation methods, which are able to form continuous lipid bilayers on substrates with nanoholes (17–19). Such pore-spanning membranes are attractive because they offer access to the liquid reservoir on both sides of the lipid bilayer (20). Also, they are advantageous for proteins with large extramembrane domains, which might lose their functionality if reconstituted into a bilayer supported directly on a flat substrate (21). The nanometer size of the pores grants a high stability of the suspended lipid membrane and opens up possibilities to use this system for certain biotechnological applications such as membrane protein arrays (22, 23) and ion channel protein biosensors (24, 25). Several different ways to form suspended lipid bilayers on various types of surfaces and methods to characterize their stability have been reported, including lipid painting methods or giant unilamellar vesicles spreading on micro-/nanofabricated silicon nitride (Si_3N_4) surfaces (23, 26) and the use of lipopolymer bilayers on porous aluminum oxide (Al_2O_3) surfaces (27, 28). The formation of suspended bilayers on top of these surfaces was characterized electrochemically by measuring the electrical resistance across the pores (21, 29). A drawback of this technique is that no kinetic or mechanistic information of suspended bilayer formation can be obtained, with poorly reproducible bilayer formation as a possible consequence. In addition, the production of suitable Si_3N_4 surfaces requires rather advanced micro-/nanofabrication procedures (30), and the anodization of aluminum to Al_2O_3 produces less controllable size and dimension of the porous surfaces (21, 31). Here, we describe a simple method termed colloidal lithography (32, 33), which allows one to create surfaces with homogenous nanoscale pits or through holes (34) and multiple surface chemistries. As an example, we describe below the fabrication of pitted surfaces with SiO_2 forming the walls of the pits and Au constituting the bottom of the pits (Fig. 1). In order to form a pit-spanning bilayer, biotin-amidocaproyl bovine serum album in (BBSA) was added to the system in a first step in order to passivate exposed Au areas at the bottom of the pits against lipid vesicles adsorption (35). Subsequently, 1-palmitoyl-2-oleoyl-*sn*-glycero-3-phosphocholine (POPC) lipid vesicles with 160 ± 40 nm average diameter were introduced to the surface. Using this strategy, phospholipid vesicles that adsorb on the SiO_2 top surface rupture and form a continuous bilayer that spans over the pits (Fig. 1f). Since we would like to follow the bilayer formation process in situ using the QCM-D technique, we produced the SiO_2-pitted surface on top of Au-coated QCM-D sensor crystals. Compared to supported bilayer formation on flat SiO_2, QCM-D revealed that the kinetics and the mechanism of vesicle adsorption and bilayer formation are altered on these nanostructured surfaces (Fig. 2). The formation of a suspended phospholipid

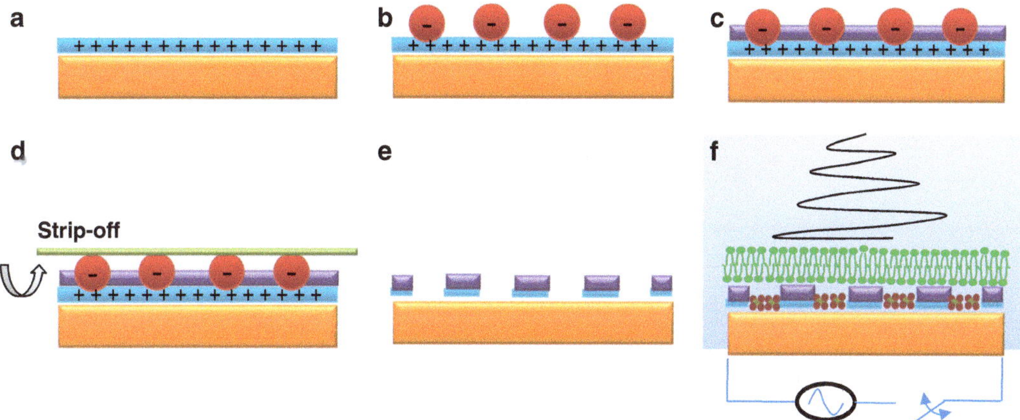

Fig. 1 Schematic illustration of the fabrication of nanostructured SiO_2 surface using colloidal lithography and the QCM-D measurement setup used to monitor the formation of a suspended phospholipid bilayer: (**a**) Deposition of polyelectrolyte triple layer on gold-coated sensor crystal. The final PDDA polyelectrolyte layer creates a positively charged surface. (**b**) Deposition of negatively charged polystyrene particle suspension on top of polyelectrolyte triple layer. (**c**) Deposition of thin films of Ti and SiO_2 using electron beam evaporation. (**d**) Polystyrene particle stripping using an adhesive tape. (**e**) The end product; a homogenous, pitted SiO_2 surface. (**f**) Adsorption of BBSA and the formation of a suspended phospholipid bilayer on the nanostructured silicon dioxide surface, as measured by the QCM-D technique

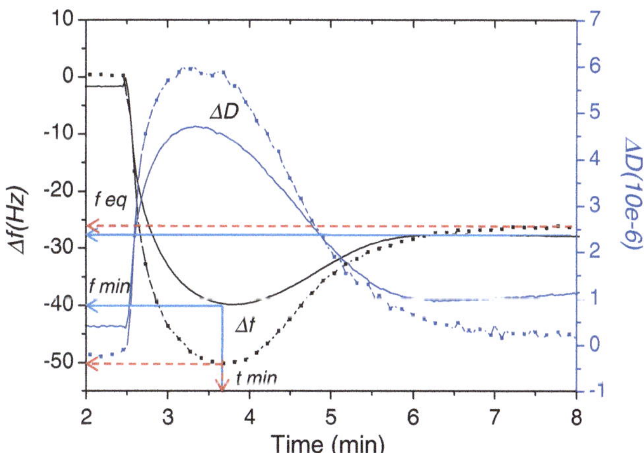

Fig. 2 Example of QCM-D graphs showing changes in frequency and dissipation (Δf, ΔD) as a function of time and monitoring the kinetics of supported phospholipid bilayer formation on a flat SiO_2 surface (*dashed lines*) and suspended phospholipid bilayer formation on a BBSA-modified nanostructured SiO_2 surface (*solid lines*). Both graphs show similar f_{eq}, indicating that a continuous phospholipid bilayer was formed on both surfaces. However, as shown by the values of f_{min}, the addition of nanotopography on the surface alters the critical vesicular coverage (the surface concentration of vesicles requires for triggering the rupture of vesicles) (flat surface, f_{min}, *dash arrow*; nanostructured surface, f_{min}, *solid arrow*), and accelerates lipid bilayer formation

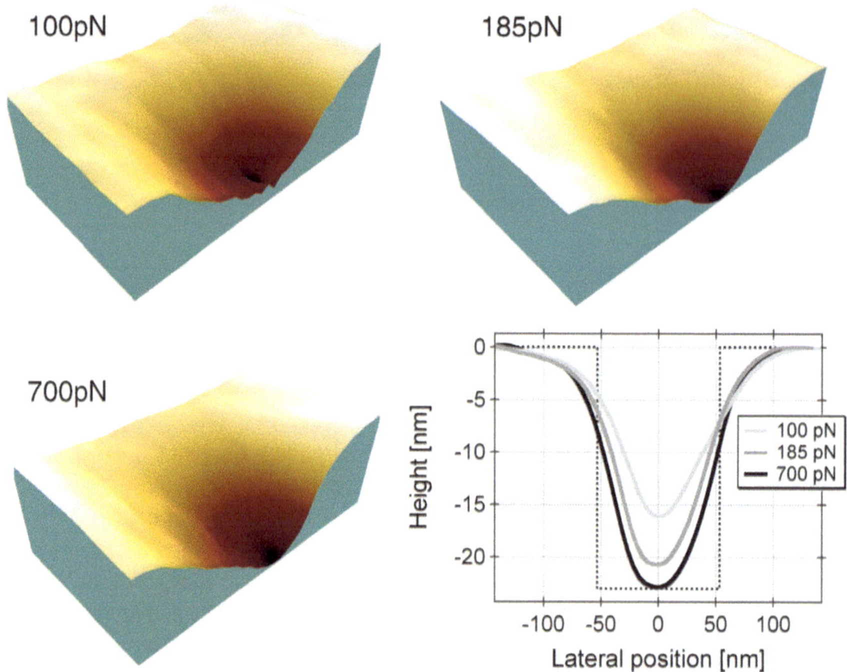

Fig. 3 3D topography images of pit-spanning bilayer as acquired at three different imaging forces using AFM. The height scale is 25 nm for all three images, and the lateral dimensions are ca. 100 nm × 175 nm. The apparent pit depth increases with increasing imaging force due to bilayer compliance, as illustrated by the cross sections in the *lower right panel*. For the largest imaging force shown here, the bilayer is pushed all the way down to the underlying BBSA layer. The theoretically expected pit diameter (as given by the colloid size) and pit depth (as given by the total evaporated film thickness minus the thickness of a BBSA layer) are indicated by the *dotted line*

bilayer was verified using AFM, which revealed a characteristic dependence of the apparent pit depth on the applied imaging force due to deformation of the pit-spanning membrane (Fig. 3). Results were highly reproducible, owing to the homogeneity of surface nanotopographies produced by colloidal lithography.

2 Materials

Purified water with an electrical resistance of 18.2 MΩ at 25 °C was used to prepare all solutions. All buffers and solutions used in the experiments were prepared and stored at room temperature unless otherwise specified. Waste solutions were handled as instructed by local waste disposal regulations.

2.1 Colloidal Lithography Components

1. A gold-coated QCM-D sensor crystal with 5 MHz resonance frequency and 13 mm diameter.
2. A pair of tweezers (see Note 1).

3. Hydrogen peroxide, 30%.
4. Aqueous ammonia solution, 28%.
5. Water.
6. A flat beaker with a diameter of around 15 cm.
7. A 100 mL glass beaker.
8. A custom-made Teflon holder used to carry the QCM-D sensor crystal (see Note 2).
9. Poly-(Diallyl-Dimethyl-Ammonium) chloride (PDDA) polyelectrolyte: 2% w/w solution in water. Weigh 1 mg of PDDA (20% w/w solution, MW = 200,000–350,000) and mix it with 49 mL of water in a 50 mL falcon tube (see Note 3).
10. Poly (sodium 4-Styrene-Sulfonate) (PSS) polyelectrolyte: 2% w/w solution in water. Weigh 1 g of PSS powder (MW = 70,000) in a 50 mL falcon tube. Add 49 mL of water to dissolve (see Note 3).
11. Polystyrene particle suspension: 0.2% v/v solution in water. Mix 200 μL of polystyrene particle suspension with average diameter of 107 nm with 9.8 mL of water in a 15 mL falcon tube to make a stock solution of 2% v/v. Store the stock solution at 4°C. Prior to use, take 1 mL of stock solution and mix it with 9 mL of water in a 15 mL falcon tube. Use immediately (see Note 4).
12. N_2 gun (see Note 5).
13. Clean room wipes.
14. Titanium (Ti) evaporation source.
15. Silicon dioxide (SiO_2) evaporation source.
16. Adhesive tape (see Note 6).
17. A 20 mL flat beaker filled with isopropyl alcohol.
18. A 20 mL flat beaker filled with water.

2.2 Phospholipid Vesicle Solution Components

1. Vesicle preparation buffer: 100 mM NaCl, 10 mM Tris, pH 8. Weigh 5.84 g NaCl and 1.21 g Tris, and transfer to a 1 L glass beaker. Add water to a volume of 900 mL. Mix and adjust the pH using HCl. Add water up to 1 L and filter through a 0.45 μM pore-size Nalgene filter. Store at room temperature and degas prior to use.
2. Phospholipid: 1-palmitolyl-2-oleoyl-*sn*-glycero-3-phosphocholine (POPC) lysophilized powder. Store at −20°C (see Note 7).
3. Chloroform.
4. A 1 L tank of liquid N_2.
5. A 1 L beaker filled with warm water around 50°C.
6. Polycarbonate membranes with 100 nm pore size.

2.3 QCM-D and AFM Measurement Components

1. Hellmanex cleaning solution: 1% v/v solution in water. Pipette 1 mL of Hellmanex solution to a 100 mL glass bottle. Add water up to 100 mL. Store at room temperature.
2. Measurement buffer: same buffer as for vesicle preparation (see item 1 in Subheading 2.2).
3. Biotin-amidocaproyl bovine serum albumin (BBSA) stock solution: 1 mg/mL concentration in water. Add 2 mL of water to a bottle containing 10 mg BBSA lysophilized powder. Gently tap the bottle to help the protein to dissolve. Transfer this 2 mL BBSA solution to a 15 mL falcon tube. Rinse the bottle with 1 mL of water, and then transfer the water to the same falcon tube. Repeat the rinsing procedure one more time before adding 6 mL of water to the falcon tube to make a total of 10 mL BBSA solution. Distribute this solution into Eppendorf tubes, each containing 50 μL protein aliquots, and store at −20°C.
4. A pair of tweezers: same as used for colloidal lithography (see item 2 in Subheading 2.1).
5. AFM tip: commercially available MSCT-AUNM Microlever-sharpened silicon nitride tip supported by a cantilever with a spring constant below 30 pN/nm (or equivalent product).

2.4 Surface Regeneration Component

1. Sodium dodecyl sulfate (SDS) solution: 10 mM. Weigh 0.72 g SDS powder and mix it with 250 mL of water in a glass beaker. Store at 25°C (see Note 8).
2. A small glass Petri dish.

3 Methods

3.1 Fabrication of Nanostructured Surface Using Colloidal Lithography

It is preferable to carry out all nanofabrication steps in a relatively dust-free environment. The use of a clean room is however not compulsory; a good laminar flow cabinet can be used as an alternative:

1. Cleaning the sensor crystal: fill a flat beaker (15 cm in diameter) around one third with water, and prepare a hot bath with a temperature of 80°C (see Note 9). Arrange the sensor crystal in the Teflon holder and put both inside a 100 mL glass beaker. Fill this glass beaker with 50 mL of water, 10 mL of hydrogen peroxide 30%, and 10 mL of ammonia solution 28%. Put this beaker in the hot bath for 10 min, take it out, and remove the sensor crystal. Rinse the sensor crystal thoroughly with water and blow it dry immediately with N_2. Let the mixture inside the beaker cool down before disposing it into a base waste container.

2. Deposition of polyelectrolyte triple layer on sensor crystal: pipette PDDA solution onto the sensor crystal and let it stay for 90 s. Make sure that the polyelectrolyte solution covers the entire surface area, and prevent the surface from drying. Rinse the sensor crystal thoroughly with water, and put the wet crystal onto a triple layer of clean room wipes with the Au surface facing upwards. Position the N_2 gun perpendicular around 2 cm above the center of the crystal, and blow the crystal dry in one shot (see Note 10). Repeat the procedure sequentially using PSS, and another layer of PDDA.
3. Deposition of polystyrene particle suspension on top of polyelectrolyte triple layer: pipette the 0.2% v/v polystyrene particle suspension onto the polyelectrolyte triple layer, which has previously been deposited, and let it stay for 60 s. Rinse with water and blow-dry the crystal in the same manner as after polyelectrolyte deposition.
4. Deposition of metal layer: load the modified sensor crystal, Ti source, and SiO_2 source into a thin film deposition system. Pump down the system to a pressure of 10^{-6} mbar (see Note 11). Set the evaporation angle to normal incidence and electron beam evaporate 1 nm of Ti, followed by 24 nm of SiO_2 at a deposition rate of 2 Å/s (see Note 12). Let the sources cool down before venting the chamber and unloading the sample.
5. Stripping-off the polystyrene particles: lay the sensor crystal on clean room wipes with the sensor surface facing upwards. Cut the adhesive tape into square-shaped pieces with a size of 2 cm × 2 cm. Apply the tape on the sensor surface and press gently. Avoid any air bubbles being trapped in between the tape and the surface. By using a pair of tweezers, strip the tape off the sensor surface at once (see Note 13).
6. Final cleaning: lay the nanostructured sensor crystal at the bottom of a small, flat beaker containing 20 mL of isopropanol, and sonicate it for 2 min. Take care so the modified surface will not touch any surface of the glass beaker or other object such as tweezers. Repeat the same procedure using water, and blow-dry the sample immediately using N_2 in the same manner as described previously (see step 2 in Subheading 3.1).

3.2 Preparation of Lipid Vesicles

1. Weigh 5 mg of POPC lysophilized powder in the round-bottom flask. Using a glass pipette, add 1 mL of chloroform to dissolve the lipid powder. Evaporate the chloroform using a constant stream of N_2 for 1–2 h (see Note 14) to form a thin lipid film on the round-bottom flask.
2. Dissolve the dried lipid film by adding 1 mL of vesicle preparation buffer and vortex it gently. Freeze the lipid solution by dipping the round-bottom flask into liquid N_2, and thaw it

immediately in the 50°C warm water bath. Repeat this procedure 4–5 times.

3. Assemble the extruder (see Note 15) by following the distributor's instructions. During extrusion, pass the lipid solution back and forth 21 times through a polycarbonate membrane with 100 nm pore diameter. Store the lipid solution under N_2 gas in a 2 mL flask at 4°C. Make sure to close the flask tightly and additionally seal it using Parafilm.

3.3 QCM-D Measurements

1. General cleaning prior to the measurement (see Note 16): put a blank sensor crystal in the measurement chamber, and pass 1 mL of 1% Hellmanex solution through all liquid handling parts of the instrument (i.e., both the temperature and measurement chamber loops if using a D300 system from Q-Sense). Stop the flow and let the solution stay inside the measurement chamber for 30 min. Resume the flow and rinse all liquid handling parts with around 100 mL of water. Take out the blank sensor crystal, and dry the chamber and tubing thoroughly.
2. Measurement: mount the clean, nanostructured sensor crystal inside the measurement chamber. Close valves that connected to all liquid handling parts. Connect a new 5 mL syringe to an inlet of QCM-D system (if using D300 system from Q-Sense), and fill it with measurement buffer. Open the valves and fill all liquid handling parts with measurement buffer (see Note 17). Set the temperature to 22°C, and start the measurement by stabilizing the frequency and dissipation baselines. Fill the inlet syringe with 1.98 mL of measurement buffer, and pipette 20 μL of 1 mg/mL BBSA stock solution into this buffer to make a 10 μg/mL BBSA solution. Introduce this solution into the measurement chamber for 20 min, and rinse immediately with measurement buffer until the frequency and dissipation shifts reach stable values. Thereafter, in the same manner as BBSA (pipette 80 μL of 5 mg/mL vesicle stock solution into 1.92 mL buffer in the syringe), introduce a vesicle solution with a concentration of 200 μg/mL to the measurement chamber. The adsorption of vesicles and the formation of a bilayer on the surface can be directly monitored as a function of time by following the changes in frequency and dissipation values until they stabilize. The measurement chamber can then be rinsed with measurement buffer in order to remove excess vesicles.

3.4 AFM Measurements

1. Clean the AFM fluid cell and tip holder (see Note 18) by immersing them in isopropyl alcohol for 20 min. Rinse all parts with water, measurement buffer, and water again before blow-drying them.
2. Clean the cantilever chip by immersing it in measurement buffer for 20 min, followed by careful rinsing with water and

blow-drying. Mount the cantilever chip in the tip holder and coarse-align the laser deflection measurement system according to your AFM supplier's instructions.

3. Transfer the sensor crystal with the freshly formed bilayer from the QCM-D instrument into the AFM fluid cell, making sure that the sensor surface is covered with measurement buffer at all times. Mount the fluid cell onto the AFM, fine-adjust the laser deflection system, and let the instrument equilibrate until temperature-induced drifts have settled to an acceptable level (see Note 19).
4. Establish contact between your tip and the sensor surface, and minimize the imaging force (see Note 20). Adjust scan size, scan speed, and feedback gains.

3.5 Surface Regeneration

1. After each measurement, remove the nanostructured sensor crystal from the AFM liquid cell and immerse it into the glass Petri dish filled with 10 mM SDS solution. Incubate the crystal for a minimum of 2 days before the next use; it may be stored in SDS for longer periods of time.
2. Prior to the next measurement, sonicate the sensor crystal for 5 min in fresh SDS solution. Remove the crystal from the SDS solution, rinse it thoroughly with water, and blow it dry with N_2 gas. Treat the crystal in a UV ozone chamber (e.g., FHR UVOH 150 LAB) two times for 10 min, followed by rinsing with water and blow-drying with N_2. Use the cleaned crystal immediately.

4 Notes

1. Stainless steel round-tip tweezers with 2 mm flat gripping point are the best option for this purpose. The sensor crystal needs to be held firmly during surface modification to prevent sliding. On the other hand one cannot apply too much force when holding the crystal in order for it not to crack or break. Therefore, it is very important that the gripping point is thin and flat.
2. Prior to surface modification, the sensor crystal has to be cleaned. The Teflon holder will be used to hold the sensor crystal during the cleaning procedure so that it will not move around and avoiding the risk that the surface of the crystal will touch the glass beaker.
3. A vortexer can be used to ensure better mixing.
4. Prior to each dilution, mix the stock solution of polystyrene particle suspension thoroughly. Such a low concentration of

solution is necessary to avoid polystyrene particle aggregation on the surface.

5. A N_2 gun is required for a very critical, quick blow-dry step.
6. "Blue tape" is used because it has an adhesion level that is strong enough to pull off the polystyrene particles but low enough to be cleanly removed from the surface of the sensor crystal without damaging it and without leaving any adhesive residue behind. So far, we did not get good results with any other type of adhesive tape.
7. Normally it is more cost-effective to order the lysophilized lipid powder in 1 g quantity and aliquot it in smaller amounts (50 mg) to avoid lipid hydration. Keep the lipid powder at −20°C.
8. Based on our experience, a 10 mM concentration of SDS solution will start to crystallize when stored below 21°C.
9. This cleaning procedure should be done inside a fume hood. To make a hot bath, use a heating plate equipped with thermocouple and magnetic stirring function. Adjust the temperature set point to 80°C, and use a magnetic stirrer inside the hot bath to homogenize the temperature.
10. Organize the working space in the laminar flow hood in such a way that the N_2 gun, the water bottle, and the falcon tubes containing 2% w/w PDDA, 2% w/w PSS, and 0.2% v/v polystyrene particle suspension are ready and easily accessible. Prepare one plastic pipette for each solution by blowing its interior with N_2. One may use the sink inside the laminar flow hood if available, or find a beaker for waste disposal. Hold the clean sensor crystal steadily using the tweezers with one hand on top of the sink, and use the other hand to pipette the polyelectrolyte solution onto the sensor crystal and to rinse the crystal with water. Make sure that the tweezers are dry before the next polyelectrolyte layer deposition step.
11. A good vacuum is required during metal evaporation in order to reduce the possibility of contamination by residual gases in the vacuum chamber.
12. A thin layer of titanium is used as adhesive layer between Au and SiO_2. It is important to sweep the electron beam across the source during SiO_2 evaporation so that the beam will not penetrate the SiO_2 source. Otherwise the evaporation rate may jump suddenly in an uncontrolled way.
13. It is very important to be able to remove the tape completely in one quick stripping. Otherwise the polystyrene particles will slightly roll and as a result the pitted surface will not be homogenous.

14. If a vacuum chamber is available, the best is to completely evaporate the chloroform using a N_2 stream for around 1 h and then to place the round-bottom flask inside the vacuum chamber for a few hours or overnight.
15. The most convenient way to make vesicles in this size range is by using a lipid extruder. The lipid extruder system can be obtained from several distributors. So far the best results were obtained when using the extruder system from Avanti Polar Lipids. The size distribution of the resulting vesicles in solution was 160 nm ± 40 nm as determined using an ALV dynamic light scattering system equipped with a krypton-ion laser. The size distribution of vesicles can be varied depending on the number of times the vesicle solution is passed through the polycarbonate membrane and the pressure used to push the vesicle solution through the membrane (in our case, 21 times and 1 kgf/cm^2). Please follow carefully the distributor's advice for maintenance and cleaning of the lipid extruder.
16. We used a Q-Sense D-300 QCM-D system for our measurements. However, the cleaning procedure can be adapted to all types of QCM-D systems. The flow operation on the D300 system is based on gravitation, while in newer models (E1 and E4) a peristaltic pump is used to control the flow rate. Please refer to www.q-sense.com for more information regarding the theoretical foundation of the QCM-D technique and modes of operation.
17. Before starting each measurement, rinse all liquid handling parts with buffer for couple of times to remove air bubbles that may trap inside the liquid handling parts.
18. We used a PicoSPM microscope with a large-area scanner from Agilent (formerly Molecular Imaging) in constant-force contact mode for our experiments, but in principle any other AFM system able to operate in liquid should work. A custom-made, O-ring-sealed fluid cell able to accommodate standard QCM sensor crystals was manufactured.
19. Variations of the imaging force due to cantilever bending induced by thermal drift must strictly be avoided due to the fragile nature of the pit-spanning membrane. If the imaging force exceeds a critical value (typically around 1–2 nN, depending on lipid composition and tip radius), the AFM tip will penetrate the lipid bilayer and faithful imaging of the pit-spanning membrane will be impossible. It is recommended to use soft cantilevers with spring constants below ≈0.05 nN/nm in order to avoid large variations of the imaging force.
20. If there is a remaining thermal drift, the imaging force needs to be readjusted during scanning for the reasons mentioned in Note 18 above.

References

1. Gennis RB (1989) Biomembranes: moleculare structure and function. Springer, New York
2. Lipowsky R, Sackmann E (eds) (1995) Morphology of vesicles, vol 1. Elsevier Science B.V, Amsterdam
3. Mathews CK, van Holde KE, Ahern KG (1999) Biochemistry. Addison Wesley Longman, San Fransisco
4. Sackmann E (1996) Supported membranes: scientific and practical applications. Science 271:43–48
5. Salafsky J, Groves JT, Boxer SG (1996) Architecture and function of membrane proteins in planar supported bilayers: a study with photosynthetic reaction centers. Biochemistry 35:14773–14781
6. Larsson C, Rodahl M, Hook F (2003) Characterization of DNA immobilization and subsequent hybridization on a 2D arrangement of streptavidin on a biotin-modified lipid bilayer supported on SiO_2. Anal Chem 75:5080–5087
7. Graneli A, Rydstrom J, Kasemo B, Hook F (2003) Formation of supported lipid bilayer membranes on SiO_2 from proteoliposomes containing transmembrane proteins. Langmuir 19:842–850
8. Larsson C, Bramfeldt H, Wingren C, Borrebaeck C, Hook F (2005) Gravimetric antigen detection utilizing antibody-modified lipid bilayers. Anal Biochem 345:72–80
9. Radler J, Strey H, Sackmann E (1995) Phenomenology and kinetics of lipid bilayer spreading on hydrophilic surfaces. Langmuir 11:4539–4548
10. Groves JT, Boxer SG (2002) Micropattern formation in supported lipid membranes. Acc Chem Res 35:149–157
11. Keller CA, Kasemo B (1998) Surface specific kinetics of lipid vesicle adsorption measured with a quartz crystal microbalance. Biophys J 75:1397–1402
12. Reimhult E, Hook F, Kasemo B (2003) Intact vesicle adsorption and supported biomembrane formation from vesicles in solution: influence of surface chemistry, vesicle size, temperature, and osmotic pressure. Langmuir 19:1681–1691
13. Franz V, Loi S, Mueller H, Bamberg E, Butt HJ (2002) Tip penetration through lipid bilayer in atomic force microscopy. Colloids Surf B Biointerfaces 23:191–2000
14. Reimhult E, Zach M, Hook F, Kasemo B (2006) A multitechnique study of liposome adsorption on Au and lipid bilayer formation on SiO_2. Langmuir 22:3313–3319
15. Dimitrievski K, Reimhult E, Kasemo B, Zhdanov VP (2004) Simulations of temperature dependence of the formation of a supported lipid bilayer via vesicle adsorption. Colloids Surf B Biointerfaces 39:77–86
16. Richter RP, Berat R, Brisson AR (2006) Formation of solid-supported lipid bilayers: an integrated view. Langmuir 22:3497–3505
17. Pfeiffer I, Petronis S, Koper I, Kasemo B, Zach M (2010) Vesicle adsorption and phospholipid bilayer formation on topographically and chemically nanostructured surfaces. J Phys Chem B 114:4623–4631
18. Sapuri-Butti AR, Butti RC, Parikh AN (2007) Characterization of supported membranes on topographically patterned polymeric elastomers and their applications to microcontact printing. Langmuir 23:12645–12654
19. Okazaki T, Morigaki K, Taguchi T (2006) Phospholipid vesicle fusion on micropatterned polymeric bilayer substrates. Biophys J 91: 1757–1766
20. White RJ, Zhang B, Daniel S, Tang JM, Ervin EN, Cremer PS, White HS (2006) Ionic conductivity of the aqueous layer separating a lipid bilayer membrane and a glass support. Langmuir 22:10777–10783
21. Schmitt EK, Vrouenraets M, Steinem C (2006) Channel activity of OmpF monitored in nano-BLMs. Biophys J 91:2163–2171
22. Reimhult E, Kumar K (2008) Membrane biosensor platforms using nano- and microporous supports. Trends Biotechnol 26:82–89
23. Han X, Studer A, Sehr H, Geissbühler I, Di Berardino M, Winkler FK, Tiefenauer LX (2007) Nanopore arrays for stable and functional free-standing lipid bilayers. Adv Mater 19:4466–4470
24. Cheng Y, Bushby RJ, Evans SD, Knowles PF, Miles RE, Ogier SD (2001) Single ion channel sensitivity in suspended bilayers on micromachined supports. Langmuir 17:1240–1242
25. White RJ, Ervin EN, Yang T, Chen X, Daniel S, Cremer PS, White HS (2007) Single ion-channel recordings using glass nanopore membranes. J Am Chem Soc 129:11766–11775
26. Kresak S, Hianik T, Naumann RLC (2009) Giga-seal solvent-free bilayer lipid membranes: from single nanopores to nanopore arrays. Soft Matter 5:4021–4032
27. Mager MD, Almquist B, Melosh NA (2008) Formation and characterization of fluid lipid bilayers on alumina. Langmuir 24: 12734–12737
28. Hennesthal C, Drexler J, Steinem C (2002) Membrane-suspended nanocompartments

based on ordered pores in alumina. Chemphyschem 10:885–889
29. Romer W, Steinem C (2004) Impedance analysis and single-channel recordings on nano-black lipid membranes based on porous alumina. Biophys J 86:955–965
30. Grant AW, Hu QH, Kasemo B (2004) Transmission electron microscopy 'windows' for nanofabricated structures. Nanotechnology 15:1175–1181
31. Weiskopf D, Schmitt EK, Kluhr MH, Dertinger SK, Steinem C (2007) Micro-BLMs on highly ordered porous silicon substrates: rupture process and lateral mobility. Langmuir 23: 9134–9139
32. Hanarp P, Sutherland DS, Gold J, Kasemo B (2003) Control of nanoparticle film structure for colloidal lithography. Colloids Surf A 214. doi:dx.doi.org/10.1016/S0927-7757(02)00367-9
33. Pfeiffer I, Seantier B, Petronis S, Sutherland D, Kasemo B, Zach M (2008) Influence of nanotopography on phospholipid bilayer formation on silicon dioxide. J Phys Chem B 112:5175–5181
34. Jonsson MP, Dahlin AB, Feuz L, Petronis S, Höök F (2010) Locally functionalized short-range ordered nanoplasmonic pores for bioanalytical sensing. Anal Chem 82:2087–2094
35. Svedhem S, Pfeiffer I, Larsson C, Wingren C, Borrebaeck C, Hook F (2003) Patterns of DNA-labeled and scFv-antibody-carrying lipid vesicles directed by material-specific immobilization of DNA and supported lipid bilayer formation on an Au/SiO_2 template. Chembiochem 4:339–343

Chapter 13

Characterization of Nanoparticle–Lipid Membrane Interactions Using QCM-D

Rickard Frost and Sofia Svedhem

Abstract

In vitro characterization of nanoparticles is becoming increasingly important due to the rapid development of novel nanoparticle formulations for applications in the field of nanomedicine and related areas. Commonly, nanoparticles are simply characterized with respect to their size and zeta potential, and additional in vitro characterization of nanoparticles is needed to develop useful nanoparticle structure–activity relationships. In this context it is highly interesting to characterize the interactions between nanoparticles and model interfaces, such as lipid membranes. Here, we describe a methodology to study such interactions using the quartz crystal microbalance with dissipation monitoring technique (QCM-D). In order to mimic some aspects of the native cell membrane, a supported lipid membrane is formed on the QCM-D sensor surface. Subsequently the membrane is exposed to nanoparticles, and the nanoparticle–lipid membrane interactions are monitored in real time. The outcome of such analysis provides information on the adsorption process (importantly kinetics and adsorbed amounts) as well as on the integrity of both the nanoparticles and the lipid membrane upon interaction. QCM-D analyses are suitable for screening of nanoparticle–lipid membrane interactions due to the fair throughput of the technique, which can be complemented, when needed, by additional analyses by other surface-sensitive analytical techniques.

Key words Nanoparticles, Lipid membrane, Extrusion, Liposomes, QCM-D

1 Introduction

Due to the great complexity of native cell membranes, model systems are commonly used to learn more about the membrane properties and interactions. The most prominent and widely used model systems are based on supported lipid membranes (1). Other common types of model systems include liposomes in suspension (2) and Langmuir–Blodgett films at the liquid–air interface (3). All of these systems have been used for nanoparticle–lipid membrane interaction studies.

The main advantage of supported lipid membranes is the confinement of the lipid membrane to a solid surface, which is useful for the application of various surface-sensitive analytical techniques.

Volkmar Weissig et al. (eds.), *Cellular and Subcellular Nanotechnology: Methods and Protocols*, Methods in Molecular Biology, vol. 991, DOI 10.1007/978-1-62703-336-7_13, © Springer Science+Business Media New York 2013

Common substrate materials for such studies include silica and titania, and the lipid membrane formation onto these substrates has been characterized in great detail by, e.g., the quartz crystal microbalance with dissipation monitoring (QCM-D) technique. The key feature of QCM-D, which is sensitive to small mass changes on the sensor surface and to the viscoelastic (nanomechanical) properties of the adsorbed material (4), is the ability of the technique to distinguish between supported membrane and adsorbed, intact liposomes (5, 6). The QCM-D analysis is based on the piezoelectric properties of thin, single crystalline quartz discs (sensors) and gives two different responses: the shift in resonance frequency (Δf) of the quartz crystal, which is related to the amount of mass adsorbed to the sensor surface, and the shift in dissipation (ΔD), which originates from energy losses during the measurement. A soft material on the sensor surface, as, e.g., adsorbed intact liposomes, efficiently damps the oscillatory motion of the quartz crystal and gives rise to a large dissipation response. This is opposite to the small dissipation shift that is a characteristic for a supported lipid membrane. In particular, the combination of (QCM-D) and atomic force microscopy (AFM) has proven valuable to understand the mechanisms of the process by which the lipid membrane forms on the solid support (5, 7, 8), a process which depends on, e.g., the substrate material (8), the ion content, and the ionic strength of the buffer used (9), as well as lipid composition, size, and charge of the liposomes (8, 10). QCM-D is an acoustic technique, sensitive to all mass that is acoustically coupled to the oscillatory motion of the sensor and thereby complementary to, e.g., the more well-established optical sensor techniques based on surface plasmon resonance. Importantly, optical techniques are generally not able to distinguish different mechanisms for the formation of supported lipid membranes from each other (7, 11).

Here, we describe a methodology developed in our laboratory to study the interactions between nanoparticles and supported lipid membranes using QCM-D (Fig. 1). The described approach is general and can be applied to all water-soluble nanoparticles. More specifically, nanoparticles developed for drug delivery (nanodrugs) can be characterized with respect to their membrane interactions and also with respect to other properties such as pH responsiveness, structural stability, or effect of surface modifications (10, 12). We have also demonstrated lipid exchange between liposomes and supported lipid membranes (13). Studies of inorganic nanoparticles are in progress. The ability of this screening platform to predict, e.g., cytotoxicity remains to be proven. It is however tempting to suggest that conditions for the experiments may be tuned such that disruption of the model membrane becomes indicative for cytotoxic effects. In such cases, the QCM-D analysis can be combined with AFM to visualize nanoparticle-induced membrane damages.

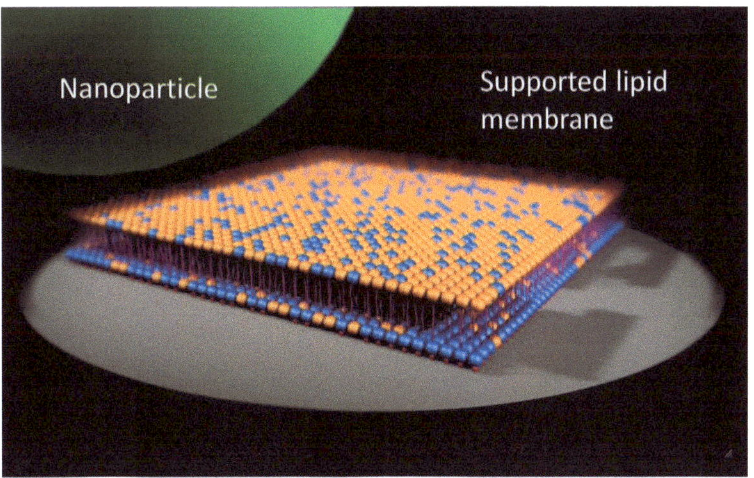

Fig. 1 Schematic illustration of a nanoparticle approaching a supported lipid membrane

The procedures below describe the protocol that we commonly follow for the preparation of liposomes (by extrusion), the spontaneous formation of supported lipid membranes on SiO_2-coated QCM-D sensors, and the subsequent real-time interaction analysis. In addition to the use of a standard QCM-D flow module, a protocol using an open QCM-D module suitable for the verification of manual sample preparation procedures (commonly required in, e.g., AFM analysis) is described. For modeling of the data, we refer to other sources.

2 Materials

Chemicals and solvents should be of analytical grade. Water should be deionized (typically to a resistivity above 18 MΩ cm) and filtered, using, e.g., a MilliQ unit, Millipore, France.

2.1 Extrusion Components

1. Buffer for extrusion of liposomes. Here, phosphate buffered saline (PBS): 0.0015 M KH_2PO_4, 0.0081 M Na_2HPO_4, 0.0027 M KCl, 0.137 M NaCl, pH 7.4 (see Note 1), which can be conveniently prepared from tablets available from Sigma.
2. Extruder. Here, mini extruder, including syringes and stand, from Avanti Polar Lipids, USA (Fig. 2a). Polycarbonate membranes with suitable pore sizes (here, 100 and 30 nm) (see Note 2) and membrane supports (Whatman, USA).
3. Lipid molecules (see Note 3). Here, e.g., 1-palmitoyl-2-oleyl-*sn* glycero 3 phosphocholine (POPC), 1 palmitoyl 2 oleyl *sn*-glycero-3-phospho-L-serine (POPS), and 1-palmitoyl-2-oleyl-*sn*-glycero-3-ethylphosphocholine (POEPC) (Avanti Polar Lipids, USA).

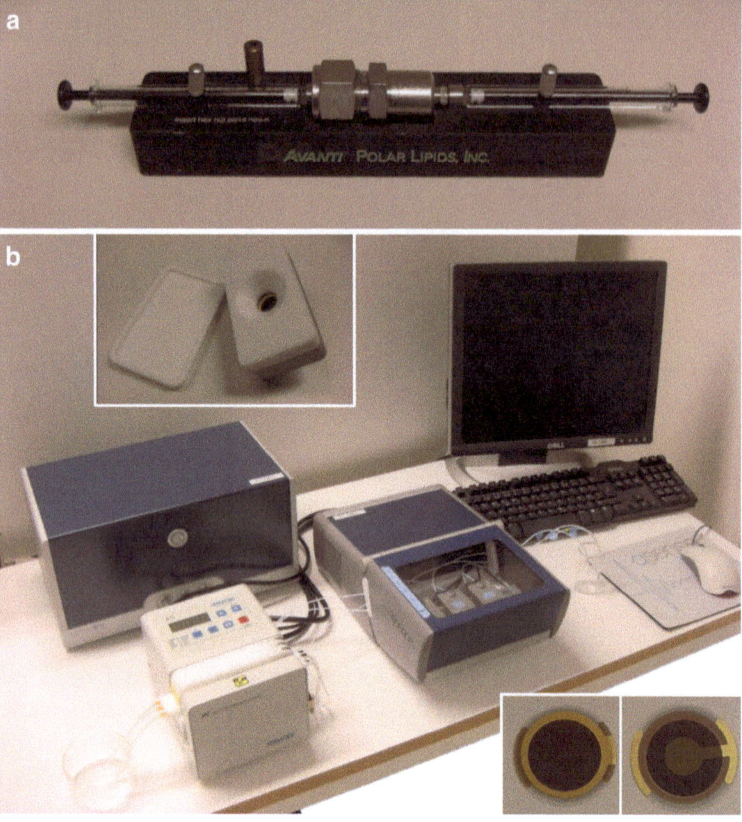

Fig. 2 The main equipment needed in the protocol. (**a**) Mini extruder for production of small unilamellar liposomes. (**b**) QCM-D E4 instrument with associated electronic unit, pump, and computer. An open QCM-D module is shown in the *upper inset*, and the two faces of a QCM-D sensor are shown in the inset at the *bottom*

4. Chloroform (or other suitable solvents for the lipid stock solution).
5. Round-bottomed flask.
6. Nitrogen gas.
7. Glass syringes for transfer of lipid stock solutions.
8. Ethanol for cleaning.
9. Optional: Equipment for determination of liposome size and zeta potential.

2.2 QCM-D Components

1. QCM equipment that measures both frequency and dissipation responses. Here, QCM-D E4 instrument (Q-Sense, Sweden) (Fig. 2b), including standard flow modules. Open QCM-D module (Q-Sense, Sweden) (Fig. 2b).
2. SiO_2-coated QCM-D sensors (Q-Sense, Sweden) (Fig. 2b).

3. UV–ozone cleaner (or oxygen plasma cleaner).
4. Degassed and filtered PBS buffer (see Subheading 2.1).
5. Equipment for degassing of buffers (e.g., sonication under reduced pressure).
6. Equipment for filtering buffer (e.g., plastic syringes and 0.2 μm syringe filters).
7. Cleaning solutions. Here, 10 mM SDS in water, 2% Hellmanex (Hellma, Germany), and Cobas cleaner (Roche, Germany).
8. Nanoparticles to be studied.
9. Buffer for dilution of nanoparticles.
10. Optional: Equipment for determination of nanoparticle size and zeta potential.

3 Methods

3.1 Preparation of Liposomes by Extrusion

There are several methods for the preparation of liposomes. Extrusion was first suggested by MacDonald et al. in 1991 (14), and a similar protocol is available from Avanti Polar Lipids:

1. If the supplier, provides the lipids in powder form, prepare stock solutions of the lipids in chloroform (or another suitable solvent, e.g., methanol). For example, weigh 40 mg of lipid in a glass vial with a cap resistant to chloroform, and add 4.0 mL of chloroform, to reach a final lipid concentration of 10 mg/mL (see Note 4).
2. Add the desired amount of lipid solution(s) to a round-bottomed flask (see Note 4). Keep the total amount of lipid to 5–10 mg per extrusion.
3. Evaporate the solvent in a fume hood using a low flow of nitrogen gas. Especially towards the end of the procedure, rotate the flask while evaporating the last solvent such that a thin lipid film forms on the inner walls of the flask.
4. Connect the flask to a vacuum pump to remove any residual chloroform. Keep the low pressure for >1 h.
5. Rehydrate the lipids by adding buffer (see Note 1) to reach a final lipid concentration of 5 mg/mL. Swirl the flask to wet the lipid film or vortex to speed up the process. Upon dissolution of the lipid film, the solution becomes turbid.
6. Extract the lipid solution into one of the two syringes belonging to the mini extruder.
7. Mount a polycarbonate membrane with a pore size of 100 nm in the extruder. Use two membrane supports on each side of the membrane to avoid rupture of the membrane during extrusion.

8. Press the lipid solution through the membrane 11 times (see Note 5). In this process the lipid solution becomes less turbid as small vesicles are formed (see Note 6).
9. Optionally, the liposomes can be reextruded 11 times through a membrane with 30 nm pore size, to further improve on the liposome size distribution.
10. Place the liposome solution in a glass vial, purge with nitrogen gas, and store at 4°C.
11. Clean the mini extruder by rinsing the different parts in ethanol and water.

3.2 Formation of Supported Lipid Membranes on QCM-D Sensor Surfaces and Exposure to Nanoparticles

1. Before conducting QCM-D experiments investigating nanoparticle–lipid membrane interactions, it is preferred to ensure that the nanoparticle suspension is well defined. The nanoparticle size distribution, determined by, e.g., dynamic light scattering, should be homogenous (i.e., be represented by a single peak, PDI < 0.2).
2. Clean the QCM-D instrument prior to use. Depending on previous usage of the instrument, thorough cleaning (disassembly of the flow modules) may be required. It is often sufficient to rinse the system with 2% Hellmanex (>10 min) and water (>5 mL/module). See also Note 12. Blow-dry the flow modules with nitrogen gas.
3. Clean SiO_2-coated QCM-D sensors in UV–ozone for >30 min.
4. Mount the sensors, using a clean pair of tweezers, in the flow module for the QCM-D E4 instrument and connect the tubings to the pump (see Note 7).
5. Fill the module with buffer (here, PBS) at 100 μL/min and turn off the flow (see Note 8).
6. Activate the temperature control, find the resonance frequencies of the sensors (use several harmonics, e.g., the 3rd, 5th, 7th, 9th, 11th, and 13th overtone), and start the data acquisition. Wait until the baselines (both f and D) are stable.
7. Restart the data acquisition, in order to reset both QCM-D responses (Δf and ΔD) to zero, and restart the flow (100 μL/min).
8. After 10 min of stable baselines, stop the flow (to avoid air bubbles in the system) and change solution from the buffer to a solution of liposomes (0.1 mg/mL) (see Note 9). Restart the flow (100 μL/min). The supported lipid membrane will form spontaneously within approximately 10 min (Fig. 3).
9. Evaluate the Δf and ΔD responses. For a supported lipid membrane of good quality, characteristic QCM-D responses,

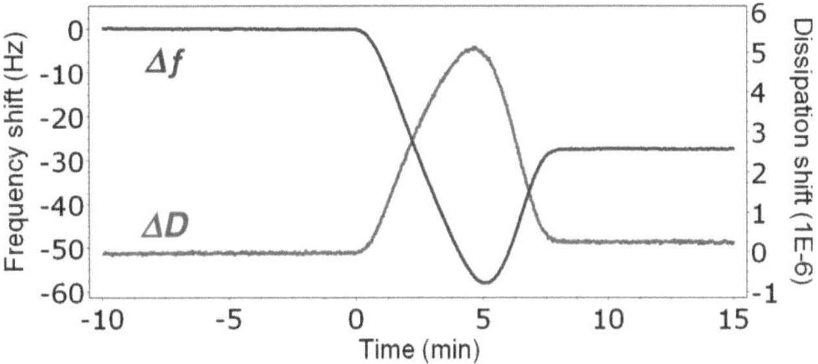

Fig. 3 QCM-D responses (7th overtone) during the formation of a POPC membrane. At $t = 0$ min the SiO_2 surface is exposed to POPC liposomes in PBS. The liposomes first adsorb intact, which generates a large decrease in frequency (corresponding to mass uptake) and a large increase in dissipation (corresponding to the formation of a soft layer) and later rupture and fuse into a supported lipid membrane. When the liposomes rupture, the enclosed liquid is released from the liposomes, a process that is accompanied by an increase in frequency and a decrease in dissipation, and the result is a supported lipid membrane

$\Delta f \approx -26$ Hz and $\Delta D < 0.5 \times 10^{-6}$, should be obtained (see Note 10).

10. If necessary, exchange the buffer in the QCM-D system to a buffer suitable for the nanoparticles to be studied (see Note 11). Do not forget to stop the flow while changing liquids.
11. Dilute the nanoparticle suspension to a suitable concentration, e.g. ,10–100 μg/mL, shortly before use.
12. Pass the nanoparticle suspension over the formed lipid membranes until the QCM-D responses have reached a steady state.
13. If applicable, the ambient medium (buffer) can be exchanged at this point to evaluate the responsiveness of the adsorbed nanoparticles, e.g., by changing the pH or the ionic strength.
14. When the experiment is finished, rinse the system with an SDS solution and water before emptying the tubing by pumping air. Dismount the flow modules and replace the sensors with ones that no longer are in use (maintenance sensors). Dry the sensors with nitrogen gas, and store until next use. Alternatively, to reduce surface contamination when reusing sensors frequently, the sensors can be stored in 10 mM SDS until next use.
15. Mount the maintenance sensors and clean the system with appropriate cleaning solutions, e.g., 2% Hellmanex and Cobas cleaner (see Note 12). Rinse extensively with water. Depending on the system under study, more thorough cleaning may be needed.
16. Dry the flow modules and the maintenance sensors with nitrogen gas

3.3 Validation of Manual Preparation Procedures Using an Open QCM-D Module

1. Clean the Teflon part of the open module and the o-ring by sonication in ethanol (5 min) and water (5 min).
2. Clean a SiO_2-coated QCM-D sensor in UV–ozone for >30 min.
3. Mount the sensor, using a clean pair of tweezers, in the open module, and place the module on the E4 (or E1) instrument.
4. Add 300 μL of buffer (e.g., PBS), put on the lid, start the temperature control, find the resonances of the quartz crystal, and start the data acquisition (see Note 13).
5. When the baselines are stable, form the supported lipid membrane by adding the liposome solution (0.1 mg/mL). When adding a new solution, partially exchange the liquid by adding 300 μL of solution, mixing with the pipette, and withdrawing 300 μL of the solution (see Note 14). Repeat this step 3–5 times (see Note 15).
6. Proceed with the same steps as when performing the experiment in a flow module (see Subheading 3.2), i.e., evaluate the quality of the supported lipid membrane, exchange buffer (if necessary), add nanoparticles, and rinse with buffer. The pipetting may induce some small spikes in the QCM-D response.
7. When the experiment is finished, empty and dismount the module. Clean the Teflon part of the module and the o-ring by sonication in ethanol (5 min) and water (5 min). Dry the clean parts.

4 Notes

1. It is also common to prepare liposomes in buffers based on, e.g., Tris or Hepes. If possible, it is preferred to use the same buffer as the nanoparticles to be studied are dispersed in.
2. The membrane pore size that is used affects the final size of the extruded liposomes (15). However, the hydrodynamic diameter (typically measured with DLS) does not necessarily correspond to the membrane pore size. Liposomes extruded through a 100 nm membrane followed by a 30 nm membrane typically measure 80–90 nm in diameter. The applied pressure is another parameter affecting the final size of the liposomes (15, 16).
3. Liposomes can be prepared with a wide variety of different lipid compositions, yielding, e.g., liposomes with different net charge. However, it should be noted that the lipid composition of the liposomes affects both their stability in suspension and their ability to spontaneously rupture and fuse into a supported lipid membrane on a SiO_2 surface.

4. Use glassware, e.g., glass syringes, when handling chloroform. Ordinary plastic pipette tips may release contaminants that alter the properties of the lipid solution.
5. The more passages through the membrane that are performed, the more homogeneous the liposome suspension becomes. Often, 11 passages have been used to produce liposomes with a narrow size distribution (low polydispersity). Use an odd number of passages through the membrane to clear the liposome solution from larger size contaminations.
6. Depending on the phase transition temperature of the lipids that are used, the extrusion may need to be performed at elevated temperatures. The extrusion must be performed at a temperature exceeding the phase transition temperature of the lipids. If necessary, the equipment can be preheated in an oven and placed on a hot plate during the extrusion.
7. Before filling the modules with liquid, it is suggested to check that it is possible to find the resonance frequencies of all sensors that are to be used.
8. When starting a new measurement, it normally takes some time (<20 min) before the baselines have stabilized (drift <1 Hz during 10 min). During this time it is preferred not to have any flow over the sensor surface to avoid contamination.
9. Dilute the liposome stock solution (5 mg/mL) in PBS to reach a final concentration of 0.1 mg/mL. If negatively charged lipids are used, e.g., POPS or POPG, it is preferable to add 5 mM of $MgCl_2$ to the buffer. Divalent cations, e.g., Mg^{2+} or Ca^{2+}, aid the formation of negatively charged lipid membranes. Note that Ca^{2+} should not be added to phosphate buffers (to avoid precipitation).
10. If the expected shifts in frequency and dissipation are not met, i.e., if $\Delta f < -26$ Hz and $\Delta D > 0.5 \times 10^{-6}$, a lipid membrane of good quality was not formed. Ideally ΔD should be close to zero. Any increase in dissipation suggests that there are still intact liposomes adsorbed to the surface. The most common reason for not reaching the desired criteria is that the sensor surface is contaminated before it is exposed to liposomes, or simply needs to be replaced. A second reason for why the characteristic QCM-D responses not always are obtained regards the lipid composition. Liposomes of some lipid compositions do not spontaneously fuse and form a supported lipid membrane on a SiO_2 support (8).
11. When changing buffer solution in the system, there will often be a shift in Δf and ΔD due to the change in bulk composition.

12. The procedure necessary to clean the instrument after use depends on what samples that have been used in the experiment. One general cleaning protocol is as follows. Fill the system with 2% Hellmanex (basic), turn off the flow, and leave for 30 min. Rinse briefly with water and fill the system with Cobas cleaner (acidic), turn off the flow, and leave for 10 min. Rinse extensively with water before drying the modules.
13. The lid to the open module prevents contamination of the liquid from the surroundings and slows down the evaporation process. However, if the lid is used, it will generate reversible shifts in the QCM-D responses when it is removed. When the lid is put back, the responses returns to their original values.
14. Partially exchanging the liquid in the open module with a pipette is performed to avoid exposing the surface to air. However, this is a rather uncontrolled process. The rate of which the new solution is added potentially influences the outcome of the analysis. If the addition rate is critical for a specific sample is easily evaluated in the open module. This information is important when manually preparing samples for other types of analyses, without real-time data acquisition. When adding the liposome solution to form a supported lipid membrane, the pipetting procedure is not crucial.
15. The number of liquid exchanges that is needed depends on the two liquid compositions. Since the new solution is diluted twofold in every liquid exchange, the liquid will never be completely exchanged. The dilution can be calculated according to $2n$, where n is the number of liquid exchanges. This fact is particularly important when removing one component from the system. In such case, perhaps 15–20 liquid exchanges are needed to sufficiently lower the concentration of that component. When adding a component (e.g., the nanoparticles), the dilution factor is not as crucial since the sample concentration often can be adjusted beforehand. However, another factor may be important in these cases. Since the analysis is performed in batch mode, the sample could be depleted close to the sensor surface if adsorption occurs. In such cases additional liquid exchanges may be needed after some time.

Acknowledgment

The Swedish Environmental Research Council FORMAS is acknowledged for financial support.

References

1. Sackmann E (1996) Supported membranes: scientific and practical applications. Science 271:43–48
2. Sessa G, Weissmann G (1968) Phospholipid spherules (liposomes) as a model for biological membranes. J Lipid Res 9:310–318
3. Zasadzinski JA, Viswanathan R, Madsen L, Garnaes J, Schwartz DK (1994) Langmuir-Blodgett films. Science 263:1726–1733
4. Rodahl M, Höök F, Fredriksson C, Keller CA, Krozer A, Brzezinski P, Voinova M, Kasemo B (1997) Simultaneous frequency and dissipation factor QCM measurements of biomolecular adsorption and cell adhesion. Faraday Discuss 107:229–246
5. Keller CA, Kasemo B (1998) Surface specific kinetics of lipid vesicle adsorption measured with a quartz crystal microbalance. Biophys J 75:1397–1402
6. Cho NJ, Frank CW, Kasemo B, Höök F (2010) Quartz crystal microbalance with dissipation monitoring of supported lipid bilayers on various substrates. Nat Protoc 5:1096–1106
7. Reimhult E, Zäch M, Höök F, Kasemo B (2006) A multitechnique study of liposome adsorption on Au and lipid bilayer formation on SiO_2. Langmuir 22:3313–3319
8. Richter RP, Bérat R, Brisson AR (2006) Formation of solid-supported lipid bilayers: an integrated view. Langmuir 22: 3497–3505
9. Rossetti FF, Textor M, Reviakine I (2006) Asymmetric distribution of phosphatidyl serine in supported phospholipid bilayers on titanium dioxide. Langmuir 22:3467–3473
10. Frost R, Grandfils C, Cerda B, Kasemo B, Svedhem S (2011) Structural rearrangements of polymeric insulin-loaded nanoparticles interacting with surface-supported model lipid membranes. J Biomater Nanobiotechnol 2: 180–192
11. Edvardsson M, Svedhem S, Wang G, Richter R, Rodahl M, Kasemo B (2009) QCM-D and reflectometry instrument: applications to supported lipid structures and their biomolecular interactions. Anal Chem 81:349–361
12. Frost R, Coué G, Engbersen JFJ, Zäch M, Kasemo B, Svedhem S (2011) Bioreducible insulin-loaded nanoparticles and their interaction with model lipid membranes. J Colloid Interface Sci 362(2):575–583
13. Kunze A, Sjövall P, Kasemo B, Svedhem S (2009) In situ preparation and modification of supported lipid layers by lipid transfer from vesicles studied by QCM-D and TOF-SIMS. J Am Chem Soc 131:2450–2451
14. MacDonald RC, MacDonald RI, Menco BPM, Takeshita K, Subbarao NK, Hu L-R (1991) Small-volume extrusion apparatus for preparation of large, unilamellar vesicles. Biochim Biophys Acta 1061:297–303
15. Frisken BJ, Asman C, Patty PJ (2000) Studies of vesicle extrusion. Langmuir 16:928–933
16. Patty PJ, Frisken BJ (2003) The pressure-dependence of the size of extruded vesicles. Biophys J 85:996–1004

Chapter 14

Single-Cell Nanosurgery

Maxwell B. Zeigler and Daniel T. Chiu

Abstract

This chapter explains the steps necessary to perform laser surgery upon single adherent mammalian cells, where individual organelles are extracted from the cells by optical tweezers and the cells are monitored post-surgery to check their viability. Single-cell laser nanosurgery is used in an increasing range of methodologies because it offers great flexibility. Its main advantages are (a) there is not any physical contact with the cells so they remain in a sterile environment, (b) high spatial selectivity so that single organelles can be extracted from specific areas of individual cells, (c) the method can be conducted in the cell's native media, and (d) in comparison to other techniques that target single cells, such as micromanipulators, laser nanosurgery has a comparatively high throughput.

Key words Laser nanosurgery, Optical tweezers, Apoptosis, Opticution, Necrosis

1 Introduction

The effects of focused laser light on living bodies, ranging from single cells to entire organisms, have been thoroughly researched (1–3). At the scale of entire organisms, lasers have been used in neuroscience applications, like the ablation of individual synapses in *Caenorhabditis elegans* (4, 5), or in developmental biology with model organisms such as *Drosophila melanogaster* (6) or zebra fish (7). Laser surgery is also particularly advantageous for single-cell or subcellular studies because mammalian cells are highly heterogeneous and segregated into biologically relevant units (8), such as organelles, which are comparable in size to the diffraction-limited focal spot of a laser.

High numerical aperture (NA) objective lenses can focus laser light down to spots on the order of hundreds of nanometers, which provide excellent spatial resolution without the need to physically manipulate the sample. When these diffraction-limited spots for spatially localized ablation are paired with a single-beam gradient optical trap, also known as optical tweezers, cell surgery can add exogenous material (9) or remove organelles from individual cells (1).

Volkmar Weissig et al. (eds.), *Cellular and Subcellular Nanotechnology: Methods and Protocols*, Methods in Molecular Biology, vol. 991, DOI 10.1007/978-1-62703-336-7_14, © Springer Science+Business Media New York 2013

Optical tweezers are not limited to biological samples and can be used to trap any object with a higher refractive index than its surroundings, assuming it is of an appropriate size and the trapping laser has sufficient power. A setup with optical tweezers and an ablation laser is amenable to a wide array of complementary biophysical laboratory techniques, such as fluorescence microscopy, and they have generated great interest for some time (10, 11).

Post-surgery, the cell may do one of three things: It may recover and remain a viable cell; it may undergo laser-induced necrosis, known as "optication"; or it may undergo apoptosis over a period of many hours. Our area of interest has been in the viability of different cell lines following cell surgery (1) and the effect of different membrane disruption techniques on cell survival (3). The method described here explains how to monitor the long-term viability of a single cell after the removal of an organelle by laser nanosurgery.

2 Materials

2.1 Polydimethylsiloxane Wells Used for Cell Surgery and Staining

1. Sylgard elastomer kit (Dow Corning, Midland, MI).
2. Photoetched glass coverslips (Bellco, Vineland, NJ).
3. Scientific oven.
4. Plasma cleaner (Harrick Plasma, Ithaca, NY).
5. Vacuum pump.
6. Vacuum chamber.

2.2 Cell Culture Equipment and Reagents

1. All cell types were grown in 75-cm^2 culture flasks and stored in an incubator with 5% CO_2 atmosphere at 37°C.
   (a) NG108-15 cells (ATCC, Manassas, VA) grown in Dulbecco's modified Eagle's medium (DMEM), supplemented with 8% fetal bovine serum (FBS) and 1% penicillin/streptomycin.
   (b) CHO-M1 cells (ATCC) grown in DMEM with 10% FBS and 1% penicillin/streptomycin.
   (c) ES-D3 (ATCC) grown in DMEM with 10% FBS and leukocyte inhibitory factor to prevent differentiation.
2. 0.25% trypsin-EDTA solution.
3. Polydimethylsiloxane (PDMS) wells for cell surgery.
4. Milli-Q water.

2.3 Optical Setup

1. Microscope: TE2000 microscope (Nikon, Melville, NY) with epi-illumination port, camera port, turret for optical filter cubes, bright-field light source, and condenser.

2. Objective: 100× magnification 1.3 NA S Fluor oil immersion objective (Nikon, Melville, NY).
3. Differential Interference Contrast (DIC) with de Sénarmont optics for bright-field imaging to improve contrast.
4. Video capture: CCD camera (Cohu, San Diego, CA).
5. Ablation laser: Nanosecond pulse 337-nm N_2 Ultraviolet laser (Laser Sciences, Inc., Franklin, MA).
6. Laser for optical tweezers: continuous wave (CW) Nd:YAG laser with a TEM_{00} Gaussian beam profile (Spectra-Physics, Mountain View, CA).
7. Laser power meter (Ophir Optronics, Jerusalem, Israel).
8. Assorted neutral density (ND) filters, dichroics, lenses, optical filters, mirrors, and optomechanics (Omega Optical, Burlington, VT; Thorlabs, Newton, NJ; and Chroma Tech Corp, Rockingham, VT).

2.4 Cell Viability Assay

1. 2-μM ethidium homodimer.
2. 200-nM MitoTracker Red (Invitrogen, Grand Island, NY).
3. Annexin V Alexa Fluor 488 (Invitrogen, Grand Island, NY).
4. 488-nm CW sapphire laser (Coherent, Santa Clara, CA).

3 Methods

3.1 Making PDMS Wells

1. Mix the Sylgard elastomer kit base and catalyst thoroughly in a 10:1 ratio in a disposable cup; 100 g of base and 10 g of catalyst are enough to make ~12 wells.
2. Place the uncured silicone mixture in the vacuum chamber and pump air out of the chamber to remove air bubbles.
3. Pour the PDMS into a large petri dish. We have found that placing evenly spaced glass vials with a diameter slightly smaller than the "1" photoetched coverslips makes suitable molds to shape the PDMS as it cures. To ensure that the vials do not move through the PDMS before it hardens, keep the vials in place using a petri dish lid with holes drilled such that the vials go through the holes and thus are held in place by the holes in the lid.
4. Bake at 60°C for at least 1 h to cure.
5. Pull out the glass vials from the PDMS, remove the PDMS from the petri dish, and cut out the individual wells.
6. Expose PDMS and photoetched glass coverslips to oxygen plasma for 60 s.
7. Seal the PDMS to the coverslip by bringing both together; apply light pressure to expunge any air pockets between the glass and PDMS.

3.2 Membrane Disruption

1. Split cells into a PDMS well 24 h before surgery using the cell's native media. The cells should be split at a low density to aid in identification during surgery and subsequent observation (see Note 1). Just prior to surgery, supplement the media in the PDMS well with ~25% Milli-Q water to induce cell swelling and help organelle extraction.
2. For the pulsed N_2 laser to disrupt the membrane, use a beam expander or lens pair to collimate and expand the beam to fill the back aperture of the microscope objective. This will produce a submicron focal spot size. We would recommend mounting at least one lens base on a rail to facilitate alignment of the focal spot in the axial direction.
3. Each laser pulse needs ~800 nJ of energy per pulse when measured before the objective. Our particular laser has a 10% pulse-to-pulse variability in pulse energy at 2 Hz at a 3-ns duration, and the objective has a roughly 30% transmission efficiency. This power is sufficient to disrupt the plasma membrane of an NG-108 cell with a single pulse to allow organelle extraction 50% of the time. Greater laser powers could lead to excessive cell damage and large cavitation bubbles; lower powers would have no visible effect. We modulate the pulsed laser's power using a neutral density filter and select the laser pulses using a shutter controlled by a transistor–transistor logic (TTL) switch.
4. Laser pulses should target the cell's plasma membrane to maximize membrane disruption and minimize cell damage (see Note 2). Focusing on the coverslip leads to ablation of the coverslip and immediate cell lysis; focusing on the cell cytoplasm creates large cavitation bubbles and excessive cell damage. Regardless of where the laser focus is directed, morphological changes will be evident for all cells following surgery.
5. If post-surgery cell viability is important and the cell line is particularly susceptible to death after surgery, a femtosecond near-IR pulse laser, instead of a nanosecond UV pulse laser, may improve cell viability. The mechanism of membrane disruption greatly affects cell viability after surgery (3). Membrane disruption by multiphoton absorption from a femtosecond laser produces a smaller focal volume, smaller cavitation bubbles, and other effects that help cell viability (2). The drawbacks of using a femtosecond laser for surgery are added cost and complexity.

3.3 Optical Trap Construction

This section provides brief guidance on the construction of a simple, robust, single-beam, gradient-force, three-dimensional optical trap with commercially available components. If more detail or explanation of optical alignment is needed, more comprehensive guides are available (12). With additional modification, trap position sensing (13), vortex-shaped optical traps (14), multiple traps (15, 16), force measurements (17, 18), and a myriad of other techniques are possible (19) but are beyond the scope of this chapter.

1. An inverted microscope is a stable and flexible platform for optical trapping. The back epifluorescence port is convenient for directing the trapping beam into the objective. Most research-grade inverted microscopes also come with a turret where multiple beam splitter cubes may be installed, which makes it convenient to switch between cell surgery and epifluorescence imaging modes. It is possible to build a platform for optical trapping without a microscope, but this requires a high degree of expertise in microscopy.
2. The objective lens is probably the most critical element of the optical trapping setup. For stable trapping, a high NA objective lens is necessary to generate the tight focus (20), such as the Nikon S Fluor 1.3 NA 100× oil immersion objective lens.
3. CW lasers emitting in the NIR, like the Nd:YAG laser, are used for optical trapping because they produce sufficient power at a wavelength that is relatively transparent to cells. Wavelengths shorter than ~700 nm are more readily absorbed by cells and should be avoided if possible because the high intensity in the focal volume results in excessive photodamage. Also, the laser should have an $M^2 < 1.1$ and emit the TEM_{00} mode to produce a light gradient steep enough to form a stable optical trap. For the sake of safety, a shutter should be installed at the point immediately after the beam exits the laser (see Note 3).
4. Place a dichroic mirror in the microscope's laser filter cube so that it will reflect IR and UV light but allow visible light to pass through. After alignment of the IR laser, an emission filter used should also block back-reflected IR and UV light from the optical tweezers and the ablation laser (see Note 4).
5. The power of the trapping beam should be minimized, but still be great enough to stably trap the organelle of interest. Most Nd:YAG lasers emit significantly more power than is necessary. The power is best modulated by using a $\lambda/2$ wave plate installed in a rotating mount followed by a polarizing beam splitter and a beam block that absorbs the excess laser power. With this configuration, the laser's power can be attenuated by rotating the wave plate. Controlling the laser power by external means instead of adjusting the drive current of the laser improves the laser power stability.
6. After the power attenuator, insert a pair of lenses into the beam path that expand the beam to fill or slightly overfill the back aperture of the objective. Beam expanders are commercially available for this purpose.
7. The most convenient way to steer the optical trap is by using a combination lens pair and a mirror to create a 4f system. Mount

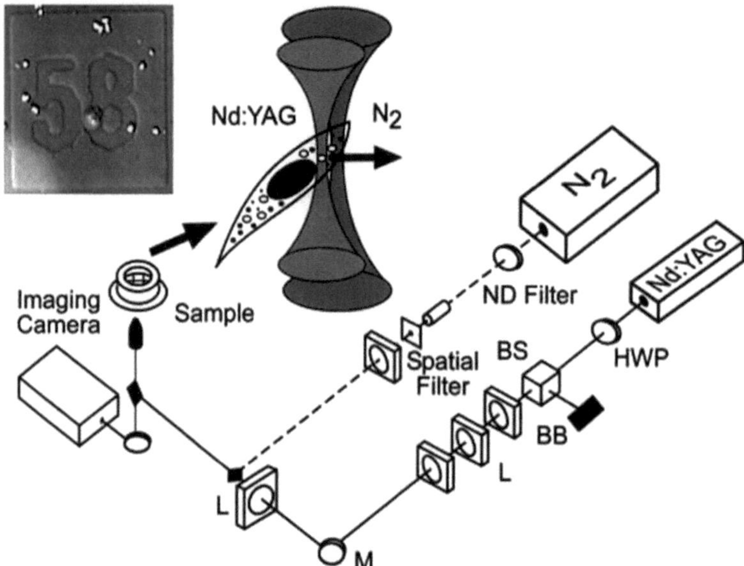

Fig. 1 Schematic of optical setup. The outputs of a pulsed N_2 laser (337 nm; 3-ns pulse width) and a CW Nd:YAG laser (1,064 nm) were aligned collinearly and sent into the microscope. The two beams were directed into the objective (NA = 1.3) by a dichroic mirror and focused onto the cell by the objective lens. The microscope was equipped with Nomarski optics, but the components are omitted in the schematic for clarity. The pulsed N_2 laser (~0.7 μJ) was used to ablate a small hole (~1–3 μm) in the cell membrane, and the YAG-trapping laser (200 mW) was employed to extract a subcellular organelle (*arrow* pointing away from focal volume). To track individual cells post-surgery, we cultured cells in PDMS wells formed by sealing PDMS to photoetched coverslips (*upper left inset*). *HWP* half-wave plate, *BS* polarizing beam splitter, *BB* beam blocker, *ND* neutral density filter, *M* mirror, *L* lens. Reprinted with permission from (1)

the lens nearest the microscope on an *x*, *y*, *z* movable lens mount and arrange the lens pair so that each lens is the same distance from the central focal spot as its focal length. The lenses closest to the objective can be used to fine-tune the alignment of the optical tweezers. The movement of the final lens along the optical axis can be used to adjust the height of the optical trap above the objective. This adjustment will ensure that the plane of the optical trap and the image are parfocalized (see Note 5). The mirror is installed in a kinematic mirror mount at the conjugate focal plane of the objective and can be used to steer the optical trap (Fig. 1).

3.4 Organelle Trapping

1. Organelle trapping is performed using the optical tweezers described in the previous section. For our setup, a power of ~200 mW at the nosepiece of the microscope generates a sufficient gradient force to trap individual organelles. The power transmitted through the microscope's objective will vary depending on the % transmission through the objective and whether or not the trapping beam is larger than the back aperture of the objective (13).

2. Organelles near the spot of membrane disruption can be extracted from the cell after steering of the trapping beam (see Note 6). Adding Milli-Q water to the media improves the ease of organelle extraction without significantly reducing cell viability, but a small amount of cytoplasm is still often extracted with the targeted organelle (Fig. 2). The extracted organelles are often lysosomes or mitochondria. These organelles can be differentiated with the appropriate fluorescent stain.
3. We have observed that the high fluence of the IR laser leads to fast photobleaching of organelle stains and possible photodamage to the organelle. If the extracted organelle is intended for further analysis, such as single mitochondria capillary electrophoresis (21), photobleaching and photodamage to the organelle can be reduced by using an optical vortex trap (14) instead of the Gaussian optical trap described above.

3.5 Cell Viability Monitoring

1. After surgery, incubate the cells at regular intervals in pre-warmed media for 10 min containing the viability reagents (see Subheading 2.4). The particular dyes are chosen to determine cell viability because (a) they can all be excited with a single 488-nm laser, (b) their fluorescence is spatially and spectrally distinct so they are easily resolved, and (c) in addition to cell necrosis, they also suggest apoptosis prior to the morphological signs of apoptosis, which may take many hours to become apparent. Ethidium homodimer is a membrane-impermeable dye that will brightly stain the cell nucleus if the membrane loses integrity. MitoTracker Red is a mitochondrial stain that will have a significant decrease in brightness if the mitochondria lose the ability to maintain a proton gradient across their membrane. Annexin V Alexa Fluor 488 is a membrane-impermeable antibody that binds to phosphatidylserine, a phospholipid that can be found on the external leaflet of a cell's lipid bilayer only during apoptosis.
2. After 10 min, remove the cells from the incubator and wash with fresh pre-warmed media. Cell viability can be determined by observing the fluorescent stains and cell morphology. Some morphological indicators that a cell is dying, such as membrane blebs, a granular appearance in their cytoplasm, more spherical shape, and loss of adherence to the coverslip, can be viewed under bright-field microscopy.
3. Cell apoptosis can be differentiated with bright-field DIC and epifluorescence illumination (Fig. 3). The viable cell (a1, a2) has a normal morphology under bright field and displays bright-orange fluorescence under epifluorescent illumination from the MitoTracker stain. The cell undergoing apoptosis (b1, b2) has a more granular appearance under bright field. Under epifluorescence, it has decreased emission intensity from MitoTracker but a brightly fluorescent nucleus from the ethidium homodimer and a green fluorescent plasma membrane as Annexin V Alexa 488 binds to the lipid bilayer.

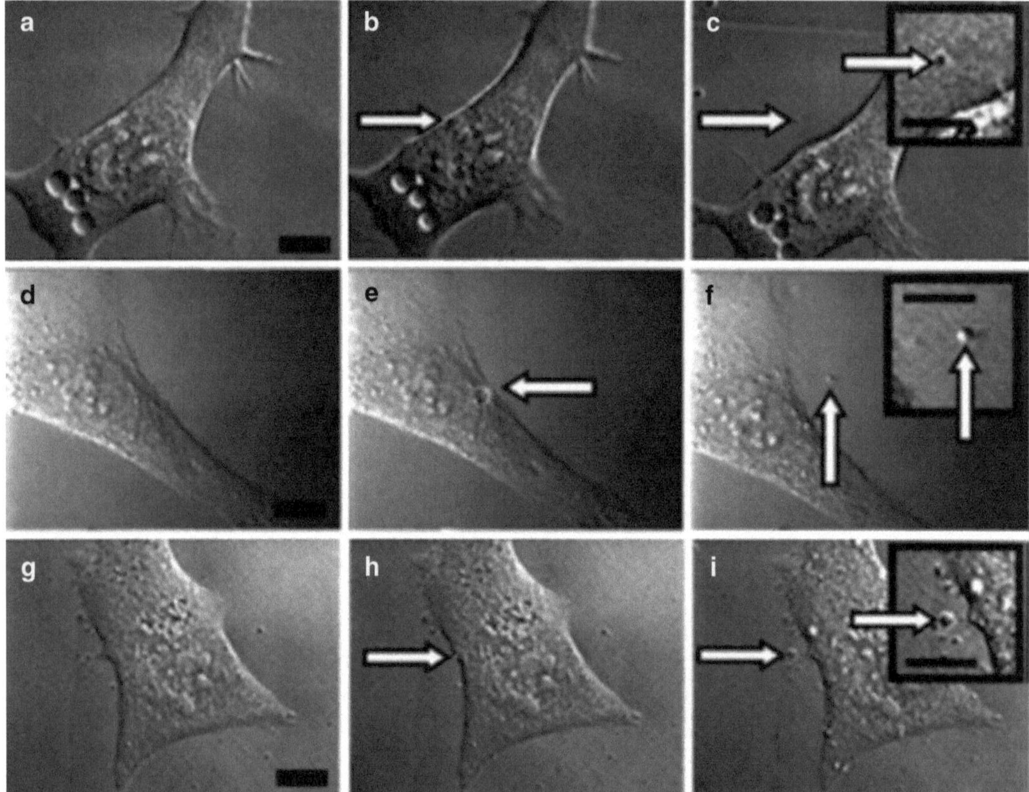

Fig. 2 Organelle extraction in successful surgeries. An NG108-15 cell (**a**) before, (**b**) during, and (**c**) after surgery; the *arrows* point to the organelle that was extracted. A similar sequence showing a CHO cell (**d**) before, (**e**) during, and (**f**) after surgery. A small cavitation bubble is visible in (**e**); (**f**) shows the organelle (*arrow*) that was completely removed from the cell. An ES-D3 stem cell (**g**) before, (**h**) during, and (**i**) after surgery. Scale bars = 10 μm. Reprinted with permission from ref. 1

4 Notes

1. Cells are known to migrate over time, a property dependent on the cell line. If long-term observation is necessary, be sure to either split the cells at a lower density or take images regularly to track cell movement. Photoetched glass coverslips also help to track individual cells.
2. When performing cell surgery, watch the surgery with a CCD camera. Do not use the microscope's oculars. IR light can leak through the microscope's oculars so place a light blocker over them when appropriate to prevent accidental exposure.
3. Proper laser safety is crucial and should include the appropriate laser goggles, beam blocks, shut-off keys, IR laser display cards, and so on. Safety measures are especially important in cell laser surgery because both the trapping and ablative lasers are out-

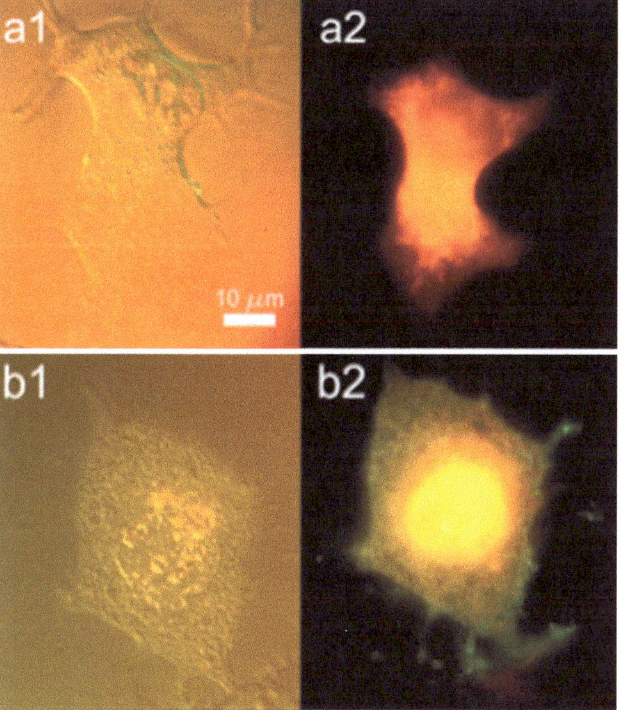

Fig. 3 Apoptosis assay. A healthy cell is observed using (**a1**) DIC microscopy and then with (**a2**) epifluorescence microscopy. Another cell undergoes apoptosis, as observed using (**b1**) DIC microscopy and (**b2**) epifluorescence microscopy. In panel (**a2**), the intense fluorescence from MitoTracker Red indicates that the mitochondria are able to maintain a potential gradient. In panel (**b2**), the cell is undergoing apoptosis so the MitoTracker Red fluorescence is decreased. In addition, ethidium has penetrated the cell's membrane and intercalated with the cell's DNA and Annexin V Alexa 488 has bound to the cell's membrane. Reprinted with permission from ref. 3

side the visible range and scattered light from either laser may be intense enough to cause eye damage.

4. Back reflection from optical tweezers can be intense enough to damage the camera so make sure its power is minimized during alignment and that the proper IR emission filter is in place.

5. The distance from the focal spot where an object is trapped by the optical tweezers can change with the object's size and composition. Some tweaking of the trap's focal spot may be necessary when trapping different samples to keep the objects in the same plane as the imaging plane.

6. If the emission filter blocks 100% of the back-reflected IR and UV light, then the trapping beam and ablation laser are invisible to the camera. Organelle extraction is easier if you find a method for tracking the position of the trapping beam and the

ablative pulsed laser so that they do not get lost between experiments or are steered outside the field of view of the camera. There are many ways to track the beams depending on the application: It can be as complex as building a trapping positioning system or as simple as marking the laser beam focal points on the monitor with a dry-erase marker.

Acknowledgments

We are grateful to NIH (NS062725, NS052637, GM085485) and NSF (CHE0844688 and CHE0924320) for support of this work.

References

1. Shelby JP, Edgar JS, Chiu DT (2005) Monitoring cell survival after extraction of a single subcellular organelle using optical trapping and pulsed-nitrogen laser ablation. Photochem Photobiol 81(4):994–1001
2. Niemz MH (2004) Laser-tissue interactions. Springer, Heidelberg
3. Zeigler MB, Chiu DT (2009) Laser selection significantly affects cell viability following single-cell nanosurgery. Photochem Photobiol 85(5):1218–1224
4. Allen PB, Sgro AE, Chao DL et al (2008) Single-synapse ablation and long-term imaging in live C. elegans. J Neurosci 173(1):20–26
5. Chung SH, Mazur E (2009) Femtosecond laser ablation of neurons in C. elegans for behavioral studies. Appl Phys A 96:335–341
6. Chang TN, Keshishian H (1996) Laser ablation of drosophilia embryonic motoneurons causes ectopic innervation of target muscle fibers. J Neurosci 16(18):5715–5726
7. Kohli V, Elezzabi AY (2008) Laser surgery of zebrafish (Danio rerio) embryos using femtosecond laser pulses: optimal parameters for exogenous material delivery, and the laser's effect on short- and long-term development. BMC Biotechnol. doi:10.1186/1472-6750-8-7
8. Heilmann I, Pidkowich MS (2004) Switching desaturase enzyme specificity by alternate subcellular targeting. Proc Natl Acad Sci USA 101(28):10266–10271
9. Stevenson DJ, Gunn-Moore FJ, Campbell P et al (2010) Single cell optical transfection. J R Soc Interface 7(47):863–871
10. Zhang H, Liu KK (2008) Optical tweezers for single cells. J R Soc Interface 5(24):671–690
11. Berns MW, Aist J, Edwards J et al (1981) Laser microsurgery in cell and developmental biology. Science 13(4507):505–513
12. Block SM (1998) Construction of optical tweezers. In: LS David, DG Robert, AL Leslie (ed) Cells: a laboratory manual, vol 2. Cold Spring Harbor Laboratory Press, Woodbury
13. Lee WM, Reece PJ, Marchington RF et al (2007) Construction and calibration of an optical trap on a fluorescence optical microscope. Nat Protoc 2(12):3226–3238
14. Jeffries GDM, Edgar JS, Zhao Y et al (2007) Using polarization-shaped optical vortex traps for single-cell nanosurgery. Nano Lett 7(2):415–420
15. Curtis JE, Koss BA, Grier DG (2002) Dynamic holographic optical tweezers. Opt Commun 207:169–175
16. Martín-Badosa E, Montes-Usategui M, Carnicer A et al (2007) Design strategies for optimizing holographic optical tweezers set-ups. J Opt A Pure Appl Opt 9: S267–S277
17. Mejean CO, Schaefer AW, Millman EA et al (2009) Multiplexed force measurements on live cells with holographic optical tweezers. Opt Express 17(8):6209–6217
18. Rohrbach A, Tischer C, Neumayer D et al (2004) Trapping and tracking a local probe with a photonic force microscope. Rev Sci Instrum 75(6):2197–2210
19. Moffitt JR, Chemla YR, Smith SB et al (2008) Recent advances in optical tweezers. Annu Rev Biochem 77:205–228
20. Ashkin A (1970) Acceleration and trapping of particles by radiation pressure. Phys Rev Lett 24(4):156–159
21. Allen PB, Doepker BR, Chiu DT (2009) High-throughput capillary-electrophoresis analysis of the contents of a single mitochondria. Anal Chem 81(10):3784–3791

Chapter 15

Single Quantum Dot Imaging in Living Cells

Jerry C. Chang and Sandra J. Rosenthal

Abstract

Direct visualization of biological processes at single-molecule level provides a detailed perspective which conventional bulk measurements are hard to achieve. Among various classes of fluorescent tags used in single-molecule tracking, quantum dots are particularly useful due to their unique photophysical properties. In this chapter, we describe the principles, methodologies, and experimental protocols for qdot-based single-molecule imaging. The first half provides an overview of fluorescent microscopy and advances in single-molecule tracking using quantum dots. The remainder of this chapter describes methods to carry out qdot-based single-molecule experiments. Detailed protocols including qdot labeling, microscopy setup, and single-molecule analysis using appropriate computational programs are given.

Key words Quantum dot, Biological labeling, Protein trafficking, Single-molecule, Fluorescence microscopy

1 Introduction

In recent years, microscopy techniques in cell biology have seen remarkable progress. Various new methods, such as near-field scanning optical microscopy (NSOM) (1), single-molecule high-resolution imaging with photobleaching (SHRIMP) (2), stimulated emission depletion (STED) microscopy (3), and stochastic optical reconstruction microscopy (STORM) (4), were developed to achieve a significantly higher spatial resolution, which permits discovery of subcellular structures in great detail. Despite the fact that these imaging methods are designed to break the optical diffraction barrier, they are not very accessible to cell biology researchers. The major shortcoming of these methods is that most of them are still under development in experimental physics laboratories and are not readily transferable to a standard laboratory microscope. Also, the data analyses are usually extremely complicated, requiring well-trained experts in the field.

Single-molecule fluorescent microscopy, derived from high-speed fluorescence microscopy, is probably the most accessible

Volkmar Weissig et al. (eds.), *Cellular and Subcellular Nanotechnology: Methods and Protocols*, Methods in Molecular Biology, vol. 991, DOI 10.1007/978-1-62703-336-7_15, © Springer Science+Business Media New York 2013

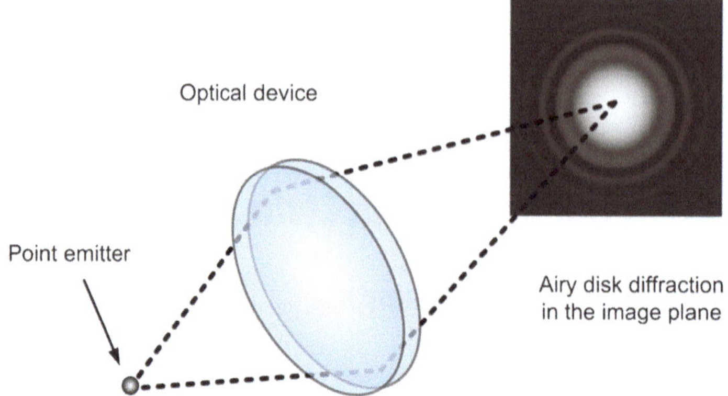

Fig. 1 Schematic representation of the diffraction pattern from a point emitter passed through an optical device. Owing to diffraction, the smallest distance to which an imaging system can optically resolve separate light sources at is limited by the size of the Airy disk

method for cell biologists to investigate single-molecule dynamics in living cells. The basis of this method is to follow the real-time movement of individual molecules by using fluorescent microscope, resulting in a map of the dynamics upon observing many individual events. Based on the above definition, an ultimate sensitivity is required, which allows *individual single molecules* to be monitored in a picoliter- to femtoliter-sized microscope sampling volume. However, even if performed on an optimal designed microscope equipped with a camera offering single photon sensitivity per frame, optical detection of a single molecule is still diffraction-limited. To achieve a sub-pixel localization in single molecule imaging, a 2D Gaussian approximation of point-spread function (PSF) approach is typically performed (5). Single emitters viewed through a microscope produce an Airy disc diffraction pattern in the image plane (Fig. 1). In the 2D Gaussian approximation approach, the diffraction pattern of a single emitter is analyzed in terms of PSF where the localized centroid is determined by fitting the PSF to a 2D Gaussian profile. The theoretical 2D paraxial PSF of the wide-field fluorescence microscope can be calculated as (5)

$$\mathrm{PSF} = \left[2 \times \frac{J_1(k_{\mathrm{em}} \cdot \mathrm{NA} \cdot r)}{k_{\mathrm{em}} \cdot \mathrm{NA} \cdot r} \right]^2, \tag{1}$$

where r is the radial distance to the optical axis, NA is the numerical aperture, J_1 is the first-order Bessel function, $k_{\mathrm{em}} = \frac{2\pi}{\lambda}$ defines the emission wave number, and λ is the wavelength of light. The smallest resolvable distance between two points of the 2D plane corresponds to the first root of the PSF and is given by the Rayleigh distances (6),

$$d_{xy}^{R} = 0.61 \frac{\lambda}{\mathrm{NA}}. \tag{2}$$

As can be seen from the Eqs. 1 and 2, under the important prerequisite in which single emitters are placed in extremely low concentration to be spatially separated greater than the diffraction-limited region, it is possible to identify the localization (x, y) of a single molecule from an optical microscope image by fitting the PSF. Indeed, the fundamental idea of the modern single-molecule microscopy is reliant on PSF fitting to localize the centroid position of single point emitters. Typically, fluorescent intensity distribution from a single emitter can be fit with a 2D Gaussian (7, 8). To calculate a subpixel estimate of the position from a single emitter, the general method is to fit the intensity distribution with a 2D Gaussian function and then calculate the local maximum intensity:

$$I_{xy} = A_0 + A \cdot e^{-\frac{(x-x_0)^2+(y-y_0)^2}{w^2}}, \tag{3}$$

where I_{xy} is the intensity of the pixel, x_0 and y_0 is the designated local maximum coordinates of the Gaussian, A is the amplitude of the signal with local background A_0, and w is the width of the Gaussian curve. The smallest distance at which two emitters can be recognized and separated is roughly equal to the full width at half maximum (FWHM) of the w:

$$w_{\mathrm{FWHM}} = w \cdot \sqrt{\ln(4)} = 1.177w. \tag{4}$$

The coordinate (x_0, y_0) acquired by fitting function Eq. 3 using chi-squared minimization is not a true position, but only an estimate. The accuracy is strongly dependent upon the respective signal-to-noise ratio (SNR), which is defined as (8)

$$\mathrm{SNR} = \frac{I_0}{\sqrt{\sigma_{\mathrm{bg}}^2 + \sigma_{I_0}^2}}, \tag{5}$$

where I_0 is the maximum signal intensity above background, σ_{bg}^2 is the variance of the background intensity values, and $\sigma_{I_0}^2$ is the true variance of the maximum signal intensity above the background. Since w width is approximately equal to the wide-field diffraction limit (for visible light is about 250 nm), the uncertainty of the fitted coordinate ($\Delta\sigma$) is approximately given by

$$\Delta\sigma \approx \frac{250}{\mathrm{SNR}} (\mathrm{nm}). \tag{6}$$

However, common organic fluorophores used in single-molecule imaging have very limited photon yield which makes them difficult to produce a sufficient SNR for single-molecule imaging. It is also important to note that organic fluorescent dyes also suffer from narrow Stokes shift (difference in excitation and emission wavelength). This drawback can increase background signals and further reduce SNR. In addition, photobleaching associated with organic fluorophores prevents long-term monitoring in single-molecule microscopy. Hence, the shortcomings associated with organic fluorophores place a high demand for new fluorescent materials for single-molecule tracking.

Over the past decade, quantum dots (qdots) have shown tremendous potential for in vitro and in vivo biological imaging (9–14). Among the fluorophores developed for single-molecule tracking, a fluorescent qdot is, perhaps, the ideal material for its enhanced brightness, exceptionally high quantum yield, narrow emission profile, large Stokes shift, and, most importantly, excellent photostability. Another interesting property associated with qdots is the fluorescent intermittency, or blinking, phenomenon (15), whereby a time trace of fluorescent intensity from a single qdot can switch between two distinct on/off states. This blinking phenomenon is sometimes considered a minor drawback since it might cause temporary trajectory data loss in single-molecule tracking (16). On the other hand, blinking is often used to identify single molecules, providing a practical benefit (17, 18).

Using qdot-based single-molecule tracking, Dahan and coworkers first reported the diffusion dynamics of individual glycine receptors in living neurons (19). By conjugating Fab fragments of antibody to single qdots, they were able to track the movement of individual glycine receptors for more than 20 min. Similar approaches were then employed for various protein targets including nerve growth factor (NGF) (20), gamma-aminobutyric acid A receptor ($GABA_A$ R) (21), glycosyl-phosphatidylinositol (GPI)-anchored protein (22), serotonin transporter (23), and GM1 ganglioside (24). In 2008, Murcia et al. pushed the temporal resolution of single qdot tracking up to an amazing 1,000 frames/s by using an extremely sensitive double micro-channel plate (MCP) image intensifiers, cooled intensified CCD camera (dual MCP ICCD) (25). Through demonstration of single qdot tracking in living mice, Tyda et al. further suggested to use single qdots as tumor-targeting nanocarriers for in vivo drug delivery study (26). Recently, Zhang et al. utilized single qdots as artificial cargo of synaptic vesicles to validate the “kiss-and-run” model of neurotransmitter secretion (17). Overall, qdot-based single-molecule imaging holds the promise to reveal molecular biological mechanisms well beyond ensemble biochemical techniques.

The process of qdot-based single-molecule microscopy is typically divided into three steps (Fig. 2). The first step is to acquire time-lapse imaging after single target-specific labeling has been made. Single fluorescent molecules should be able to produce diffraction-limited blurred spots in each flame. The second step involves imaging data processing and single-molecule localization

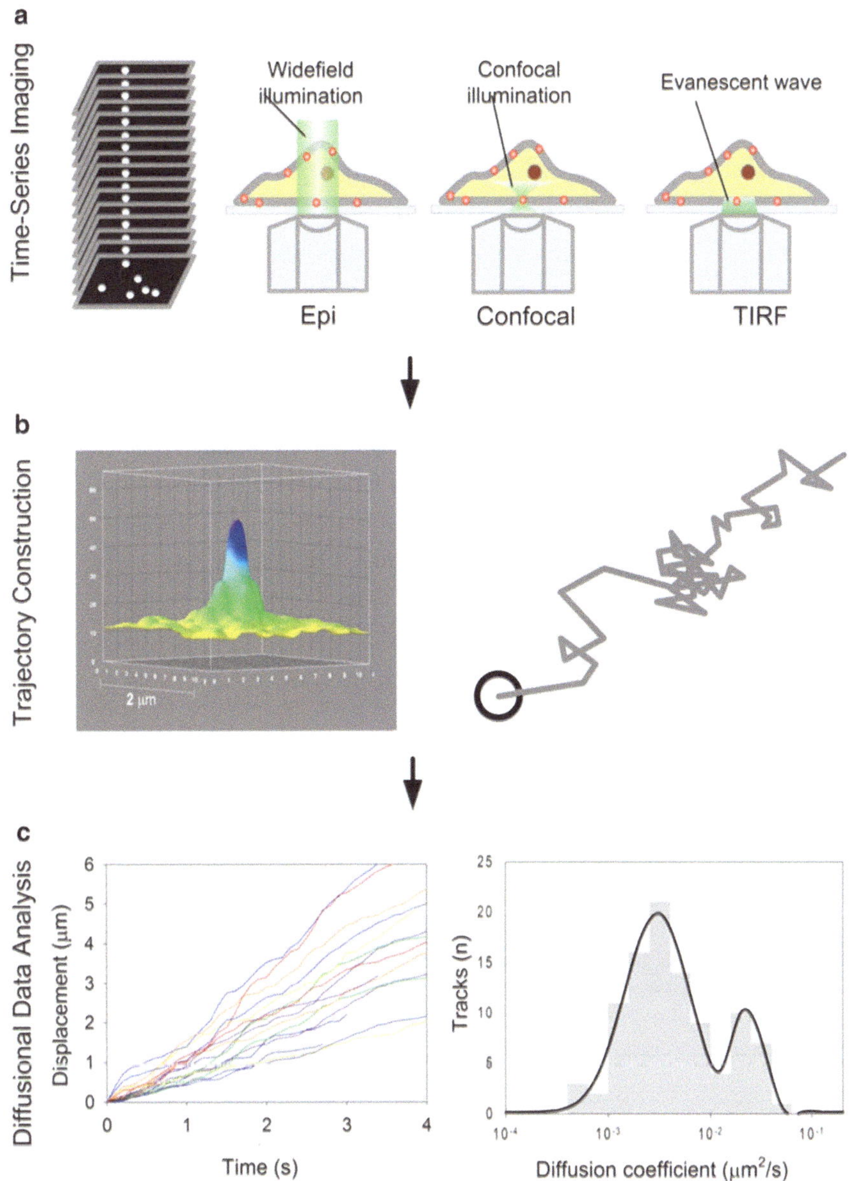

Fig. 2 Approach to single-molecule microscopy. (**a**) Time-lapse images of single fluorophore-tagged biomolecules in living cells acquired from an optical fluorescent microscope system (epifluorescence, confocal, or total internal reflection fluorescence (TIRF) microscope). (**b**) Estimation of the positions of single quantum dots with subpixel accuracy is accomplished by fitting the individual spot intensity values with a two-dimensional Gaussian distribution. After the positions of single quantum dots are identified, a trajectory of the target protein (*gray line*) can be subsequently derived from the time-series imaging data. (**c**) The final step is to analyze the diffusional properties (displacement, velocity, and diffusion coefficient) from single-molecule trajectories

from time-lapse images and is therefore normally anticipated as a computationally demanding step. In the third step, the diffusion dynamics can be analyzed from the trajectories of individual molecules, e.g., Brownian motion.

2 Materials

2.1 Reagents

1. Biotinylated small molecule or antibody against an extracellular epitope.
2. Qdot streptavidin conjugate (1 μM, Invitrogen Corporation, Carlsbad, CA).
3. 35-mm cell culture dishes with cover glass bottom (MatTek Corporation, Ashland, MA).
4. HeLa cells.
5. Dulbecco's Modified Eagle Medium (DMEM) (GIBCO, Invitrogen Corporation, Carlsbad, CA).
6. Phenol red-free Dulbecco's Modified Eagle Medium (DMEM) (GIBCO, Invitrogen Corporation, Carlsbad, CA).
7. Penicillin-streptomycin antibiotic mixture (100×) (GIBCO, Invitrogen Corporation, Carlsbad, CA).
8. Fetal bovine serum (FBS) (GIBCO, Invitrogen Corporation, Carlsbad, CA).
9. L-glutamine (GIBCO, Invitrogen Corporation, Carlsbad, CA).

2.2 Equipment, Software, and Accessories

1. Fluorescent microscope (see Subheading 3.1).
2. Microscope mounted heating chamber.
3. Microscope image acquisition and analysis software (MetaMorph 7.6, Molecular Devices, Sunnyvale, CA).
4. Technical computing software for numerical analysis (Matlab R2008b, MathWorks, Natick, MA).
5. Spin coater.

3 Methods

3.1 Imaging System Calibration Using Spin-Coated Single Quantum Dots

A general strategy to identify whether the optical system is able to detect individual qdots is to carry out time-lapse imaging of a very dilute qdot solution to avoid multiple qdots overlapping within diffraction-limited distance. Individual qdots are characterized by their blinking properties (8, 18). The emission intensity of a single qdot is shown in Fig. 3, where a single qdot blinks completely on and off during a time-lapse sequence of 80 s at a 10-Hz frame rate. When the fluorescence is produced by an aggregate structure consisting of several qdots, such blinking effects are completely canceled out. The protocol described below is based on a custom-built Zeiss Axiovert 200-M inverted fluorescence microscope coupled with a charge-coupled device (CCD) camera (Cool-Snap$_{HQ2}$, Roper Scientific, Trenton, NJ). To track single qdots in real time, the acquisition rate

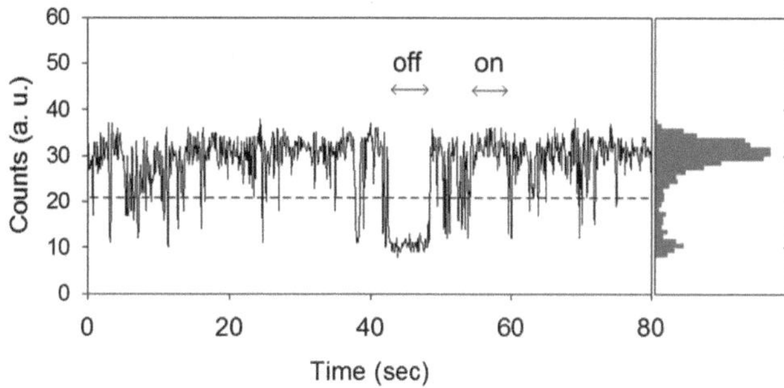

Fig. 3 A typical intensity over time plot from a single blinking qdot. As displayed in the *left panel*, the intensity trajectory of a single qdot displays two dominant states: an "on" state and an "off" state, termed blinking. A predefined intensity threshold is shown by the *dashed line*. *Right panel* displays the probability density distribution

should be set at 10 Hz or higher. However, the imaging rate is usually limited by the frame readout time of the camera. This particular CCD is chosen due to its adequate 60% quantum efficiency (QE) throughout the entire visible spectrum range (450–650 mm) with a frame rate >20 at 512 × 512 pixels (see Note 1). And again, as we introduced in the background section, a more advanced back-illuminated Electron Multiplying CCD (EMCCD) with sub-msec temporal resolution will be a much better choice. Imaging should be performed with a high-resolution (63× or 100×) oil-immersion objective lens with numerical aperture of 1.30 or greater.

The sample preparation steps are:

1. Prepare a clean microscope glass slide coverslip (or 35-mm culture dish with a coverslip at the bottom).
2. Add one drop (20 mL) of 100-pM Qdot® ITK™ carboxyl quantum dots solution onto the coverslip.
3. Spin cast the qdot solution on the coverslip for 30 s at 500 rpm (~30 × g for common compact spin coater) sufficient to disperse qdot particles uniformly across the coverslip (see Note 2).
4. Mount the coverslip on the microscope stage.
5. Qdots are excited using a xenon arc lamp (excitation filter 480/40 BP) and detected with CCD camera through appropriate emission filter (600/40 BP for Qdot 605). Acquire time-lapse images (100 ms per frame, 60 s).

3.2 Single Quantum Dot Labeling in Living Cells

Single quantum dot labeling can be prepared through either a direct labeling (one-step) procedure or an indirect (two-step) protocol. In the direct labeling procedure, the target-specific probe (small molecule organic ligand, peptide, or antibody) is directly conjugated to the qdots surface to make ligand-qdot nanoconjugates. Therefore, the cellular labeling strategy could be performed in one

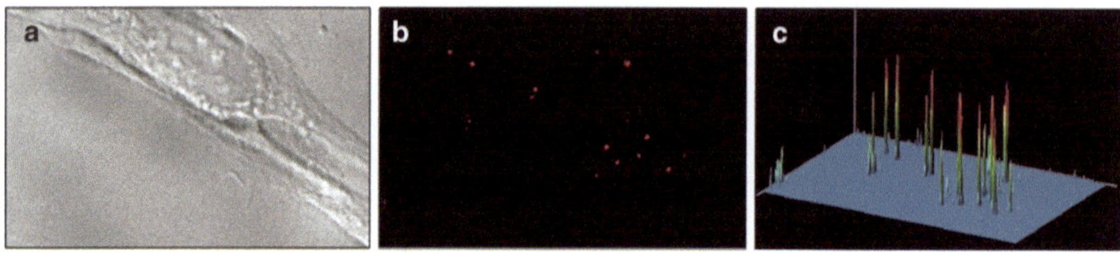

Fig. 4 Example of live-cell imaging of membrane proteins labeled with single qdots (**a**: bright-field image, **b**: fluorescence image, and **c**: surface intensity plot of **b**)

step in which the live-cell sample is incubated with a target-specific nanoconjugate prior to fluorescent imaging. In the two-step procedure, the cell sample is first incubated with biotinylated ligand to yield the desired specific ligand-protein binding. After an appropriate washing step, strep-qdots are added as the fluorescent tag of the biotinylated ligand-protein complex for the single-molecule imaging.

We provide a general protocol for single qdot labeling of adherent cells which is applicable to most mammalian cell lines. The standard protocol given below should be followed:

1. Prepare a 35-mm coverslip-buttoned culture dish with cells that have reached about 50% confluence.
2. Wash the cells gently three times with phenol red-free culture medium by repeatedly pipetting out.
3. Incubate cells with a biotinylated small molecule probe (0.5 nM to 0.5 mM dependent upon the biological affinity) or antibody (1–10 mg/mL) in red-free DMEM for 20 min at 37°C. (For one-step labeling protocol, incubate cells with 10–50 pM ligand-qdot nanoconjugates and skip steps 4 and 5) (see Note 3).
4. Wash cells gently three times with phenol red-free culture medium.
5. Incubate the cells with Qdot streptavidin conjugate (0.1–0.5 nM) in phenol red-free culture medium for 5 min at 37°C.
6. Wash the cells at least three times with phenol red-free culture medium.
7. Place the culture dish on the microscope stage with mounted heating chamber and heat to 37°C.
8. The labeling quality can be observed under fluorescent microscope. Punctate qdot staining should be visible through the eyepiece or CCD detector (Fig. 4). Single qdots can be identified by their blinking property.
9. Acquire time-lapse images at 37°C. In our experiments, acquisition procedure typically lasts for 60 s at 10 Hz rate.

3.3 Tracking Programs for Single-Molecule Analysis

Tracking and trajectory construction is a computationally demanding step of following single Qdot-labeled biological targets through successive images. One of the most important determinants of modern single-molecule tracking techniques is the nanometer accuracy, which is heavily weighted by the PSF fitting to localize the centroid position for subpixel resolution, normally demonstrated as a fitting of a 2D Gaussian function to a PSF. For practical application, estimated background-corrected intensities of an image are normally filtered out, a necessary step for the calculation of centroid position (x_0, y_0). After locations of single molecules are identified in each frame, the next step is to link the detected single-molecule positions. However, single qdot blinking brings additional difficulties for the trajectory generation, as the spots can temporarily disappear. A practical and most frequently used approach is to define a tolerance limit of blinking frames (usually 10 frames) and process an additional association step in the trajectory generation algorithm to merge multiple trajectories into one (27). This procedure allows tracking to continue and thus compensates for the transient data loss caused by qdot blinking. In addition, qdot blinking frequency is dependent on the excitation power; hence, it is generally recommended to perform single qdot tracking experiments with low-power excitation if signal intensity is sufficiently high.

An ImageJ plug-in for single-molecule/single-particle tracking offers several user-friendly features including an easily understandable interface, free on-line tutorial, and computationally efficient process. The program is free to download at ImageJ website: http://rsbweb.nih.gov/ij/plugins/index.html (see Note 4). In addition, particle tracking using IDL, developed by Crocker and Grier, provides a total solution including 2D Gaussian fit for spot localization, trajectory generation, as well as MSD calculation. The algorithms with detailed tutorial are freely available at http://www.physics.emory.edu/~weeks/idl/index.html. Matlab version of these routines can be found at http://physics.georgetown.edu/matlab/ (Fig. 5). With recent advances in computing power and numerical software, the development of tracking algorithms has evolved rapidly during the past few years in supporting better correspondence for motion detection (28), high computational efficiency (29), or 3D motion segmentation and localization (30). All the tracking algorithms mentioned above may, however, require technical training to operate since they all established under technical computing environments such as Matlab, IDL, or C^{++}.

Below is a general tracking procedure using ParticleTracker:

1. Use the *File/Open* commend in the ImageJ to import the prerecorded TIF stack or uncompressed avi file (Fig. 6a). If the avi file contains multichannel imaging data, use the *Image/Color/RGB Split* to qdot data channel extraction.
2. Next, click the *Plugins/Particle Detector & Tracker* commend. If RGB images or images with greater than 8 bits are loaded for

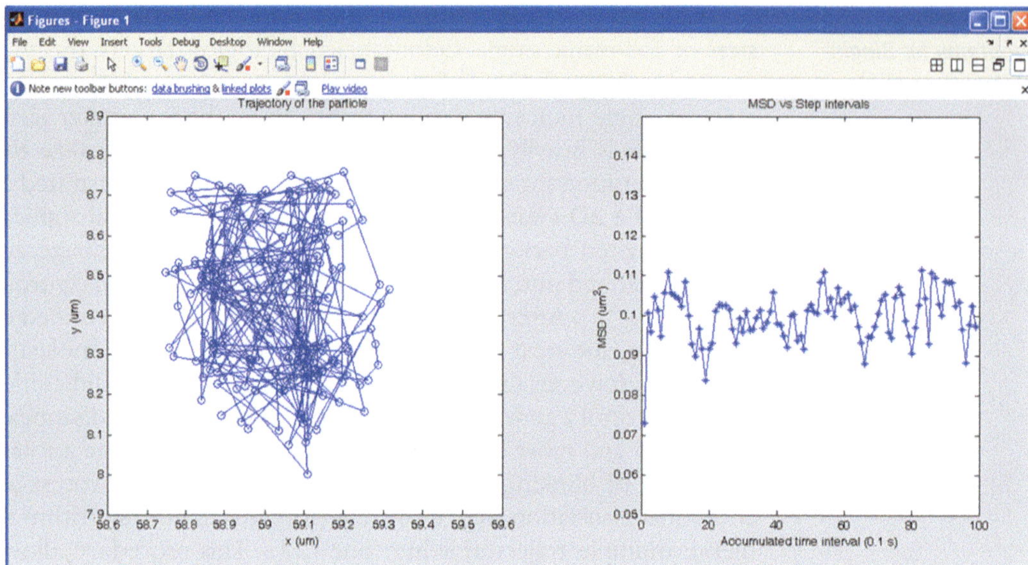

Fig. 5 Snapshot of the interface of Matlab-based particle tracking program originally developed from *particle tracking using IDL* algorithm. Data obtained from 600 frames of single Qdot imaging. *Left panel* indicates the 2D trajectory, and *right panel* shows the MSD over time

tracking, a checkable menu item will show up to ask whether the images are converted to 8 bits. If running on a computer with fewer than 2 GB of memory installed, it is strongly recommended to convert to 8 bits to reduce the memory consumption.

3. As indicated in Fig. 6b, three basic parameters for particle detection are given. *Radius*: Approximate radius of the particles in the images in units of pixels. *Cutoff*: The score cutoff for the non-particle discrimination. *Percentile*: The percentile (r) that determines which bright pixels are accepted as particles. Click on preview detected and then the successfully detected spots will be circulated. Here, we recommend to use *Radius* = 3, *Cutoff* = 0, and *Percentile* = 0.1 as initial guess, but these values might vary based on the images. Start with our recommended parameter and change these values until most of the visible particles are detected after clicking the preview button.

4. After setting the parameters for the detection, set up the particle-linking parameters (*Displacement* & *Link Range*) in the bottom of the dialog window (Fig. 6b). Here, the *Displacement* parameter means the maximum number of pixels a particle is allowed to move between two succeeding frames. The *Link Range* parameter is used to specify the number of subsequent frames that is taken into account to determine the optimal correspondence matching. We recommend to use *Displacement* = 2 and *Link Range* = 10 as initial guess, and again, these parameters can also be modified after viewing the initial results.

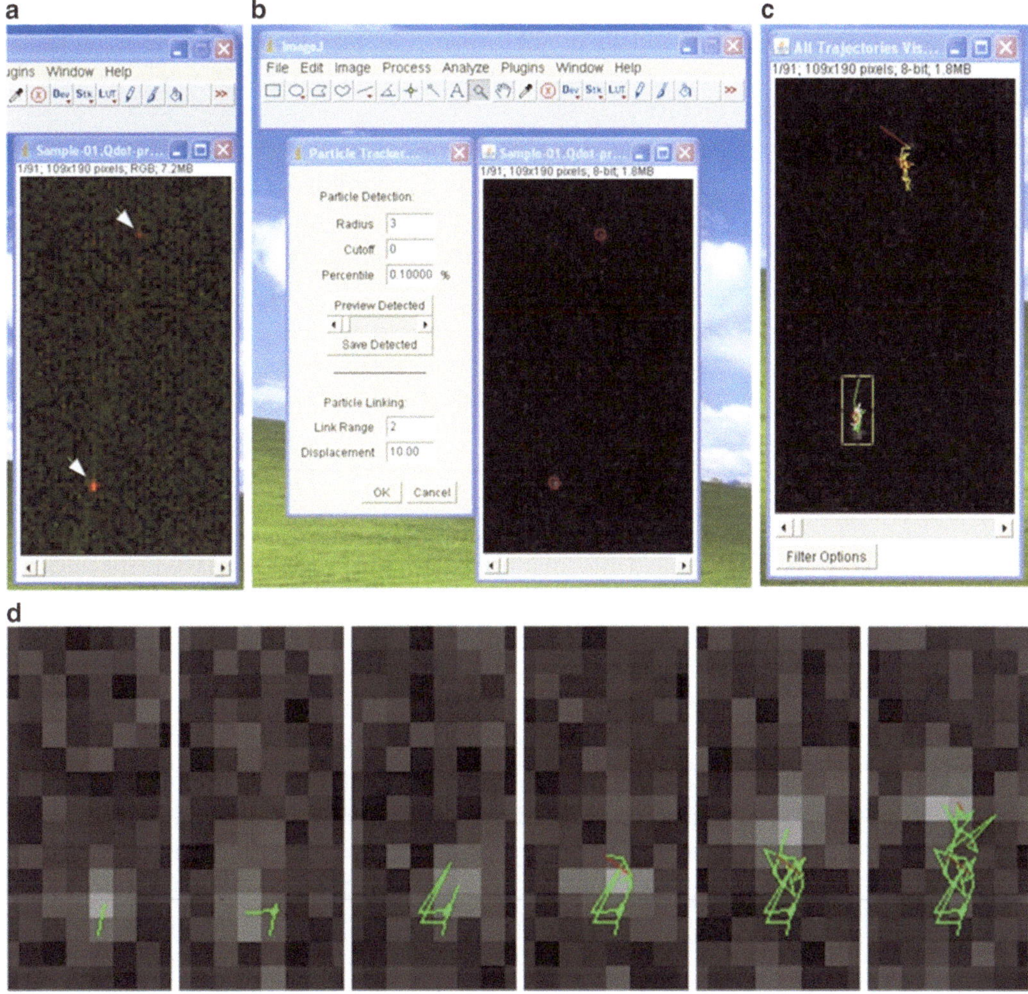

Fig. 6 Steps of single qdot tracking using ParticleTracker—an ImageJ plug-in. (**a**) A typical raw frame from a time-lapse single Qdot labeling movie. *White arrows* indicate the Qdot-labeled target proteins. (**b**) Image conversion and preview detection from the raw frame. Image conversion to Gray 8 is preferred to increase computational efficiency. *Red circle* masks the successfully targeted spots for tracking after executing the Preview Detected function. (**c**) Visualization of all trajectories after executing the Show Detected function. Particular area of interest can be selected and zoomed in as indicated in *yellow box*. (**d**) Time-lapse trajectory from the selected area of interest. *Red line* drawn indicates the "Gaps" in the trajectory

5. Next, push the OK button and the result window will then be displayed (Fig. 6c) after seconds to minutes computational calculation (for 600 frames of 128 × 128 pixel images take less than 1 min on a regular dual core PC). With the *Filter Options* button given on the dialog window, you can filter out trajectories under a given length. Particular trajectory of interest can be selected by clicking it once with the mouse left button (*see* yellow box of Fig. 6c).
6. The selected trajectory can be displayed in a separate window by clicking on the *Focus on Selected Trajectory* button. The

visualization of the selected trajectory can then be saved individually in .gif format (Fig. 6d). Detected time-series trajectory coordinates can also be exported in a single .txt file for further analyses.

7. After exporting trajectory coordinates, MSD of a specific trajectory can be obtained according to the formula below:

$$\mathrm{MSD}(n\Delta t) = (N-n)^{-1}\sum_{i=1}^{N-n}[(x_{i+n}-x_i)^2+(y_{i+n}-y_i)^2], \quad (7)$$

where x_i and y_i are the position of particle on the frame i, Δt is the time resolution, N is total number of frames, $n\Delta t$ is the time interval over which the MSD is calculated.

4 Notes

1. Detailed information regarding Photometrics CCD specification can be found at http://www.photomet.com.
2. The spinning force is not critical in this step. In most cases, a rotational speed as low as 500 rpm (~30 × *g* for common compact spin coater) is sufficient to achieve a uniform spread when using common compact spin coater.
3. The labeling concentration/cell type relationship should be adjusted for the surface protein expression level. In our experiments, we choose low concentrations for transfected cells. For labeling endogenously expressing membrane proteins in living cells, higher concentrations may be needed.
4. The algorithm used in the ParticleTracker program can easily cause false linking of different molecules/particles between frames. This could lead to incorrect trajectory construction. It may be improved by manual relinking with visual inspection. However, potential problems and limitations can still be associated with such manual relinking. Due to its respectable efficiency, ParticleTracker program is suitable for preliminary screening tests. For serious and in-depth analysis, we recommend to use particle tracking using IDL or the Matlab-based tracking routines.

Acknowledgments

The authors thank Drs. David Piston and Sam Wells for helpful advice with single quantum dot tracking experimental setup. We thank colleagues in the group, especially to Dr. James McBride and Oleg Kovtun, for helpful discussions and suggestions. This work was supported by grants from National Institutes of Health (R01EB003778).

References

1. Dunn RC (1999) Near-field scanning optical microscopy. Chem Rev 99:2891–2928
2. Gordon MP, Ha T, Selvin PR (2004) Single-molecule high-resolution imaging with photobleaching. Proc Natl Acad Sci USA 101: 6462–6465
3. Willig KI, Rizzoli SO, Westphal V, Jahn R, Hell SW (2006) STED microscopy reveals that synaptotagmin remains clustered after synaptic vesicle exocytosis. Nature 440:935–939
4. Huang B, Wang WQ, Bates M, Zhuang XW (2008) Three-dimensional super-resolution imaging by stochastic optical reconstruction microscopy. Science 319:810–813
5. Zhang B, Zerubia J, Olivo-Marin JC (2007) Gaussian approximations of fluorescence microscope point-spread function models. Appl Opt 46:1819–1829
6. Thomann D, Rines DR, Sorger PK, Danuser G (2002) Automatic fluorescent tag detection in 3D with super-resolution: application to the analysis of chromosome movement. J Microsc 208:49–64
7. Cheezum MK, Walker WF, Guilford WH (2001) Quantitative comparison of algorithms for tracking single fluorescent particles. Biophys J 81:2378–2388
8. Kubitscheck U, Kückmann O, Kues T, Peters R (2000) Imaging and tracking of single GFP molecules in solution. Biophys J 78:2170–2179
9. Alivisatos P (2004) The use of nanocrystals in biological detection. Nat Biotechnol 22: 47–52
10. Chan WC, Nie S (1998) Quantum dot bioconjugates for ultrasensitive nonisotopic detection. Science 281:2016–2018
11. Bruchez M, Moronne M, Gin P, Weiss S, Alivisatos AP (1998) Semiconductor nanocrystals as fluorescent biological labels. Science 281:2013–2016
12. Rosenthal SJ, Tomlinson I, Adkins EM, Schroeter S, Adams S, Swafford L, McBride J, Wang Y, DeFelice LJ, Blakely RD (2002) Targeting cell surface receptors with ligand-conjugated nanocrystals. J Am Chem Soc 124:4586–4594
13. Kim S, Lim YT, Soltesz EG, De Grand AM, Lee J, Nakayama A, Parker JA, Mihaljevic T, Laurence RG, Dor DM, Cohn LH, Bawendi MG, Frangioni JV (2004) Near-infrared fluorescent type II quantum dots for sentinel lymph node mapping. Nat Biotechnol 22: 93–97
14. Chang JC, Tomlinson ID, Warnement MR, Iwamoto H, DeFelice LJ, Blakely RD, Rosenthal SJ (2011) A fluorescence displacement assay for antidepressant drug discovery based on ligand-conjugated quantum dots. J Am Chem Soc 133:17528–17531
15. Nirmal M, Dabbousi BO, Bawendi MG, Macklin JJ, Trautman JK, Harris TD, Brus LE (1996) Fluorescence intermittency in single cadmium selenide nanocrystals. Nature 383: 802–804
16. Wang X, Ren X, Kahen K, Hahn MA, Rajeswaran M, Maccagnano-Zacher S, Silcox J, Cragg GE, Efros AL, Krauss TD (2009) Non-blinking semiconductor nanocrystals. Nature 459:686–689
17. Zhang Q, Li Y, Tsien RW (2009) The dynamic control of kiss-and-run and vesicular reuse probed with single nanoparticles. Science 323:1448–1453
18. Thompson MA, Lew MD, Badieirostami M, Moerner WE (2009) Localizing and tracking single nanoscale emitters in three dimensions with high spatiotemporal resolution using a double-helix point spread function. Nano Lett 10:211–218
19. Dahan M, Levi S, Luccardini C, Rostaing P, Riveau B, Triller A (2003) Diffusion dynamics of glycine receptors revealed by single-quantum dot tracking. Science 302:442–445
20. Cui B, Wu C, Chen L, Ramirez A, Bearer EL, Li W-P, Mobley WC, Chu S (2007) One at a time, live tracking of NGF axonal transport using quantum dots. Proc Natl Acad Sci USA 104:13666–13671
21. Bouzigues C, Morel M, Triller A, Dahan M (2007) Asymmetric redistribution of GABA receptors during GABA gradient sensing by nerve growth cones analyzed by single quantum dot imaging. Proc Natl Acad Sci USA 104:11251–11256
22. Fabien P, Xavier M, Gopal I, Emmanuel M, Hsiao-Ping M, Shimon W (2009) Dynamic partitioning of a glycosyl-phosphatidylinositol-anchored protein in glycosphingolipid-rich microdomains imaged by single-quantum dot tracking. Traffic 10:691–712
23. Chang JC, Tomlinson ID, Warnement MR, Ustione A, Carneiro AM, Piston DW, Blakely RD, Rosenthal SJ (2012) Single molecule analysis of serotonin transporter regulation using antagonist-conjugated quantum dots reveals restricted, p38 MAPK-dependent mobilization underlying uptake activation. J Neurosci 32:8919–8929
24. Chang JC, Rosenthal SJ (2012) Visualization of lipid raft membrane compartmentalization

in living RN46A neuronal cells using single quantum dot tracking. ACS Chem Neurosci 3:737–743

25. Murcia MJ, Minner DE, Mustata G-M, Ritchie K, Naumann CA (2008) Design of quantum dot-conjugated lipids for long-term, high-speed tracking experiments on cell surfaces. J Am Chem Soc 130:15054–15062
26. Tada H, Higuchi H, Wanatabe TM, Ohuchi N (2007) In vivo real-time tracking of single quantum dots conjugated with monoclonal anti-HER2 antibody in tumors of mice. Cancer Res 67:1138–1144
27. Ehrensperger M-V, Hanus C, Vannier C, Triller A, Dahan M (2007) Multiple association states between glycine receptors and gephyrin identified by SPT analysis. Biophys J 92:3706–3718
28. Jaqaman K, Loerke D, Mettlen M, Kuwata H, Grinstein S, Schmid SL, Danuser G (2008) Robust single-particle tracking in live-cell time-lapse sequences. Nat Methods 5:695–702
29. Smith CS, Joseph N, Rieger B, Lidke KA (2010) Fast, single-molecule localization that achieves theoretically minimum uncertainty. Nat Methods 7:373–375
30. Ram S, Prabhat P, Chao J, Sally Ward E, Ober RJ (2008) High accuracy 3D quantum dot tracking with multifocal plane microscopy for the study of fast intracellular dynamics in live cells. Biophys J 95:6025–6043

Chapter 16

Fabrication of Fluorescent Silica Nanoparticles with Aggregation-Induced Emission Luminogens for Cell Imaging

Sijie Chen, Jacky W.Y. Lam, and Ben Zhong Tang

Abstract

Fluorescence-based techniques have found wide applications in life science. Among various luminogenic materials, fluorescent nanoparticles have attracted much attention due to their fabulous emission properties and potential applications as sensors. Here, we describe the fabrication of fluorescent silica nanoparticles (FSNPs) containing aggregation-induced emission (AIE) luminogens. By employing surfactant-free sol–gel reaction, FSNPs with uniform size and high surface charge and colloidal stability are generated. The FSNPs emit strong light upon photoexcitation, due to the AIE characteristic of the silole aggregates in the hybrid nanoparticles. The FSNPs are cytocompatible and can be utilized as fluorescent visualizer for intracellular imaging for HeLa cells.

Key words Fluorescent silica nanoparticles, Fluorescent probes, Aggregation-induced emission, Sol–gel reaction, Cell imaging

1 Introduction

Fluorescent nanoparticles have been used widely in biology as sensing, imaging or tracking probes (1, 2). Among them, semiconductor quantum dots (QDs) have received particular interest because of their high fluorescence and resistance to photobleaching. Unfortunately, they are made of heavy metals and are thus inherently toxic. Although scientists have tried to solve the problem by various techniques, their cytotoxicity is still a considerable issue (3).

Silica nanoparticles (SNPs), on the contrary, are cytophilic, transparent but nonfluorescent and hence are ideal host materials for the fabrication of fluorescent silica nanoparticles (FSNPs) for imaging application (4). FSNPs can be prepared by incorporating fluorophores into silica networks via physical processes or chemical reactions. The silica matrix acts as a protective shield, reducing the possibilities of penetrations of oxygen and other harmful species that may cause photobleaching of the embedded fluorophores.

Volkmar Weissig et al. (eds.), *Cellular and Subcellular Nanotechnology: Methods and Protocols*, Methods in Molecular Biology, vol. 991, DOI 10.1007/978-1-62703-336-7_16, © Springer Science+Business Media New York 2013

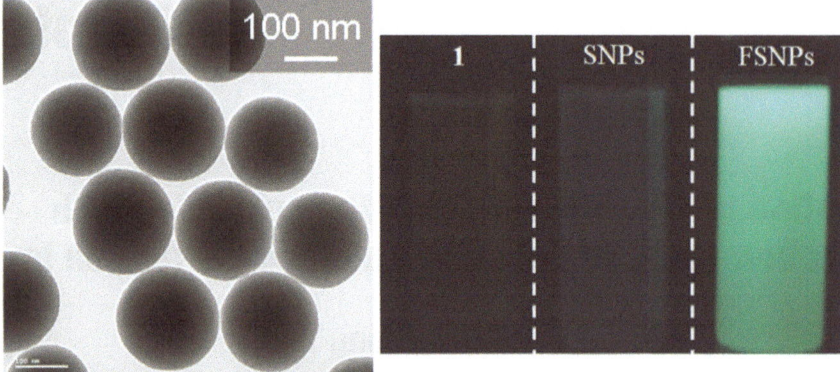

Fig. 1 *(Left)* TEM image of the FSNPs. *(Right)*: photographs of solutions and suspensions of **1**, SNPs, and FSNPs in ethanol taken under 365 nm UV irradiation. Reproduced from ref. 11 with the permission of Wiley

However, most dye molecules emit intensely in the solution state but become weakly fluorescent when aggregated in poor solvent or in the solid state. Such phenomenon is known as aggregation caused quenching (ACQ) effect (5) and has been an obstacle for the development of highly emissive FSNPs.

In 2001, Tang discovered a phenomenon of aggregation-induced emission (AIE) that is exactly opposite to the ACQ effect: nonemissive, propeller-like luminogens such as hexaphenylsilole are induced to emit efficiently by aggregate formation (6). The AIE effect dramatically boosts the fluorescence quantum yields of the luminogens, turning them from faint fluorophores to strong emitters. Mechanistic investigations reveal that the AIE effect is caused by the restriction of the intramolecular rotation in the aggregate state, which blocks the nonradiative relaxation channel and populates the radiative decay (7–10).

In this chapter, we describe a methodology of hybridizing AIE luminogens with silica nanoparticles to generate highly emissive FSNPs. The luminogens are chemically bound to the silica networks, which can prevent their leakage under harsh conditions. The FSNPs are uniformly sized, surface-charged, and colloidally stable. Whereas the solution of the AIE luminogens and the suspension of the SNPs are invisible under the UV irradiation, intense green light is emitted from the FSNPs (see Fig. 1). The FSNPs can be function as a fluorescent visualizer for selectively imaging the cytoplasm of HeLa cells (see Fig. 2) (11, 12).

2 Materials

2.1 Chemicals for Synthesis

1. THF was purchased and distilled from sodium benzophenone ketyl under nitrogen immediately prior to use.
2. Tetraethoxysilane (TEOS).

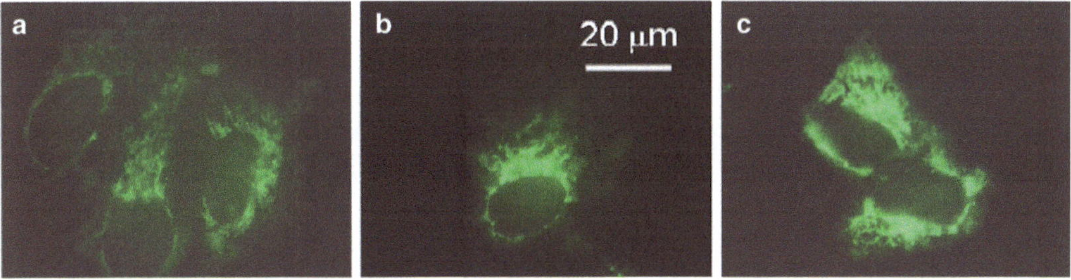

Fig. 2 Fluorescent images of HeLa cells stained by FSNPs with different luminogen loadings. Concentration of **1** (μM): (**a**) 2, (**b**) 4, (**c**) 8. Reproduced from (11) with the permission of Wiley

3. Dimethylsulfoxide (DMSO).
4. (3-Aminopropyl)triethoxysilane (APS).
5. Phenylacetylene.
6. *n*-Butyllithium (*n*-BuLi).
7. Naphthalene.
8. Decylmethyldichlorosilane.
9. Lithium.
10. Dichloro(*N*,*N*,*N*′,*N*′-tetramethylethylenediamine)zinc.
11. 4-Iodophenol.
12. 1,2-Dibromoethane.
13. Dichlorobis(triphenylphosphine)palladium(II).
14. Deionized water was used in the experiment.

2.2 Cell Culture

The following materials are used for culturing the HeLa cells.

1. Minimum Essential Medium, Fetal Bovine Serum, penicillin, and streptomycin were purchased from Gibco, Invitrogen.
2. 4-(2-Hydroxyethyl)-1-piperazineethanesulfonic acid (HEPES) was purchased from Sigma.
3. 35 mm culture dishes were purchased from Corning.
4. Phosphate buffered saline (PBS).
5. Cover slides.
6. Cell culture CO_2 incubator.

2.3 Instrument

1. NMR analysis was conducted on a Bruker ARX 400 spectrometer with tetramethylsilane (TMS; $\delta = 0$) as internal standard.
2. Mass spectroscopy was carried out on a Finnigan TSQ 7000 triple quadrupole spectrometer operating in a MALDI–TOF mode.
3. The morphologies of the FSNPs were investigated using JOEL 2010 TEM and JOEL 6700F SEM at an accelerating voltage of 200 kV.

4. Copper 400-mesh carrier grids covered with carbon-coated formvar films were used for the TEM and SEM measurements.
5. Fluorescence spectra were taken on a Perkin–Elmer LS 50B spectrofluorometer with a Xenon discharge lamp excitation.
6. Zeta potentials and particle sizes of the FSNPs were determined at room temperature by a ZetaPlus Potential Analyzer (Brookhaven Instruments Corporation, USA).
7. The HeLa cells were imaged under an inverted fluorescence microscope (Nikon Eclipse TE2000-U; λ_{ex} = 330–380 nm, diachronic mirror = 400 nm).

3 Methods

3.1 Synthesis (See Scheme 1)

3.1.1 Synthesis of 1-(2-Bromoethoxy)-4-Iodobenzene (4)

1. To a mixture of 4-iodophenol (**2**, 5.0 g, 22.7 mmol) and potassium carbonate (4.7 g, 34.1 mmol) in acetone was added 12.8 g (68.2 mmol) of 1,2-dibromoethane (**3**). The mixture was stirred and heated to reflux for 24 h.
2. After filtration and solvent evaporation, the crude product was purified by silica gel chromatography using chloroform/hexane (1:4 v/v) as eluent.

3.1.2 Synthesis of Bis(phenylethynyl)decylmethylsilane (7)

1. To a THF solution of phenylacetylene (**5**, 4.0 mL, 36.4 mmol) was added 25.0 mL (40.1 mmol) of 1.6 M *n*-butyllithium solution in hexane at −78°C. After stirring at the same temperature for 2 h, decylmethyldichlorosilane (**6**, 4.8 mL, 18.2 mmol) was added at −78°C. The mixture was warmed to room temperature and stirred overnight.
2. The solvent was removed under reduced pressure. The mixture was dissolved in dichloromethane and washed with water. The organic layer was dried over magnesium sulfate and filtered. The filtrate was evaporated and the crude product was purified by a silica gel column using hexane as eluent.

3.1.3 Synthesis of 1-Decyl-1-methyl-2,5-bis[4-(2-bromoethoxy)phenyl]-3,4-diphenylsilole (10)

1. A mixture of lithium (0.056 g, 8 mmol) and naphthalene (1.04 g, 8 mmol) in 8 mL of THF was stirred at room temperature under nitrogen for 3 h to form a deep dark green solution of lithium 1-naphthalenide.
2. The viscous solution was added dropwise to a solution of 7 (0.77 g, 2 mmol) in 5 mL of THF over 2 min at room temperature.
3. After stirring for 1 h, the mixture containing 8 was cooled to 0°C with an ice bath and diluted with 10 mL THF. $ZnCl_2$–TMEDA (2 g, 8 mmol) was then added to the mixture to give a black suspension of 9.

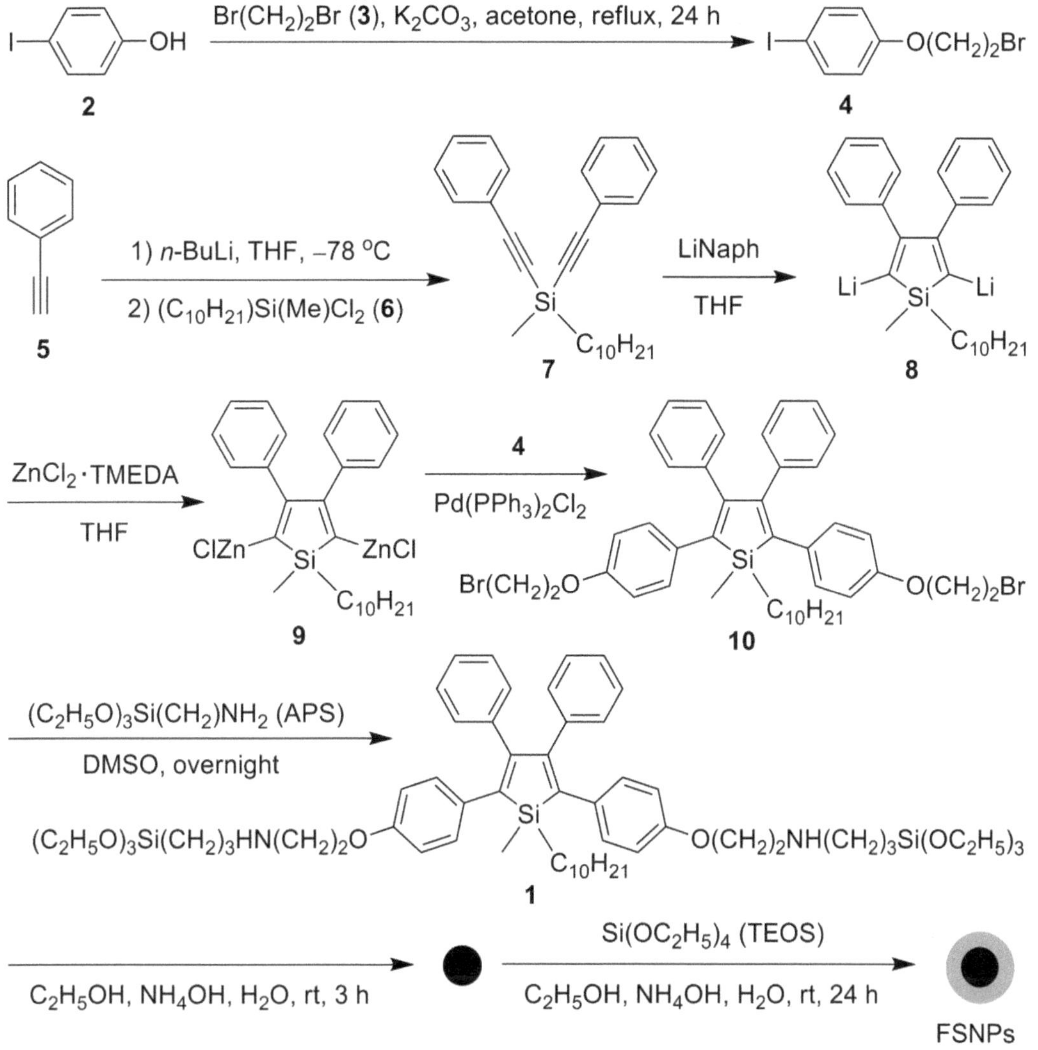

Scheme 1 Synthesis of silole derivative **1** and fabrication of fluorescent silica nanoparticles (FSNPs). *n-BuLi* n-butyllithium, *THF* tetrahydrofuran, *Naph* 1-naphthalenide, *TMEDA* N,N,N′,N′-tetramethylethylenediamine, *DMSO* dimethysulfoxide. Reproduced from ref. 11 with the permission of Wiley

4. After stirring for an additional hour at room temperature, a solution of 4 (1.63 g, 4.9 mmol) and $Pd(PPh_3)_2Cl_2$ (0.08 g, 0.1 mmol) in 10 mL of THF was added. The mixture was refluxed overnight.
5. After cooling to room temperature, 10 mL of 3 M HCl solution was added and the mixture was extracted with dichloromethane. The combined organic layer was washed with brine and dried over magnesium sulfate.

6. After solvent evaporation under reduced pressure, the residue was purified by a silica gel column using ethyl acetate/hexane (1:9 v/v) as eluent.

3.2 Fabrication of FSNPs

1. Silole–APS (1) was prepared by stirring a mixture of 4 μM of 10 and 10 μM of APS in 100 μL of DMSO overnight.
2. The reaction mixture was concentrated under vacuum.
3. 1 was added into a mixture of 64 mL of ethanol, 1.28 mL of ammonium hydroxide, and 7.8 mL of distilled water and stirred at room temperature for 3 h to prepare the silole–silica nanocores (see Note 1).
4. A mixture of 2 mL of TEOS in 8 mL of ethanol was then added dropwise (see Note 2) into the mixture of the nanocores and the reaction was stirred at room temperature for an additional 24 h (see Note 3) to coat the luminogenic nanocores with silica shells.
5. After incubation, the mixture was centrifuged and the FSNPs were redispersed in ethanol under sonication for 5 min. Such process was repeated three times and the FSNPs were finally dispersed in water for the cell imaging experiments.

3.3 Cell Imaging by FSNPs

3.3.1 Cell Culture

The HeLa cells were cultured in minimum essential medium containing 10% fetal bovine serum and antibiotics (100 units/mL penicillin and 100 μg/mL streptomycin) in a 5% carbon dioxide humidity incubator at 37°C. These cells were grown overnight on a plasma-treated 25 mm round cover slip mounted onto a 35 mm culture dish with an observation window (see Note 4).

3.3.2 Preparation of FSNPs for Imaging

1. Prepare desired concentration of FSNPs in PBS.
2. Autoclave the nanoparticles suspension at 121°C for 20 min.
3. Sonicate the suspension for 5 min.
4. Filter the suspension using a 450 nm filter in biosafety cabinet and add 250 μL of filtrate into 2 mL medium (see Note 5).
5. Remove the cell culture medium and add the medium prepared in step 4.
6. Incubate the cell in a 5% carbon dioxide humidity incubator at 37°C for 24 h.

3.3.3 Cell Imaging

1. Wash the cell with PBS for three times (see Note 6).
2. Add MEM with HEPES to the dish.
3. Observe the cell under the fluorescent microscope through the observation window.

4 Notes

1. During the fabrication of the FSNPs (Subheading 3.2, step 3), the ethanol/ammonium hydroxide/distilled water mixture should be stirred for more than 5 min before addition of 1.
2. To coat the fluorescent nanocores with silica shells (Subheading 3.2, step 4), the TEOS/ethanol mixture should be added dropwise. The rate should be well controlled and uniform. Nanoparticles with sizes of ~200 nm are generated at a rate of 3 s/drop.
3. The stirring speed is important for the particle fabrication. The rate should be high and stable to furnish FSNPs with uniform sizes.
4. The following are treated to the dishes before cell culture.
   (a) Drill a hole of around 10 mm diameter in the middle of the dish.
   (b) Stick a piece of cover slide to the dish by paraffin.
5. The FSNPs are sonicated and filtrated before use as autoclave will destabilize the nanoparticles. The FSNPS will be more uniform and well dispersed after these treatments.
6. The cells should be washed thoroughly before imaging as excess FSNPs will emit outside the cell and disturb the observation.

References

1. Velikov KP, van Blaaderen A (2001) Synthesis and characterization of monodisperse core–shell colloidal spheres of zinc sulfide and silica. Langmuir 17:4779–4786
2. Pellegrino T, Kudera S, Parak WJ (2005) On the development of colloidal nanoparticles towards multifunctional structures and their possible use for biological applications. Small 1(1):49–63
3. Hardman R (2006) A toxicologic review of quantum dots: toxicity depends on physicochemical and environmental factors. Environ Health Perspect 114(2):165–172
4. Santra S, Dutta D, Walter GA (2005) Fluorescent nanoparticle probes for cancer imaging. Technol Cancer Res Treat 4(6):593–602
5. Birks JB (1970) Photophysics of aromatic molecules. Wiley-Interscience, New York
6. Chen J, Law CW, Tang BZ (2003) Synthesis, light emission, nanoaggregation, and restricted intramolecular rotation of 1,1-substituted 2,3,4,5-tetraphenylsiloles. Chem Mater 15: 1535–1546
7. Yu G, Yin S, Luo Y (2005) Structures, electronic states, photoluminescence, and carrier transport properties of 1,1-disubstituted 2,3,4,5-tetraphenylsiloles. J Am Chem Soc 127(17):6335–6346
8. Luo J, Xie Z, Tang BZ (2001) Aggregation-induced emission of 1-methyl-1,2,3,4,5-pentaphenylsilole. Chem Commun (Camb) 18:1740–1741
9. Fan X, Sun J, Zou D (2008) Photoluminescence and electroluminescence of hexaphenylsilole are enhanced by pressurization in the solid state. Chem Commun (Camb) 26:2989–2991
10. Hong Y, Lam JW, Tang BZ (2009) Aggregation-induced emission: phenomenon, mechanism and applications. Chem Commun (Camb) 29:4332–4353
11. Faisal M, Hong Y, Tang BZ (2010) Fabrication of fluorescent silica nanoparticles hybridized with AIE luminogens and exploration of their applications as nanobiosensors in intracellular imaging. Chem Eur J 16: 4266–4272
12. Faisal M, Yu Y, Tang BZ (2011) Fabrication of silica nanoparticles with both efficient fluorescence and strong magnetization and exploration of their biological applications. Adv Funct Mater 21:1733–1740

Chapter 17

Monitoring the Degradation of Reduction-Sensitive Gene Carriers with Fluorescence Spectroscopy and Flow Cytometry

Constantin Hozsa, Miriam Breunig, and Achim Göpferich

Abstract

Polycations like poly(ethylene imine) (PEI) or poly(L-lysine) (pLL) form nanometer-sized complexes with nucleic acids (polyplexes) which can be used for gene delivery. It is known that the properties of these carriers can be greatly improved by introducing disulfide bridges on the polymers, thus making them reduction sensitive. However, little is known about how such modified carriers behave intracellularly.

Here, we describe a method that uses the reduction-sensitive fluorescent dye BODIPY FL L-cystine to label PEI and pLL. Our probe is activated under reductive conditions leading to strongly increased fluorescence intensity. Subsequently, we show how the intracellular route of polyplexes made from these labeled polymers can be monitored by flow cytometry.

Key words Poly(ethylene imine), Poly(L-lysine), Redox-sensitive gene carrier, Disulfides, Flow cytometry, BODIPY FL L-cystine, Polyplexes

1 Introduction

Most strategies in gene therapy aim at altering the cellular gene expression by either inserting therapeutic genes or inhibiting the expression of undesirable gene products (1). In both cases, nucleic acids are exogenously introduced into living cells with gene delivery vehicles. Currently, there is a growing interest in nonviral vectors because of their higher safety compared to viral vectors and their great flexibility for modification (2, 3). The two most commonly used classes of nonviral vectors are cationic lipids like DOTAP and polycations such as poly(ethylene imine) (PEI) or poly(L-lysine) (pLL) (4). Materials from both groups interact with negatively charged nucleic acids to form nanometer-sized complexes termed lipoplexes or polyplexes, respectively (2).

Such complexes must remain highly stable against ion exchange reactions (i.e., displacement of the cargo) in the bloodstream but

Volkmar Weissig et al. (eds.), *Cellular and Subcellular Nanotechnology: Methods and Protocols*, Methods in Molecular Biology, vol. 991, DOI 10.1007/978-1-62703-336-7_17, © Springer Science+Business Media New York 2013

have to release their cargo quickly after their uptake in target cells. An inefficient release of nucleic acids leads to a low gene expression level (DNA) or a low silencing effect with small interfering RNA (siRNA) (5). Therefore, it is essential to create carriers that satisfy both conflicting demands on the complex stability (6). One possibility is to create environment-responsive polyplexes by exploiting the large redox-potential gradient between extra- and intracellular compartment. In this case, disulfide bonds serve as a trigger: They remain stable outside cells but are readily cleaved at the high cytosolic glutathione concentration (7). There are numerous examples in the literature employing this strategy. For instance, both short-chained pLL and PEI were cross-linked to high molecular weight branched products. Polyplexes made from those products show an increased nucleic acid condensation capability and plasma stability as well as high transfection efficiency (6, 8, 9). Disulfide cross-linked PEI (S_2-lPEI) is also known to be less cytotoxic than its non-cleavably branched counterpart. Disulfides are also used to reversibly attach cell-targeting ligands (small molecules, peptides, or antibodies) or hydrophilic polymers as polyethylene glycol (PEG) to prevent unspecific binding on the polyplex surface (2, 7).

As beneficial as the use of disulfides might be, there are a number of facts to be considered. Among others, the choice of the cross-linking agent is important as well as the cross-linking ratio (9). Polyplexes can easily be "overstabilized", so it is indispensable to adjust the exact spatiotemporal point of the linker cleavage to the type of nucleic acid to be delivered: RNA has to be released earlier than DNA as its target is in the cytosol, not the nucleus (8). "Naked" DNA (i.e., without carrier), however, does not efficiently pass nuclear pores (3, 10).

Unfortunately, the cellular trafficking of reduction-sensitive gene carriers has hardly been investigated in detail so far. Where, when, and how the carrier's disulfides are cleaved is a matter of dispute (7). The most likely location is the cytosol, but the cell membrane, the endolysosomes, and the nucleus are also discussed (10, 11).

To gain a better understanding of those processes, we developed a fluorescent probe based on the reduction-sensitive dye BODIPY FL L-cystine (BP-Cys_2, Fig. 1). This dye consists of a cystine residue that carries two BODIPY molecules linked to its amines. In this dimeric form, it is virtually nonfluorescent due to the strong self-quenching of both fluorophores (12, 13). When the bridging cystine disulfide is cleaved, a significant fluorescence intensity increase indicates the presence of reductive conditions. Polyplexes made with our probe can address several questions such as the following: (1) Does the delivery system actually reach reductive cellular compartments? (2) At which point does the disulfide cleavage begin and how fast is it? (3) How are S_2-cross-linked polycations degraded?

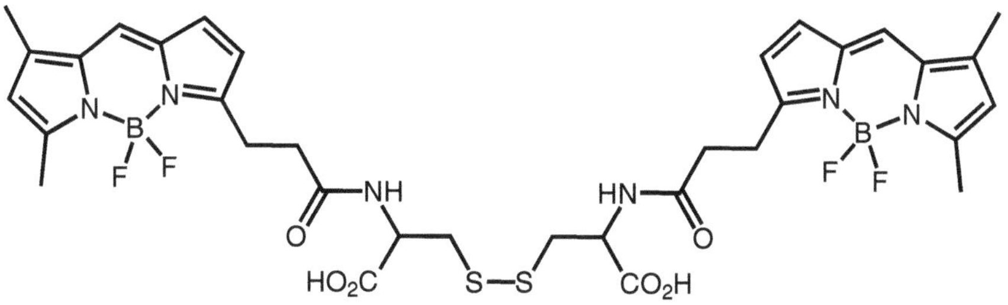

Fig. 1 Structure of the self-quenching and reduction-sensitive dye BODIPY FL L-cystine: Cleavage of the bridging disulfide bond results in an increased fluorescence intensity

In the following sections, we describe a quick and simple method to label lPEI, S_2-lPEI, and pLL with BP-Cys_2. We also explain the removal of unbound dye and the quality control of the end product. Moreover, we shortly describe the cross-linking of lPEI with cystine. Finally, we demonstrate the use of these labeled polymers in flow cytometry.

2 Materials

Ultrapure water (18 MΩ cm or 0.055 μΩ/cm) and analytical grade chemicals should be used in all experiments. Please note that BODIPY FL Cys_2, NEM, and all labeled products must be kept frozen (≤ −18°C) and protected from light (wrapped in aluminum foil). DMTMM and pLL are to be stored under argon atmosphere. Where indicated, solutions should be prepared fresh immediately before use.

2.1 BODIPY FL L-Cystine Labeling Components

1. Solvents for labeling and purification: Water, methanol, dimethyl-sulfoxide (DMSO), 0.1 M NaOH, and 10 mM HCl.
2. 50 mM borate buffer, pH = 9, for purification of labeled PEI. For preparation, add 4.77 g of $Na_2B_4O_7 \cdot 10\ H_2O$ (sodium tetraborate) to about 150 ml of water in a glass beaker (see Note 1). Stir the solution with a magnetic stir bar until the borate is completely dissolved. Remove the stir bar and adjust the pH to 9 with HCl. Add water to a volume of 250 ml. Keep refrigerated.
3. Phosphate-buffered saline (PBS buffer, see Note 2) for stock solutions of the labeled products.
4. Linear polycations: lPEI is available in different molecular weights from various manufacturers or can be synthesized by the hydrolysis of poly(2-ethyl-2-oxazoline) (see Note 3). Linear pLL in different sizes is also commercially available (see Note 4). For this work, we used 6.3 kDa lPEI and 5.2 kDa pLL·HBr.

5. Dye stock solution: 1 mg BODIPY FL L-cystine (M=788.44 g/mol, see Note 5) dissolved in 100 μl DMSO (γ=10 g/l). Store at −20°C and protected from light.
6. Coupling reagent stock solution: DMTMM (4-(4,6-Dimethoxy-1,3,5-triazin-2-yl)-4-methylmorpholinium chloride, M=276.72 g/mol) in methanol (γ=10 g/l). Must be prepared fresh prior to each use (see Note 6 and (14)).
7. Glassware: Small scale syntheses (approx. 200 mg) can easily be done in 4 ml glass vials with snap-caps. Product handling and recovery is improved with silanized glassware (see Note 7).
8. Ultrafiltration tubes with a molecular weight cutoff (MWCO) suitable for the retention of the polymer (see Note 8).
9. Smaller items: magnetic stirrer and small magnetic stir bar (ca. 5 mm), aluminum foil, pH indicator paper, syringe (1–2 ml) and corresponding hypodermic needle, syringe filter (0.22 μm pore size), TLC plates (silica gel) and separation chamber, UV lamp (254 or 366 nm), and argon.
10. Larger equipment: freeze dryer, alternatively a vacuum pump with desiccator or a SpeedVac (see Note 9).

2.2 lPEI Cross-Linking Components

The cross-linking of lPEI is quite similar to the labeling procedure. Additional components needed: t-Butyl carbamate protected cystine $(Boc\text{-}Cys\text{-}OH)_2$, 1 M HCl, and NaOH pellets. The molecular weight of the cross-linked product can be determined by size-exclusion chromatography (see Note 23).

2.3 Cell Culture and Flow Cytometry

We typically use Chinese hamster ovary cells (CHO-K1) in our transfection experiments but this method was successfully tested on other cell lines as well (13, 15). The following items are needed:

1. CHO-K1 cells grown overnight (37°C, 5% CO_2) in Ham's F12 medium (+10% fetal calf serum) in 24-well tissue-culture-treated polystyrene plates (80,000 cells per well).
2. 1.5 ml microcentrifuge tubes and 1 ml, 200 μl and 20 μl pipettes.
3. Stopwatch.
4. Trypsin and a laboratory centrifuge.
5. PBS buffer.
6. Ice for storing stock solutions and cells.
7. Cellular thiol-blocking solution: *N*-Ethylmaleimide (NEM) in PBS (γ=78.12 mg/l, c=625 μM) (see Note 10).
8. Nucleic acid stock solutions: Plasmid DNA and/or siRNA in water ($\gamma \approx 1$ g/l, see Note 11).

2.4 Spectroscopy Components

1. UV/Vis spectrophotometer (400–600 nm), fluorometer (λ_{ex} = 488 nm, λ_{em} = 500–650 nm), and suitable microtiter plates or cuvettes.
2. PBS buffer.
3. 2-Mercaptoethanol cleavage solution for a final concentration of 100 mM (see Note 12).

3 Methods

All procedures (except cell culture experiments) are performed at room temperature. The polymer amounts used here can be adapted to your needs (see Note 13).

3.1 Labeling of Polycations with BODIPY FL L-Cystine

This section describes the labeling of lPEI, Cys_2-lPEI (see Subheading 1), and pLL (see Note 14). Differences in the work up procedures are indicated.

1. Dissolve about 30 mg (see Note 15) of the dried polymer in 400 μl H_2O and 200 μl DMSO. Depending on the polymer, the dissolution process might take several minutes. Add more water if the polymer does not fully dissolve (200–400 μl), i.e., the solution is not clear.
2. Add 20–30 μl (200–300 μg/0.25–0.38 μmol) of the BODIPY FL Cys_2 stock solution (see Note 16). The solution should now be orange and show an intense fluorescence under UV.
3. Dissolve 10 mg of DMTMM·H_2O in 1 ml of methanol and then add 63–96 μl (n(DMTMM) = 2–3 μmol) to the polymer solution. This amount corresponds to four equivalents of DMTMM for each BODIPY FL Cys_2 carboxylic group.
4. Protect the reaction setup from light and stir it for at least 4 h.
5. Transfer the solution into an ultrafiltration tube and rinse the reaction vessel with 10% DMSO (in water). Fill the tube with the same solvent to its maximum volume and start the centrifugation (see Note 17).
6. Discard the permeate, add more DMSO solution, and repeat the centrifugation until the permeate is free from unbound dye (see Note 18).
7. Repeat step 6 with borate buffer (pLL: 0.1 M NaOH). Then use water until the permeate is neutral and switch to 10 mM HCl (see Note 19).
8. Filter the retentate through a syringe filter (see Note 20) and freeze-dry it in a silanized glass vial (see Notes 7 and 9). The final product is a green fluorescent, brittle foam.

3.2 Cross-Linking of Linear PEI with $(Boc\text{-}Cys\text{-}OH)_2$

1. Dissolve 200 mg lPEI (32 μmol) and 40 mg of $(Boc\text{-}Cys\text{-}OH)_2$ (91 μmol; 2.8 equivalents per PEI molecule) in 800 μl and 850 μl, respectively, in methanol (see Note 21). Combine both solutions then add 115 mg of DMTMM·H_2O (361 μmol; two equivalents per carboxylic group) in 500 μl methanol and stir the reaction for at least 4 h.
2. Evaporate the methanol by carefully heating it up (50–60°C) while stirring.
3. To remove the BOC-protecting group, add 1 M HCl (2 ml) and stir for 1 h at room temperature. Do not close the reaction vessel as CO_2 is forming.
4. Transfer the solution into a 50 ml centrifuge tube and add about 10–15 ml of water. Place the tube in an ice bath and carefully add concentrated NaOH solution (15 ml or 4–5 NaOH pellets) to form a white PEI precipitate. Fill the rest of the tube with water (do not overfill).
5. Centrifuge (4°C/12–15,000*g*) until the precipitate is completely sedimented (see Note 22). Discard the supernatant.
6. Fill the tube with water and stir up the precipitate thoroughly. Repeat steps 5 and 6 until the supernatant is neutral.
7. Dissolve the crude product in 1 M HCl (see also Note 21) and perform an ultrafiltration with the same solvent (repeat four to five times).
8. Filter and dry your final product as described under Subheading 3.1 (see Note 23).

3.3 Quality Control with UV/Vis Spectroscopy

Quality control with UV/Vis spectroscopy is a very important step because the spectroscopic properties of BODIPY FL Cys_2 tend to undergo unforeseeable changes during the labeling process resulting in nonfluorescent products. We observed this behavior also with succinimidyl ester activated BODIPY and 5(6)-Carboxytetramethylrhodamine (5(6)-CO_2H-TAMRA).

1. Dissolve the labeled polymer hydrochloride in PBS. A concentration between 2 and 10 mg/ml is usually sufficient.
2. Measure the absorbance between 400 and 600 nm with pure PBS as blank.
3. Check for the characteristic BODIPY FL absorbance maximum (Fig. 2a) at about 504 nm. A typical product spectrum is shown in Fig. 2b. A hypsochromic shift (Fig. 2c, dotted line) or a strong sideband next to the main absorbance band (Fig. 2c, black line) indicates the formation of dye aggregates. In that case, you can try lowering the labeling density or use another polymer type (see Note 14).

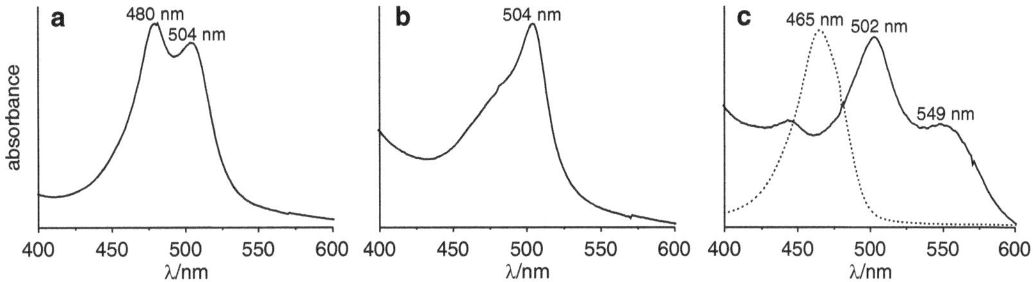

Fig. 2 Absorbance spectra of BP-Cys$_2$ (**a**) and BP-Cys$_2$-labeled polymers: disulfide cross-linked lPEI (**b**), branched PEI 25 kDa (**c**, *black line*), and disulfide cross-linked poly(L-lysine) (**c**, *dotted line*) measured in PBS. Note that the second absorbance band (480 nm) of BP-Cys$_2$ is only present when measuring in PBS. There is no difference between the absorbance of the cleaved and uncleaved probe

3.4 Quality Control with Fluorescence Spectroscopy

Fluorescence spectroscopy of the final product allows estimating the overall fluorescence intensity and the increase of fluorescence intensity after disulfide cleavage (see Note 24). Most importantly, it shows potential changes in the emission spectra that might indicate dye degradation or the formation of dye aggregates.

1. Prepare two 800 μl polymer samples (PBS, γ(polymer) ≈ 0.25–1.25 mg/ml) and add 200 μl PBS or 2-mercaptoethanol (500 mM), respectively (see Notes 12). The samples' absorbance should be below 0.1 at 504 nm to avoid self-absorbance which lowers the fluorescence intensity.
2. Measure both samples (λ_{ex} = 488 nm, λ_{em} = 500–650 nm) using PBS as blank (see Note 25).
3. Check the emission spectrum of BODIPY FL. It must not change during the labeling process (Fig. 3a). Look for any shift in the emission maximum, signal broadening (Fig. 3b, black line) or a band at around 610 nm (data not shown). In those cases, the fluorescence intensity will be too low for reasonable measurements.

3.5 Flow Cytometry

In this section, we explain how nucleic acid complexes based on the labeled polymers can be used in vitro (see Subheading 2.3). We describe the basic setup, the polyplex formation, and the data analysis in an experiment that follows the BODIPY FL fluorescence intensity during the cellular uptake of pLL-DNA nanoparticles. The use of *N*-Ethylmaleimide as a thiol-blocking agent proves the involvement of intracellular SH-groups in the probe's cleavage. NEM untreated cells show a significantly higher increase of fluorescence intensity.

1. Prepare two 24-well polystyrene tissue-culture plates (for 16 samples in triplicate) with CHO-K1 cells for the addition of pLL-DNA polyplexes (incubation times: 4 h, 2 h, 1 h, 45 min,

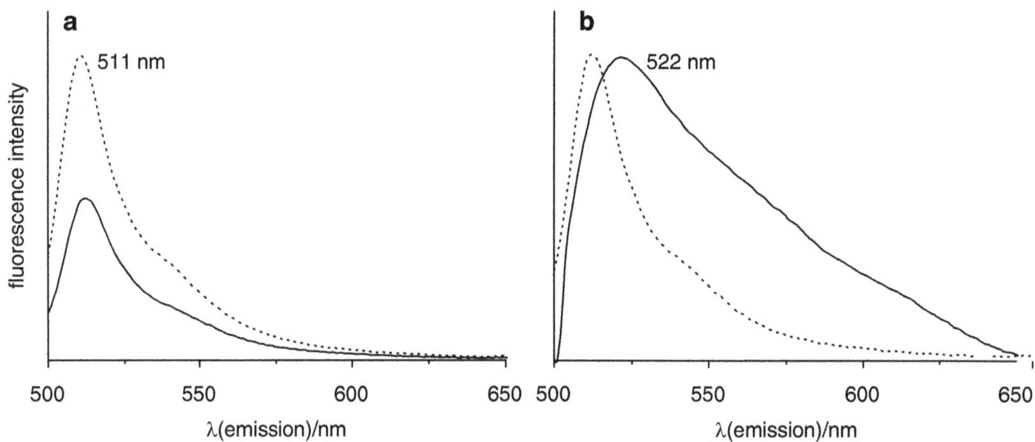

Fig. 3 Fluorescence spectra of labeled polymers (λ_{ex} = 488 nm): (**a**) disulfide cross-linked lPEI in absence (*black line*) and in presence (*dotted line*) of 100 mM 2-mercaptoethanol. The intensity increase is about twofold. (**b**) Comparison between S_2-lPEI (*dotted line*) and branched PEI 25 kDa (*black line*). Note the redshifted, broader emission band (not drawn to scale. bPEI 25 kDa has a significantly lower fluorescence intensity than S_2-lPEI)

30 min, 15 min; see Note 26) in the presence and absence of NEM (twelve samples). Cells without polyplex treatment serve as reference (four samples: 4 h, 15 min; ±NEM).

2. Before each polyplex addition, the cells are preincubated with 25 μM NEM (see Note 27) for 1 h: Replace the medium with a mixture of 240 μl Ham's F12/10% FCS and 10 μl freshly prepared NEM stock solution or PBS, respectively.
3. After 50 min, start the polyplex formation: Add 20 μl PEI stock solution to 1 μg DNA dissolved in 20 μl PBS (see Note 28). Keep it in the dark for 10 min. Wash the cells once with PBS, then add 200 μl Ham's F12 (w/o FCS) and the polyplex solution (references: 40 μl PBS). Shake the well plate gently and place it back in the incubator.
4. Repeat steps 2 and 3 with all samples at the given time points.
5. Wash all samples with PBS and detach the cells with 250 μl trypsin (see Note 29). Stop the reaction by adding 300 μl Ham's F12/10% FCS.
6. Wash all samples at least once before resuspending them in 300 μl PBS.
7. Adjust your flow cytometer as if using FITC stained samples (λ_{ex} = 488 nm; detection: "FITC-channel" (usually FL1) 530/30 nm). For finding the right PMT voltage settings, first measure the samples with the highest (4 h polyplex incubation w/o NEM) and the lowest (untreated cells) expected fluorescence intensity.

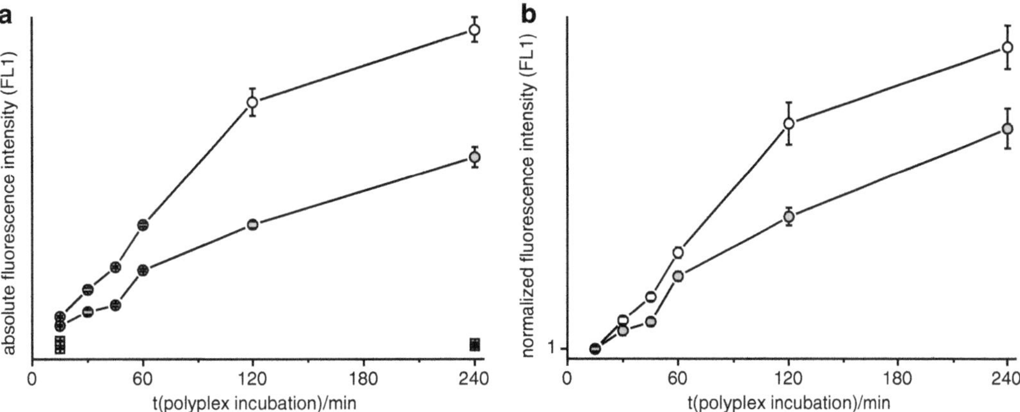

Fig. 4 Flow cytometry—fluorescence intensity of BP-Cys$_2$-pLL/DNA polyplexes in the presence (*gray symbols*) or absence (*white symbols*) of 25 μM NEM as a function of polyplex incubation time. Polyplex treated (*open circles*) and untreated/reference (*open squares*) cells: (**a**) absolute values, (**b**) normalized values

8. Data analysis: Plot the sideward scattering (SSC) intensity against the forward scattering (FSC) intensity. Gate all living, intact cells and plot their BODIPY FL fluorescence intensity (FL1) against FSC.
9. Finally, plot the mean FL1 intensity against the polyplex incubation time (Fig. 4a). Negligible changes in the intensity (15 min→4 h) of the references indicate a reliable experiment. The presence of NEM tends to shift the measured values slightly upwards so a normalization against the first time point is advisable (Fig. 4b; see Note 30).

4 Notes

1. It is not necessary to use a volumetric flask. The accuracy of the glass beaker is sufficient for this application.
2. Ready to use PBS buffer is inexpensive and can be found in any cell culture lab.
3. Selected lPEI manufacturers: Polysciences Inc., Warrington, PA, USA (http://www.polysciences.com); Polymer Chemistry Innovations Inc., Tucson, AZ, USA (http://www.polychemistry.com). PEI hydrochloride and PEI as a freebase can be labeled. There is no influence on reactivity, but PEI·HCl is insoluble in organic solvents. Please also note that commercially available PEI might contain large amounts of water. For an accurate gravimetric determination of its amount, it has to be dried in vacuum at ≈ 95°C until all water is evaporated (no bubbles in the molten polymer). For a review of lPEI syntheses, see (16) and (17).

4. Highly monodisperse linear pLL can be obtained from Alamanda Polymers, Inc., Huntsville, AL, USA (http://www.alpolymers.com). The choice of the counterion is irrelevant. Keep pLL under argon.
5. To our knowledge, BODIPY FL L-cystine is only available from Invitrogen Corporation, Carlsbad, CA, USA (http://www.invitrogen.com). Catalog Number: B-20340.
6. Solid DMTMM should be kept frozen and under an argon atmosphere. It is highly hygroscopic. Please be aware that it contains about 12% (w/w%) water as an impurity. Hence, 1 mg solid DMTMM·H_2O consists of 3.18 μmol DMTMM and 6.66 μmol H_2O.
7. Silanization can be done with Sigmacote (Sigma-Aldrich) or any comparable product. Please refer to the manufacturer's instructions.
8. We prefer ultrafiltration (centrifugal tubes) for the removal of unbound dye as it is much faster and has a higher product recovery than dialysis. As a rule of thumb from the manufacturers, select a membrane with a cutoff between 0.2 and 0.3 times the value of the molecular weight of the polymer. However, we were able to recover 50% of 3.2 kDa pLL with a 3 kDa membrane. If the MWCO is too small compared to the polymer mass, the product purification might take a very long time. Ultrafiltration tubes are available in many sizes (usually 0.5, 4, and 15 ml). For small samples (< 20 mg), use a tube volume of 4 ml. To our experience, the filtration process is faster in swing bucket than in fixed angle centrifuges.
9. If no freeze dryer is available, put the frozen product in a desiccator and use a standard laboratory vacuum pump.
10. Caution: NEM is highly toxic! Use proper safety equipment. Always prepare a fresh solution before use: Dissolve 2 mg of NEM in 639 μl PBS (c(NEM) = 25 mM) then dilute 25 μl of this solution with 975 μl PBS (c(NEM) = 625 μM). Keep this solution on ice and protected from light.
11. TE buffer can be used as well (10 mM Tris–HCl, pH = 7.0, 1 mM EDTA). RNA is highly sensitive to degradation by RNase and base hydrolysis under basic conditions. Make sure to use RNase free chemicals.

    Be careful in the choice of your gene product. The emission spectra of proteins like (E)GFP are very similar to BODIPY FL (λ_{em} = 509 nm) and will interfere in any measurement.
12. Caution: 2-Mercaptoethanol (BME) is toxic, malodorous (fume hood!), and absorbed through skin. It decomposes under the influence of water to hydrogen sulfide, so stock

solutions must always be prepared fresh. For a 500 mM stock solution, add 965 μl PBS to 35 μl of pure BME (c = 14.3 M).

13. We tested our method with the following amounts of polycations: labeling 9–50 mg, cross-linking 40–220 mg.
14. To our experience, the labeling of branched PEI (bPEI) and Cys_2-pLL leads to the formation of BODIPY aggregates with altered absorbance and emission characteristics. The resulting products show little or no fluorescence.
15. 30 mg of dried polymer correspond to about 5 and 6 μmol of 6.3 kDa lPEI and 5.2 kDa pLL·HBr, respectively.
16. A labeling ratio from 10 to 20 (n(polymer)/n(BODIPY)) is usually sufficient. Lower ratios increase the risk of forming nonfluorescent BODIPY aggregates, while higher ratios yield a low fluorescence signal of the end product.
17. At maximum speed, one centrifugation step usually takes between 20 and 30 min. The filtration devices have a dead-stop volume which prevents them from running dry.
18. Use TLC (eluant: methanol, detection: UV, R_f(BODIPY) ≈ 0.8, labeled polymer remains on baseline) to check for unbound dye. The product should be stirred up with a pipette (do not touch the filtration membrane!) before each consecutive centrifugation step. If you are unable to the finish the ultrafiltration, the product can be stored in the tube overnight.
19. PEI precipitates at higher pH values, while BODIPY FL L-cystine has an increased solubility. Do not use higher concentrated HCl (above 10 mM, pH = 2) as the dye degrades over time under more acidic conditions.
20. This filtration is necessary because the polymer hydrochloride electrostatically attracts dust.
21. The molecular weight of the cross-linked product is influenced by factors like the size of the lPEI, the amount of cross-linker, and the reaction volume. If the product turns out to be insoluble (gel-like, i.e., overly cross-linked) under acidic conditions, use less cross-linker but keep the reaction volume constant.
22. The precipitate might swim on top in the first two centrifugation steps due to the higher density of the concentrated NaOH solution. In that case, you can use a syringe with a long needle to remove the solution under the crude product.
23. The molecular weight can be determined by gel filtration chromatography: $\gamma(S_2\text{–lPEI})$ = 20 mg/ml, 150 mM NaCl + 0.1% TFA at 1 ml/min, Novema 300 Å SEC column (PSS Polymer Standards Service GmbH, Mainz, Germany) at 40°C, refractive index, and UV detector (247 nm).

Table 1
An example of how to plan a flow cytometry experiment The following parameters were used: polyplex incubation: 4 h and 15 min; NEM incubation: 1 h; polyplex formation: 10 min

t(polyplex incubation)	4 h	15 min
NEM addition	8:00	11:45
Polyplex formation	8:50	12:35
Polyplex addition	9:00	12:45
Trypsinization	13:00	13:00

24. The increase in fluorescence depends on factors like labeling density and polymer type. Therefore, no general rule can be given. It usually varies between 30 and 100% but a 30-fold increase is possible under certain conditions. You should always check if a particular polymer batch can be used for an experiment. Each instrument (e.g., spectrometer vs. cytometer) has a unique detector linearity. We even found samples with an increase as low as 30% to be very useful in flow cytometry.
25. An excitation at 504 nm would result in higher fluorescence intensity but most microscopes and flow cytometers are equipped with 488 nm light sources. In addition, due to the small Stokes shift of BODIPY FL, the excitation light might interfere with detection of the emitted light at 511 nm. It is advisable to measure the cleaved sample first as a higher emission intensity is to be expected. Adjust the measurement parameters (PMT voltage, band-passes) accordingly, and then measure the uncleaved sample and the blank under the same conditions.
26. This experiment requires thorough time planning. Frequently, different solutions have to be added at the same time. Combined with the large number of samples, this might quickly lead to confusion. It is strongly advised to create a spreadsheet that automatically calculates the time point for each work step (for an example, see Table 1).
27. NEM is cytotoxic in higher concentrations. Although in many publications concentrations around 100 μM are used, we found that this is killing virtually all CHO-K1 cells after 1 h of incubation. According to our data, 25 μM is sufficient to efficiently block cellular thiols while leaving most cells intact.
28. The ratio of positively charged polymer amine nitrogen atoms to negatively charged nucleic acid phosphorus atoms is specified by the N/P value. The higher it is the stronger polyplexes are formed; it typically lies between 8 and 24. The N/P value is calculated as follows:

n(P) estimation:

1 µg double-stranded DNA contains approximately 3.237 nmol P (RNA: 3.150 nmol)

n(N) estimation (pLL·HCl):

$$n(\mathrm{N})_{\mathrm{pLL\times HCl}} = \frac{m(\mathrm{pLL.HCl})}{M(\mathrm{pLL.HCl})} \times [N(\text{lysine residues}) + 1]$$

with $M(\mathrm{pLL.HCl}) = 18.05\ \mathrm{g/mol} + N(\text{lysine residues}) \times 164.62\ \mathrm{g/mol}$,

e.g., 1 µg $H_2N\text{-}(Lys)_{25}\text{-}OH{\cdot}HCl$ corresponds to $M(\mathrm{pLL.HCl}) = 4133.55\ \mathrm{g/mol}$, $n(\mathrm{pLL.HCl}) = 241.9\ \mathrm{pmol}$,

and $n(\mathrm{N})_{\mathrm{pLL.HCl}} = 6.290\ \mathrm{nmol}$

n(N) estimation (PEI·HCl): PEI's nitrogen content (mass fraction ω(N)) has to be determined with CHN analysis or is provided by the manufacturer,

e.g., 1 µg PEI with a nitrogen content of 17.62% corresponds to

$$n(\mathrm{N})_{\mathrm{PEI.HCl}} = \frac{m(\mathrm{PEI.HCl}) \times \omega(\mathrm{N})}{M(\mathrm{N})} = \frac{1\ \mu\mathrm{g} \times 0.1762}{14.0067\ \mathrm{g/mol}} = 12.58\ \mathrm{nmol}.$$

29. PEI or pLL can lead to an increased cell adhesion making complete trypsinization impossible. In that case, press the well plate (lid!) on a flat, smooth surface and shake it vigorously in a circular motion.

30. Calculation of normalized fluorescence intensities: $I_t^n = \frac{I_t}{I_{t_0}}$

    I_t^n: normalized intensity, I_t: absolute intensity. I_{t_0}: absolute intensity at first time point t_0.

    Corresponding propagation of uncertainty for the standard deviation σ:

$$\sigma(I_t^n) = \sqrt{\left(\frac{1}{I_{t_0}}\right)^2 \sigma^2(I_t) + \left(-\frac{I_t}{I_{t_0}^2}\right)^2 \sigma^2(I_{t_0})}$$

References

1. De Laporte L et al (2006) Design of modular non-viral gene therapy vectors. Biomaterials 27:947–954
2. Wagner E, Kloeckner J (2006) Gene delivery using polymer therapeutics. In: Satchi-Fainaro, Ronit and Duncan, Ruth (eds) Advances in polymer science: polymer therapeutics I. Springer, Berlin
3. Won YY et al (2009) Missing pieces in understanding the intracellular trafficking of polycation/DNA complexes. J Control Release 139:88–93
4. Luten J et al (2008) Biodegradable polymers as non-viral carriers for plasmid DNA delivery. J Control Release 126:97–110
5. Yadava P et al (2007) Evaluation of two cationic delivery systems for siRNA. Oligonucleotides 17:213–222
6. Miyata K et al (2004) Block catiomer polyplexes with regulated densities of charge and disulfide cross-linking directed to enhance gene expression. J Am Chem Soc 126:2355–2361

7. Bauhuber S et al (2009) Delivery of nucleic acids via disulfide-based carrier systems. Adv Mater 21:3286–3306
8. Dhananjay J et al (2009) Bioreducible polymers for efficient gene and siRNA delivery. Biomed Mater 4:025020
9. Oupicky D et al (2001) Triggered intracellular activation of disulfide crosslinked polyelectrolyte gene delivery complexes with extended systemic circulation in vivo. Gene Ther 8:713–724
10. Felgner JH et al (1994) Enhanced gene delivery and mechanism studies with a novel series of cationic lipid formulations. J Biol Chem 269:2550–2561
11. Lin C, Engbersen JF (2009) The role of the disulfide group in disulfide-based polymeric gene carriers. Expert Opin Drug Deliv 6:421–439
12. Da Poian AT et al (1998) Kinetics of intracellular viral disassembly and processing probed by Bodipy fluorescence dequenching. J Virol Methods 70:45–58
13. Lee Y et al (2007) Visualization of the degradation of a disulfide polymer, linear poly (ethylenimine sulfide), for gene delivery. Bioconjug Chem 18:13–18
14. Kunishima M et al (1999) 4-(4,6-dimethoxy-1,3,5-triazin-2-yl)-4-methyl-morpholinium chloride: an efficient condensing agent leading to the formation of amides and esters. Tetrahedron 55:13159–13170
15. Breunig M et al (2008) Mechanistic investigation of poly(ethylene imine)-based siRNA delivery: disulfide bonds boost intracellular release of the cargo. J Control Release 130:57–63
16. Lungwitz U et al (2005) Polyethylenimine-based non-viral gene delivery systems. Eur J Pharm Biopharm 60:247–266
17. Brissault B et al (2003) Synthesis of linear polyethylenimine derivatives for DNA transfection. Bioconjug Chem 14:581–587

Chapter 18

Quantification of Intracellular Mitochondrial Displacements in Response to Nanomechanical Forces

Yaron R. Silberberg and Andrew E. Pelling

Abstract

Mechanical stress affects various aspects of cell behavior, including cell growth, morphology, differentiation, and genetic expression. Here, we describe a method to quantify the intracellular mechanical response to an extracellular mechanical perturbation, specifically the displacement of mitochondria. A combined fluorescent-atomic force microscope (AFM) was used to simultaneously produce well-defined nanomechanical stimulation to a living cell while optically recording the real-time displacement of fluorescently labeled mitochondria. A single-particle tracking (SPT) approach was then applied in order to quantify the two-dimensional displacement of mitochondria in response to local forces.

Key words Atomic force microscopy, Mitochondria, Single-particle tracking, Nanomechanics, Mechanotransduction, Force transmission

1 Introduction

The atomic force microscope (AFM) (1) has become an invaluable tool for investigating biological systems. The ability to study living cells in fluid under physiological conditions has facilitated both nanoscale imaging (2) and the measurement of various mechanical and material properties of living cells, such as viscoelasticity (3, 4) and mechanical dynamics (5, 6). Recent technical developments have integrated traditional microscopy methods, such as fluorescence and laser scanning confocal microscopies, into AFM systems (7–9). This has enabled the simultaneous measurement of material properties of living cells and their biological responses and signaling pathways to be made.

Mitochondria are semiautonomous and highly dynamic organelles, which have the ability to change their shape and their location inside the living cell (10). Localization and rearrangement of mitochondria in higher eukaryotes is known to be dependent on the microtubule cytoskeleton (11, 12). More recent research suggests that actin filaments have an important role as well, such as

Volkmar Weissig et al. (eds.), *Cellular and Subcellular Nanotechnology: Methods and Protocols*, Methods in Molecular Biology, vol. 991, DOI 10.1007/978-1-62703-336-7_18, © Springer Science+Business Media New York 2013

facilitating mitochondrial organization in yeast and vertebrate neurons (13, 14) and controlling mitochondrial movement and morphology (15). Given the strong association of mitochondria with the cytoskeleton, it is predicted that forces locally applied via the AFM tip would affect their arrangement if the cytoskeleton is capable of mechanical transduction (16–18).

Feature-point tracking is a single-particle tracking (SPT) algorithm that enables an efficient, automated, two-dimensional detection and tracking analysis of particles trajectories (19). The method is suited to digital video and time-lapse fluorescence imaging, which typically generates low-intensity data. Feature-point tracking detects particle positions in a digital video or image sequence and generates particle trajectories over time. One of its main advantages is that it does not make any assumptions regarding the smoothness of the trajectories. Thus, it is extremely useful for many biological applications where the type of motion is not explicitly known in advance or when the motion is random and can change rapidly. Furthermore, by not assuming a motion model, the algorithm integrity is not biased when several modes of motion are incorporated by a single trajectory. Therefore, this method is useful for tracking both natural and force-induced motion (post-perturbation) of an organelle in the same time-lapse sequence, such as in the case of tracking mitochondrial displacements.

This protocol describes a straightforward approach for combining time-lapse fluorescent imaging while performing extracellular mechanical indentation using the AFM. The quantification of the mechanical response of mitochondria, by applying feature-point tracking approach using ImageJ's plug-in ParticleTracker, allows to correlate the effect nanomechanical forces have on the living cell and to conclude on the mechanical force transmission in the live cell. Examples for the application of this system include the comparison between basal and force-induced mitochondrial displacements (20) and the analysis of the effect of cytoskeleton disruption on force induction in the live cell (21).

2 Materials

1. Cell culture: Any standard cell type that can be stained with a live-cell mitochondria dye, such as MitoTracker Red (Life Technologies, Grand Island, NY, USA), can be used. In our case, NIH-3T3 fibroblasts were used.
2. Growth medium: GlutaMAX I media (Life Technologies, Grand Island, NY, USA) supplemented with 10% fetal bovine serum and 100 IU/ml penicillin and 100 μg/ml streptomycin.
3. Culture dishes: For epi-fluorescence imaging, either plastic or glass-bottom dishes are suitable. Note: If the AFM apparatus contain a specialized stage, then specific dish type might be needed.

4. Fluorescent dye: MitoTracker Green (Life Technologies, Grand Island, NY, USA) and MitoTracker Red stock solutions should be kept at −20°C in DMSO at concentration of 1 mM. Cells are incubated with either MitoTracker Green at 1 μM for 30 min or MitoTracker Red at 100 nM for 10 min, before replacing with fresh media.
5. AFM: Should be suitable for working in liquid and integrated onto an epi-flourescence inverted optical microscope to allow for simultaneous operation. In our case, we employed a NanoWizard I AFM (JPK Instruments); however, several manufacturers now offer similar AFMs for combined AFM and optical microscopy. Contact-mode cantilever for liquid should be used and calibrated prior to experiment. In our case, MSCT-AUWH AFM cantilevers (Veeco) with pyramidal-shaped tips were calibrated, and the spring constant was experimentally determined to be 0.05 ± 0.01 N/m (22).

3 Methods

3.1 Image Acquisition

1. Plate the cells on the day prior to experiment into suitable dishes (must be compatible with the fluorescence microscopy layout and the AFM apparatus) (see Note 1).
2. On the day of experiment, incubate the cells with MitoTracker Red at 100 nM for 10 min, before replacing with fresh media (see Note 2).
3. Leave cells to equilibrate in the incubator (37°C) for a further 30 min prior to the indentation experiment (see Note 3).
4. Calibrate the AFM (cantilever sensitivity and spring constant) by using a suitable method such as the "thermal fluctuation" method (22).
5. Put the dish containing the cells on the microscope stage (preferably temperature-controlled) (see Notes 2–6), and carefully position the AFM head on top of the stage and lower the cantilever holder into the media, paying attention not to crash the cantilever's tip into the bottom of the dish. Try to avoid having bubbles caught on the cantilever (see Note 7).
6. Optically choose live, interphase cells and position the AFM cantilever and tip above the center of the nucleus (see Note 8).
7. Set time-lapse fluorescent image capturing at the desired interval (such a one image per second).
8. After at least two images had been captured, perform indentation at the desired force (see Fig. 1 of the experimental layout).
9. Retract the AFM cantilever after the next image had been captured and stop time-lapse imaging.

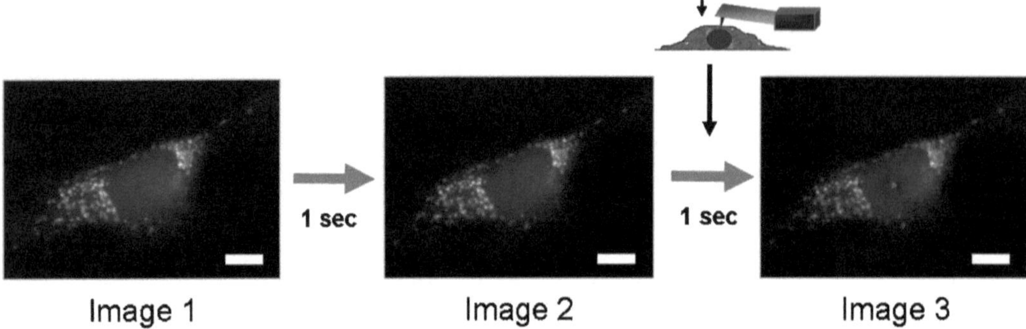

Fig. 1 Experimental layout of image acquisition. A sequence of images was taken at 1 s intervals. Three images were picked for analysis: two images taken prior to AFM indentation (images 1 and 2) and the one image that followed the indentation (image 3). Changes between image 1 and 2 reflect basal mitochondrial movement, while changes between image 2 and image 3 reflect the basal movement together with force-induced movement that resulted from the AFM indentation. The pyramidal AFM tip can be seen in the middle of the cell in image 3. Scale bars are 10 μm

3.2 Image Analysis

Mitochondria displacement analysis is carried out by the method of feature-point tracking (see Introduction). A sequence of three images is analyzed, and the comparison is made between two pairs of images: changes from image 1 to image 2, which were captured prior to AFM indentation, reflect the basal movement of mitochondria in the cell (control), and changes from image 2 to image 3, between which AFM indentation takes place, reflect the basal movement in addition to the force-induced movement (Fig. 1). Thus, this experimental layout has a built-in control that allows the normal, basal movement to be distinguished from movement that results from force application and indentation.

For analysis of mitochondria displacements, any particle-tracking software can be used; however, ImageJ, a public-domain image processing and analysis program developed at the National Institute of Health (ImageJ, http://rsbweb.nih.gov/ij/), together with the ParticleTracker plug-in (ParticleTracker, http://courses.washington.edu/me333afe/ImageJ_tutorial.html), proved to be a suitable tool for that purpose (20, 21). This plug-in is used to detect particles and calculate trajectories in an image sequence using the feature-point tracking algorithm (19). The tracking algorithm consists of two main steps: detection of feature points in every frame and linking of these points into trajectories:

1. Load the three sequential fluorescent images into ImageJ, and convert them to a stack (Image > Stacks > Images to Stack) (see Note 8).
2. Run the ParticleTracker plug-in. Since image conditions, such as the intensity and noise levels, vary between each image set, several parameters need to be adjusted manually in order to facilitate correct particle detection and avoid false detections

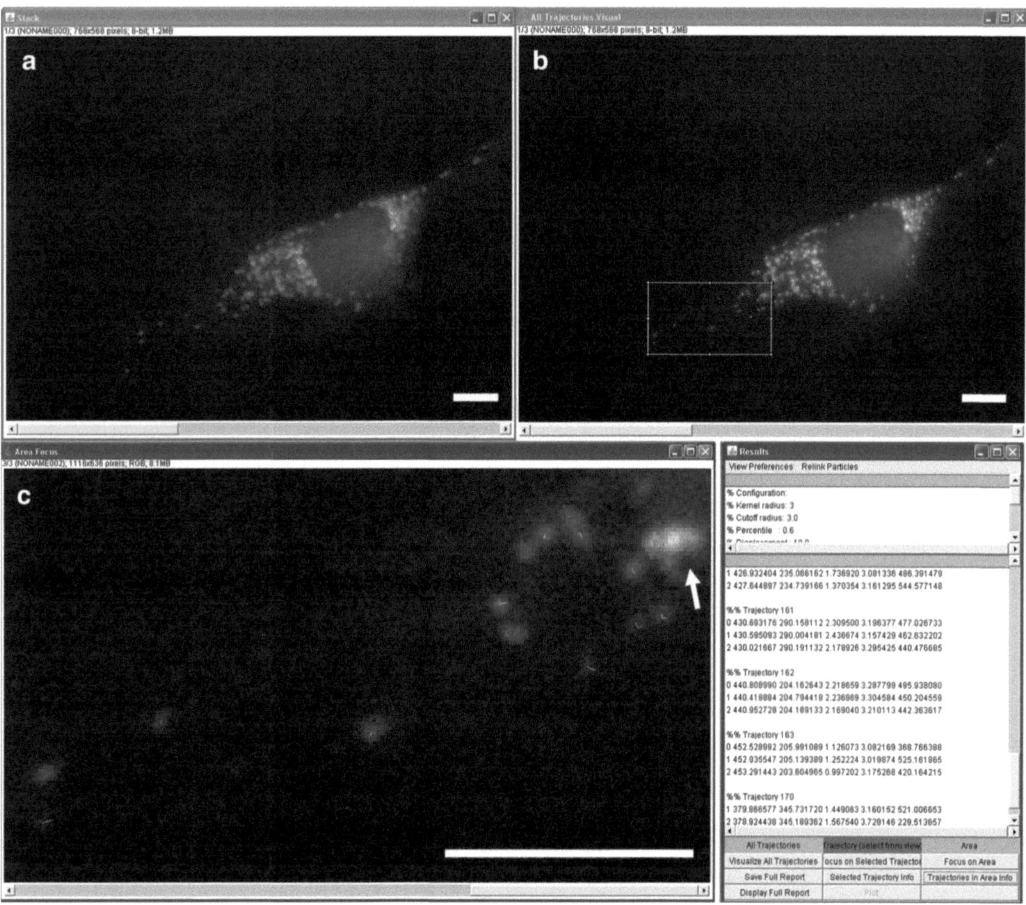

Fig. 2 Tracking mitochondrial displacements using ParticleTracker. After the three images are combined into a stack, the ParticleTracker plug-in identifies feature points according to manually entered parameters (**a**) a region for analysis is then chosen, (**b**) and the trajectories in that region are shown together with coordinates of each determined feature point at each of the three images (**c**). Before processing the results, all trajectories are manually verified and false links (**c**, *white arrow*) are removed. Scale bars are 10 μm

due to background noise. These include the approximate radius of the particle (in pixels), which should be bigger than a single-point radius but smaller than the distance between two separate points (usually set to 3), and a percentile (%) that determines the sensitivity of the algorithm to background noise when deciding the local maxima of featured points (i.e., how bright the particle needs to be in order to be accepted as a feature point). The percentile is the percentage of the upper end intensity range that will be considered as feature points, and is usually set between 0.2 and 1.5%.

3. ImageJ marks with red circles all the detected feature points (Fig. 2a). If the image is too noisy, adjustments to the above parameters should be made to make sure no "noise" was mistaken as feature points.

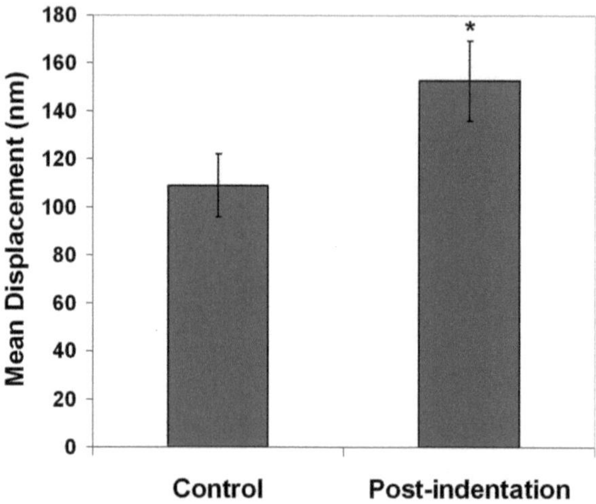

Fig. 3 Mean mitochondrial displacement calculated with the feature-point tracking algorithm. The average mitochondrial displacement increased in ~40% following AFM indentation, from 114 ± 6 nm for the natural displacement to 160 ± 10 nm post-indentation (mean ± s.e.m; * < 0.001)

4. Click on "visualize all trajectories." Then, select an appropriate region for the analysis (e.g., where single mitochondrial structures can clearly be distinguished from each other) (Fig. 2b), and click "Trajectories in area info." Data on all the trajectories in that region is extracted, including the *x* and *y* coordinates of each particle in each of the images in the stack (Fig. 2c). You can then copy and paste the list of coordinates to a separate ASCII file.
5. Manual inspection of the trajectory list is recommended, especially for more noisy images, in order to remove invalid trajectories (see Note 9).
6. The final step is the calculation of absolute displacements of each particle from the extracted trajectories, which can be done either manually or using an appropriate data processing tool, such as MatLab. Note that the second column of the output data contains the x coordinate and the third column contains the y coordinate and that every line represents a different image in the stack.

As images 1 and 2 are taken prior to indentation and image 3 taken following indentation, the absolute basal (natural) displacement (in pixels) is given by $\sqrt{(x2-x1)^2+(y2-y1)^2}$,and the force-induced movement is given by $\sqrt{(x3-x2)^2+(y3-y2)^2}$ (the numbers following *x* and *y* represent the respected images).

Figure 3 shows the results for natural and post-indentation displacement of mitochondria, calculated using the feature-point tracking algorithm (21). A total of 323 individual mitochondria were analyzed, in 42 regions for a total of 21 cells. The average natural displacement calculated to be 114 ± 6 nm, while the average post-indentation displacement was 160 ± 10 nm (mean ± s.e.m; $*P < 0.001$). That is, mitochondrial displacement increased by ~40% in response to indentation.

4 Notes

1. When investigating force transmission on a single-cell level, it is important to make sure the experiment is conducted on single cells not on a confluent layer, where intracellular mechanics are interlinked and affected by neighboring cells. However, experiments can also be devised to investigate the movement of internal cellular structures in cell monolayers.
2. It important to take into consideration medium evaporation from the dish. The AFM cantilever should be fully immersed in the fluid. In our case, 50 mm glass-bottom FluoroDish™ culture dishes (World Precision Instruments, Inc., UK) were used, containing 3 ml of media.
3. Use of HEPES buffer is recommended for maintaining appropriate pH levels in the medium. Typically, cells in a 50 mm dish are stable for about 1 h in ambient conditions. We do not recommend working beyond 2 h without medium replacement or devising a scheme to control the pH of the medium (e.g., creating a 5% CO_2 atmosphere).
4. The experiment length should be minimized, as temperature and pH are not stable under these conditions and also cell are susceptible to contamination. Ideally, the experiment should not extend longer than 1 h.
5. Following the experiment, cells should be discarded and not put back into the incubator, as contamination from the environment and/or AFM tip are prone to happen.
6. Owing to their viscoelastic properties, the cells' response to force will vary not only according to the magnitude of the force but will also be affected by loading rate; thus, tip velocity should be taken into consideration and kept constant throughout the experiments.
7. Applying a drop of ethanol or DMSO on the cantilever tip helps avoiding air bubbles being trapped underneath the tip when lowering the AFM cantilever into the media.
8. When choosing a region for tracking (in ImageJ or other software), take into consideration that when tracking a 2D movement only, it is better to choose mitochondria that are near or at the edges of the cell, where the cell is very flat and 2D motion can be assumed.

9. The output trajectories data will normally include a few invalid trajectories (sometimes up to ~10% of the total calculated trajectories, depending on image quality, particle density, and analysis parameters), which results from false linking of particles between two frames (Fig. 2c, white arrow). These false trajectories are often of high magnitude and will greatly affect the overall calculated average; thus, it is important to filter them out. However, they can be easily noticed as their magnitude of displacement (the change in x or y coordinates) will be considerably higher than the usual displacement of true trajectories.
10. As mitochondrial motion is both random Brownian motion and directed filament-based displacement, a convenient way to validate the displacement analysis is to compare between mitochondrial movement in control cells and that in cells treated with cytoskeleton-disrupting drugs such as cytochalasin and nocodazole (21).

Acknowledgments

Y.R.S. would like to thank the Japanese Society for the Promotion of Science (JSPS) for a post-doctoral fellowship grant. A.E.P. acknowledges generous support from the Canada Research Chairs program, the Province of Ontario Early Researcher Award, and the Natural Sciences and Engineering Research Council. The authors would like to gratefully acknowledge the tremendous support and mentorship of Professor Michael Horton (1948–2010) and his inspiration for this work.

References

1. Binnig G, Quate CF, Gerber C (1986) Atomic force microscope. Phys Rev Lett 56:930–933
2. Putman CA, van der Werf KO, de Grooth BG, van Hulst NF, Greve J (1994) Viscoelasticity of living cells allows high resolution imaging by tapping mode atomic force microscopy. Biophys J 67:1749–1753
3. Radmacher M, Tillmann RW, Fritz M, Gaub HE (1992) From molecules to cells—imaging soft samples with the atomic force microscope. Science 257:1900–1905
4. Charras G, Horton MA (2001) Cellular mechanotransduction and its modulation: an atomic force microscopy study. Biophys J 80: 305A–306A
5. Rotsch C, Jacobson K, Radmacher M (1999) Dimensional and mechanical dynamics of active and stable edges in motile fibroblasts investigated by using atomic force microscopy. Proc Natl Acad Sci USA 96: 921–926
6. Pelling AE, Sehati S, Gralla EB, Valentine JS, Gimzewski JK (2004) Local nanomechanical motion of the cell wall of Saccharomyces cerevisiae. Science 305:1147–1150
7. Haupt BJ, Pelling AE, Horton MA (2006) Integrated confocal and scanning probe microscopy for biomedical research. ScientificWorldJournal 6:1609–1618
8. Horton M, Charras G, Ballestrem C, Lehenkari P (2000) Integration of atomic force and confocal microscopy. Single Mol 1:135–137
9. Lehenkari PP, Charras GT, Nykänen A, Horton MA (2000) Adapting atomic force microscopy for cell biology. Ultramicroscopy 82:289–295
10. Bereiterhahn J, Voth M (1994) Dynamics of mitochondria in living cells—shape changes, dislocations, fusion, and fission of mitochondria. Microsc Res Tech 27:198–219
11. Brady S, Lasek R, Allen R (1982) Fast axonal transport in extruded axoplasm from squid giant axon. Science 218:1129–1131

12. Heggeness MH, Simon M, Singer SJ (1978) Association of mitochondria with microtubules in cultured cells. Proc Natl Acad Sci USA 75:3863–3866
13. Morris R, Hollenbeck P (1995) Axonal transport of mitochondria along microtubules and F-actin in living vertebrate neurons. J Cell Biol 131:1315–1326
14. Drubin D, Jones H, Wertman K (1993) Actin structure and function: roles in mitochondrial organization and morphogenesis in budding yeast and identification of the phalloidin-binding site. Mol Biol Cell 4:1277–1294
15. Rudiger Suelmann RF (2000) Mitochondrial movement and morphology depend on an intact actin cytoskeleton in Aspergillus nidulans. Cell Motil Cytoskeleton 45:42–50
16. Wang N, Butler JP, Ingber DE (1993) Mechanotransduction across the cell-surface and through the cytoskeleton. Science 260: 1124–1127
17. Alenghat FJ, Ingber DE (2002) Mechanotransduction: all signals point to cytoskeleton, matrix, and integrins. Sci STKE 2002:pe6
18. Blumenfeld R (2006) Isostaticity and controlled force transmission in the cytoskeleton: a model awaiting experimental evidence. Biophys J 91:1970–1983
19. Sbalzarini IF, Koumoutsakos P (2005) Feature point tracking and trajectory analysis for video imaging in cell biology. J Struct Biol 151:182–195
20. Silberberg YR, Pelling AE, Yakubov GE, Crum WR, Hawkes DJ, Horton MA (2008) Tracking displacements of intracellular organelles in response to nanomechanical forces, presented at Biomedical imaging: from nano to macro, 2008. ISBI 2008. 5th IEEE international symposium on 14–17 May 2008, Paris, France. pp 1335–1338
21. Silberberg YR, Pelling AE, Yakubov GE, Crum WR, Hawkes DJ, Horton MA (2008) Mitochondrial displacements in response to nanomechanical forces. J Mol Recognit 21:30–36
22. Levy R, Maaloum M (2002) Measuring the spring constant of atomic force microscope cantilevers: thermal fluctuations and other methods. Nanotechnology 13:33–37

Chapter 19

Imaging Select Mammalian Organelles Using Fluorescent Microscopy: Application to Drug Delivery

Paul D.R. Dyer, Arun K. Kotha, Marie W. Pettit, and Simon C.W. Richardson

Abstract

The microscopic imaging of specific organelles has become a staple of the single-cell assay and has helped define the molecular regulation of many physiological processes. This definition has been made possible by utilizing different criteria to identify specific subpopulations of organelles. These criteria can be biochemical, immunological, or physiological, and in many cases, markers regulate fusion to the organelle they define (e.g., Rab-GTPase proteins). Single-cell imaging technology allows, within the context of drug delivery, an evaluation of the intracellular trafficking of both biological and synthetic macromolecules. However, it should be remembered that there are many limitations associated with this type of study and quantitation is not easy. The temporal dissection of novel and default trafficking of both macromolecular "drugs" and macromolecular drug delivery systems is possible. These methodologies are detailed herein.

Key words Endocytosis, Drug delivery, Organelle, Fluorescent microscopy, Live-cell imaging, Polymers

1 Introduction

Why do we care about cell imaging? Despite being subject to many limitations, including the misinterpretation of fixation artifacts, the interpretation of static depictions of dynamic systems, a high degree of difficulty interpreting quantitative information, microscopy is still a very useful tool to both the cell biologist and the drug delivery scientist. However, false assumptions based upon propter hoc arguments, i.e., "because a protein is visible on a specific compartment, it functions there," must be identified and rejected. Further, cognizance of the limitations of the systems is also critical, as essentially ex vivo systems, i.e., cells grown in culture, may or may not behave the same way as cells in a functioning organism. Fluorescent quenching (as an effect of fluorophore concentration) and changes in fluorescent yield due to vesicular acidification make qualitative

Volkmar Weissig et al. (eds.), *Cellular and Subcellular Nanotechnology: Methods and Protocols*, Methods in Molecular Biology, vol. 991, DOI 10.1007/978-1-62703-336-7_19, © Springer Science+Business Media New York 2013

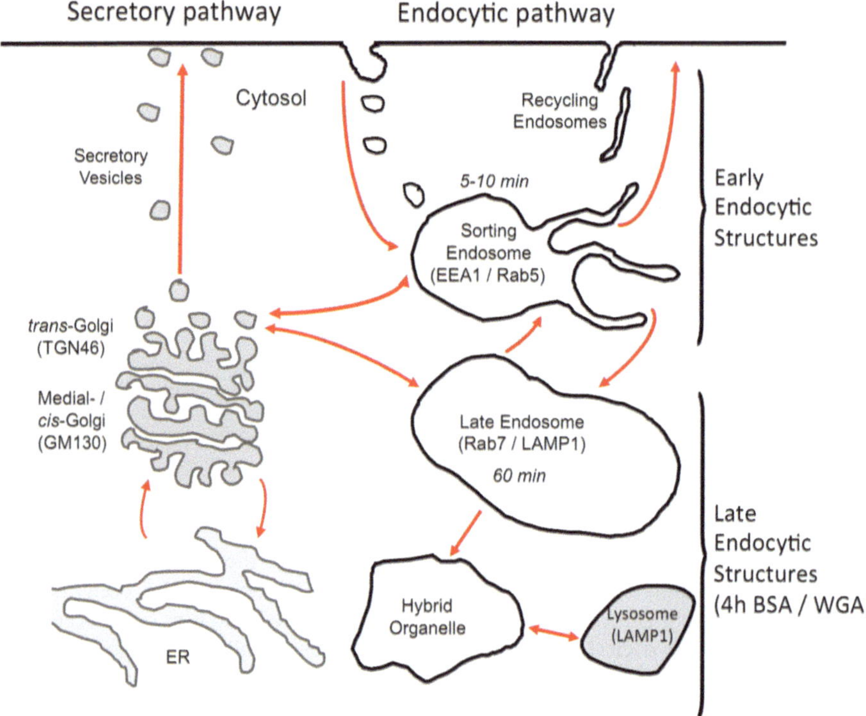

Fig. 1 Shows a cartoon representing the trafficking of material between the organelles of the secretory and endocytic system of mammalian cells

and temporal data easier to interpret reliably than quantitative data defining drug uptake. However, that said, quantitation is not impossible but beyond the scope of this chapter (1). The useful application of microscopy to drug delivery is defined by experimental design, and can vary from affirming the cellular uptake and internalization of fluorescently labeled material, to examining the kinetics associated with the movement of material between intracellular compartments (see Fig. 1) (1–4). Mitigating the caveats outlined above is the philosophical view that some data is infinitely more than no data and that in an imperfect world, we need to start somewhere.

In this chapter, we will document protocols that have been used to examine the trafficking of macromolecular material to a select panel of intracellular compartments in a qualitative or temporal manner. This is explored in both live and fixed cells in order to try to control for fixation artifacts. Particular attention has been paid to sample preparation as, given the current technology base, imaging is viewed as routine and not rate limiting. Discussion pertaining to specific markers delimiting intracellular compartments (within the context of drug delivery) can be found in an earlier review (4, 5).

This chapter has three objectives and these are to describe methodologies that:

(a) Establish subcellular markers in fixed cells (see Fig. 1).

(b) Localize “hard to fix” materials, i.e., biocompatible polymers.

(c) Facilitate the visualization of material in live cells.

Examples of select markers are shown in both non-transfected (see Fig. 2) and transfected (see Fig. 3) squamous epithelial (Vero) cells.

2 Materials

2.1 Materials, Equipment, and Cells

Transient transfections were performed using Lipofectamine® (Invitrogen, Paisley, UK). Vero cells (designated CCL-81) were from the American Type Culture Collection (ATCC) (Middlesex, UK). PD10 columns (GE Healthcare, Fairfield, CT, USA) were used for conjugate-fluorophore separation. Cells were imaged using either a Nikon 90i overhead epifluorescent microscope attached to a Nikon digital camera (DS-Qi1Nc) and a computer running Nikon NIS-Elements Advanced Research software. The principal objective used for fluorescent imaging was an oil immersion CFI Plan Apochromat VC 60X N2 (NA 1.4, WD 0.13 mm).

Live-cell imaging was performed using a Nikon Eclipse Ti inverted microscope with a heated stage (set to 37°C) attached to a Nikon (DS-Fi1) digital camera also connected to the Nikon NIS-Elements Advanced Research software. The principal objective used for inverted fluorescent imaging was a CFI Super Plan Fluor 60× (dry) (NA 0.70, WD 2.61–1.79 correction collar 0.1–1.3 mm).

2.2 Solutions and Culture Media

1. *Phosphate-buffered saline (PBS)*: This was prepared as a 10× stock by adding sodium chloride (80 g), potassium chloride (2 g), disodium hydrogen phosphate (14.4 g), and potassium dihydrogen phosphate (2.4 g) to 800 ml of double distilled deionized (dddi) H_2O. The final volume was adjusted to 1,000 ml with dddi H_2O. This was diluted appropriately using dddi H_2O (see Note 1).
2. *Richardson Piper (RP) media*: Magnesium acetate (1 mM), calcium chloride (1 mM), glucose (5 mM), glutamate (5 mM), fetal bovine serum (10% v/v), were made up to their stated concentration in 1× PBS. This was then filter sterilized through a 0.2 µm filter (see Note 2).
3. *Buffered formalin*: First 10× PBS (5 ml) was added to a glass beaker. Dddi H_2O was added to bring the volume to 30 ml. The preparation was heated to approximately 80°C. Paraformaldehyde (PAF) (1 g) was added as well as 5 N sodium

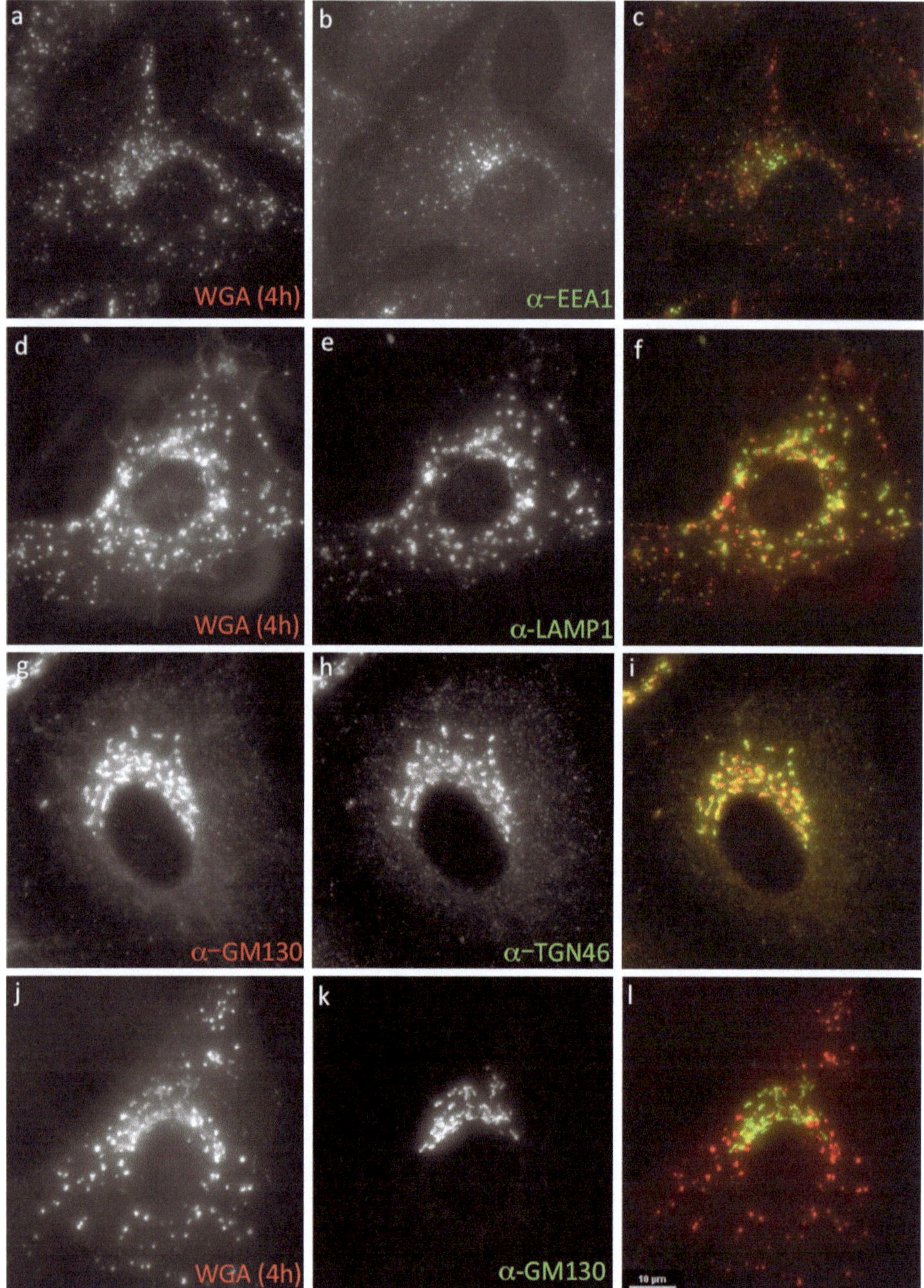

Fig. 2 Panels (**a**–**c**) cells fixed in aldehyde, having been incubated for 1 h with 25 μg/ml WGA-TxR (with 200 μM leupeptin), were washed with PBS and incubated for a further 4 h in complete media supplemented with 200 μM leupeptin. Little co-localization between the early endocytic structures labeled by the anti-early endosomal antigen 1 (EEA1) antibody and the WGA-TxR was evident. This was not surprising as the majority

hydroxide (1 ml). The reagent was subject to aspiration to dissolve the PAF. Hydrochloric acid (5 N) was then added (900 μl). More hydrochloric acid was added drop-wise to bring the pH to 7.0 before dddi H_2O was added to bring the final volume to 50 ml (see Note 3).

4. *Saponin extraction buffer*: PIPES buffer (10 ml of 400 mM pH 6.8 in water, stored at 4°C) was added to a 50 ml sterile plastic conical tube. EGTA (5 ml of a 50 mM stock) was also added. Saponin (500 μl of a 5% w/v solution in water, store at −20°C) and 5 μl magnesium chloride solution (1 mM in H_20) were also added. Dddi H_2O was added to a final volume of 50 ml (see Note 4).
5. *Triton permeabilization buffer*: PBS (5 ml of a 10× stock solution) was added to a 50 ml sterile plastic tube. Glycine was added to bring its final concentration to 50 mM. Triton-X-100 was added to a final concentration of 0.2% v/v (see Note 5).
6. *Blocking buffer*: PBS (5 ml of a 10× stock) was added to a 50 ml sterile plastic tube. Serum (1 ml) was added and the final volume adjusted to 50 ml using dddi H_2O (see Note 6).
7. *Mounting media* (see Note 7): *N*-propyl gallate (10 mg) was added to a sterile 1.5 ml Eppendorf tube. PBS (100 μl of a 10× stock) was added as well as dddi H_2O (400 μl) (see Note 8). Glycerol (500 μl) was then added and the preparation gently heated to (60–70°C) to facilitate the dissolution of the *N*-propyl gallate. The preparation was then allowed to cool to room temperature.

3 Methods

3.1 Establishing Subcellular Markers in Fixed Cells

As reported (4–6), robust markers for subcellular endocytic and secretory organelles are often proteins that are integral to maintaining organelle identity ("gate keepers"). Examples include tethering proteins, receptors, or overexpressed GTPases fused in frame

Fig. 2 (continued) of the WGR-TxR was documented (**d–f**) as occupying late endocytic structures (Fig. 1). Panels (**d–f**) shows cells fixed in methanol having been treated with WGA-TxR as above. Here a high degree of co-localization was evident between WGA-TxR and antibodies specific for lysosomal-associated membrane glycoprotein 1 (LAMP1), LAMP 1 being a marker for later endocytic structures (Fig. 1). Panels (**g–i**), cells subject to aldehyde fixation were immunostained with a monoclonal antibody specific for Golgi matrix (GM) protein of 130 kDa (GM130) and a polyclonal (sheep) anti-Trans-Golgi network protein of 46 kDa (TGN46) antibodies. Here a high degree of co-localization was evident, though TGN46 signal may also appear upon puncta surrounding the medial-Golgi. The anti-GM130 antibody may be seen decorating a reticular structure that corresponded to the Golgi ribbon (medial-Golgi). GM130 is known to facilitate tethering to the cis-Golgi. This reticular structure is visible below (**k**). Panels (**j–l**) show aldehyde fixed cells that have been treated with WGA (as above) and immunostained using an anti-GM130 antibody. Here little co-localization of signal from the late endocytic structures labeled with WGA-TxR and the Golgi labeled with the anti-GM130 antibody was evident

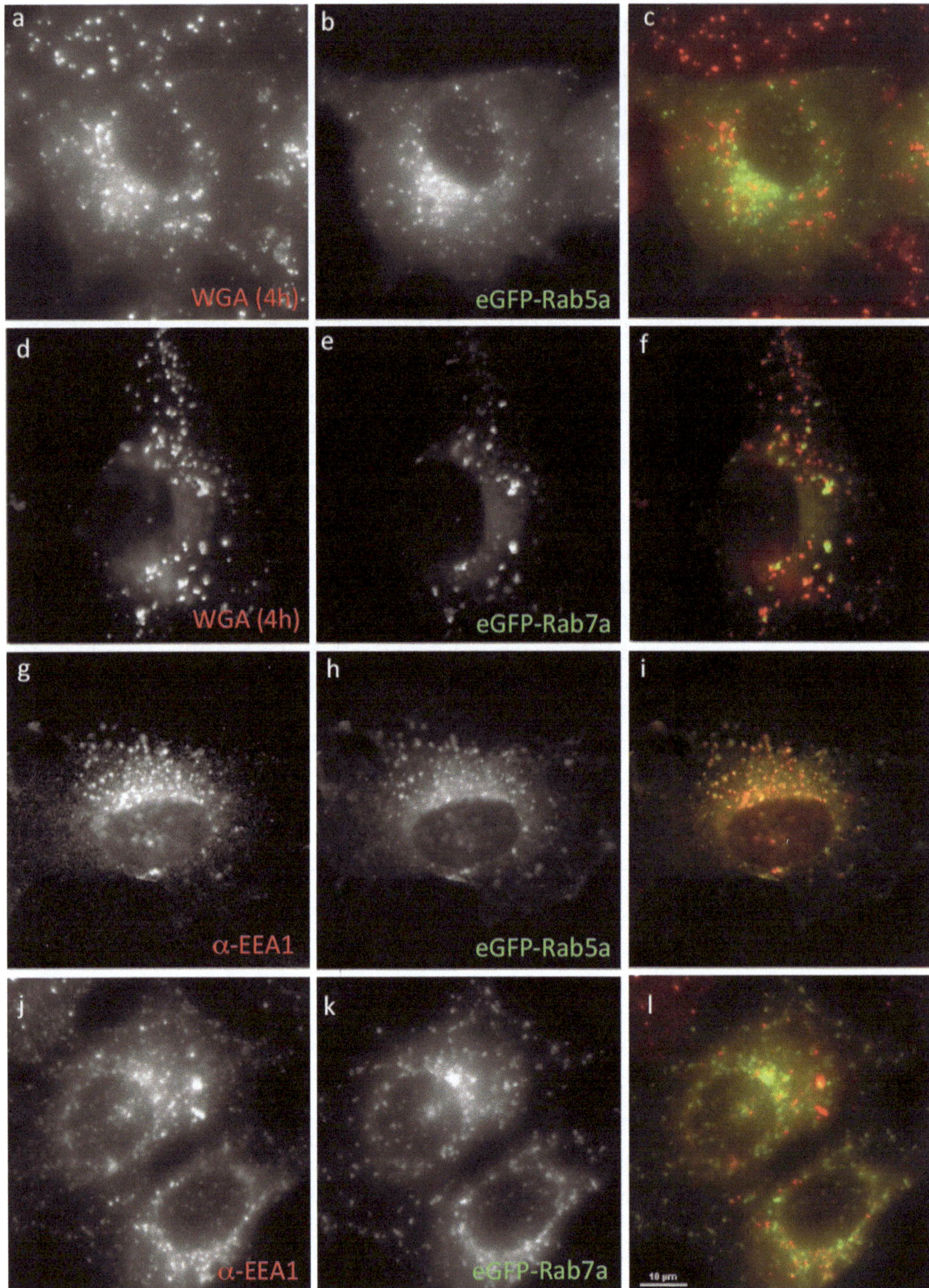

Fig. 3 Transfected cells transiently expressing either enhanced green fluorescent protein (eGFP)-rat sarcoma ((Ras)-related in brain) (Rab) 5a (**b**, **h**) or eGFP-Rab7a (**e**, **k**), 48 h after transfection. Panels (**a**, **d**) show cells that have been treated with WGA as described in Fig. 1 after the cells were transfected. A very small degree of co-localization was evident between eGFP-Rab5 and WGA-TxR (**a–c**) and eGFP-Rab7a and the anti-EEA1

to eGFP. In order to image these proteins, either immunostaining or transfection may be employed. The trafficking of fluorescent physiological probes subject to endocytic capture may also be characterized (temporally) relative to these immunological or eGFP-tagged markers (see Note 9). To this end:

1. A glass coverslip (22 mm × 22mm × 0.5mm) was placed in a small glass beaker and the coverslip completely submerged in absolute ethanol. The beaker was then placed in a class II cell culture hood containing a lit spirit burner.
2. Using a set of metal micro-forceps, the glass coverslip was removed from the alcohol and the excess liquid removed. Taking care not to ignite the reservoir of ethanol in the beaker, the ethanol still on the coverslip was ignited. The coverslip was then sterile (see Note 10).
3. Within the class II hood, a sterile glass coverslip was placed in each well of either a sterile 6-well TC coated plate or a 35-mm-diameter sterile TC-coated dish.
4. Squamous epithelial cells suspended in complete media were then added to each well at a density of 5×10^5 cells/well for transfection experiments or 1×10^5 for non-transfection immunofluorescence (IF) experiments (see Notes 11 and 12).
5. The final volume of media in each well was then adjusted to 2 ml, and the cells left in a cell culture incubator overnight, i.e., at 37°C (5% v/v CO_2).
6. If the transient expression (via transfection) of marker proteins was required, plasmid DNA (1 μg/well) was placed in 0.5 ml/well of serum- and antibiotic-free cell culture media. This preparation was sterile (see Note 13).
7. The DNA preparation (step 6 above) was then mixed with an equal volume of serum-free and antibiotic-free media containing a 5× excess (relative to DNA by weight) of Lipofectamine®. This preparation was sterile.
8. The transfection reagent (from the previous step) was left at room temperature for 30 min to come to equilibrium.
9. After 30 min., 1 ml of the lipofection reagent was applied per well, to cells that had been washed 3× with sterile PBS pH 7.4. The cells were then left under standard incubation conditions for 4–6 h.
10. After 4–6 h, the lipofection reagent was removed, and the cells were washed three times with sterile PBS before being left in complete media under standard incubation conditions (see Note 14) prior to examination.

Fig. 3 (continued) antibody (**j–l**). A high degree of co-localization between cells expressing eGFP-Rab7a and WGA-TxR (**d–f**) was evident. Separately, cells immunostained with the anti-EEA1 antibody, and expressing the eGFP-Rab5a transgene (**g–i**) also showed a high degree of co-localization. All cells were fixed with aldehyde

3.2 Immunostaining

The following steps were performed in rapid succession, having made the necessary reagents immediately beforehand (see Note 15).

1. Cells were set up as described above (Subheading 3.1, steps 1–5).
2. Following an appropriate incubation time, the cell monolayer was washed three times with PBS (see Notes 16 and 17). Fixation may be undertaken with formalin (aldehyde) or cold solvent (4).
3. Fixation with formalin (see Note 18): *Buffered formalin* (2% w/v) (Subheading 2.2, item 3) was added to the cell monolayer immediately after the final PBS wash had been removed. The cells were left in the buffered formalin for 20 min at room temperature.
4. Fixation by solvent extraction: Immediately after the final PBS wash was removed, prechilled absolute methanol (−20°C) was added (2 ml/well). The dish was then left at −20°C for 5 min (see Note 19).
5. Following fixation, the cells were washed three times with PBS, and the final wash removed.
6. The *Triton permeabilization buffer* (Subheading 2.2, item 5) was added to the cell monolayer after fixation in aldehyde. The cells were incubated in this buffer for 5–20 min at room temperature (see Note 20).
7. Blocking nonspecific antibody binding: After a further 3× PBS washes, *blocking buffer* (1–2 ml) (Subheading 2.2, item 6) was added to each well. The cells were left in *blocking buffer* at room temperature for 60 min.
8. Primary antibody hybridization: Antibody was diluted to an appropriate concentration using 1% v/v serum solution made up in PBS. The diluted antibody (40 μl) was placed on a strip of Parafilm™ within a humidified sealable container (see Note 21). Following this incubation, the cells were washed three times with PBS.
9. Secondary antibody hybridization: A dilution of 1:100 to 1:300 of fluorophore-labeled secondary antibody diluted in 1% v/v serum in PBS was made. Hybridizations were performed as before (step 8 above) with the exception that the coverslips were left in the dark for 30–45 min. After this incubation, the coverslip(s) were washed three times with PBS.
10. *Mounting media* (~30 μl) (Subheading 2.2, item 7) was placed on a glass slide, and the edge of one coverslip was lowered into it, cells facing the glass slide. *Mounting media* was then smeared across the microscope slide, ensuring that the area that the coverslip would eventually occupy was covered. The angle between the coverslip and the glass slide was then reduced

until the two were in contact ensuring there were no trapped air bubbles.

11. The coverslip was then positioned and excess mounting media removed. A folded tissue was then used to absorb any excess liquid while gently pushing the coverslip into the slide (see Note 22).
12. The edges of the coverslip were sealed to the microscope slide with nail varnish and the slide stored at −20°C (see Note 23).

3.3 Localizing "Hard to Fix" Materials, i.e., Biocompatible Polymers

This can be achieved by localizing material relative to a well-characterized endocytic probe (see Note 24) such as *Texas red-conjugated BSA* (see Note 25) or *Texas red-conjugated WGA*. Having established the temporal kinetics of the probe relative to immunological markers, it was possible to use an aforementioned physiological probe in conjunction with an unknown (i.e., a fluorophore-labeled polymer) to delineate the intracellular trafficking of material that will not fix well but is subject to compartmentalization. That said, the test material to be evaluated must have been conjugated to a fluorophore. This approach allows the omission of a permeabilization step, which would allow the postfixation movement of material. Although *Texas red-conjugated WGA* can be obtained commercially, *Texas red-conjugated BSA* must be made and characterized by the researcher.

1. Bovine serum albumin (BSA) (100mg) was dissolved in 9.5 ml of PBS and placed in a glass, lightproof container.
2. Texas red®-X, succinimidyl ester (TxR-NHS) was dissolved in 1 ml of dimethyl sulfoxide (DMSO).
3. TxR-NHS (0.5 ml) was then added to the BSA solution and left at room temperature for 30 min. Residual TxR-NHS was frozen @ −20°C for use at a later date.
4. PD10 columns (GE Healthcare, Fairfield, CT, USA) (x4) were equilibrated with 30 ml of PBS.
5. Free fluorophore was separated from the conjugate by loading 2.5 ml of the reaction onto each (of four) PD10 columns and eluted in a volume of 3.5 ml/column. Characterization of the conjugates was performed as per the manufacturer's instructions (TxR-NHS) (2) (see Note 26).
6. Vero cells were plated onto either 35 mm diameter or 6-well sterile TC coated places containing a sterile coverslip and left under optimal culture conditions for 24 h (Subheading 3.1, steps 1–5).
7. The cell culture media was removed, and cells were washed three times with sterile PBS.
8. Wheat germ agglutinin (WGA)-Texas red (TxR) (5–50 μg/ml) or BSA-conjugated Texas red (BSA-TxR) (5–10 mg/ml)

was then added to the cells, co-incubated with leupeptin (200 μM). These reagents were typically administered in a final volume of 0.5 ml cell culture media.

9. Media containing WGA-TxR was left on the cells at 37°C in 5% v/v CO_2 for 1 h and media containing BSA-TxR for 4 h at 37°C in an atmosphere of 5% v/v CO_2.
10. Both physiological probes were then "chased" (temporally) into late endocytic structures, typically over a 4–48 h time span. The chase phase was performed after removing the media containing the fluorescent marker, washing the cells 3× with sterile PBS (pH 7.4), and replacing the PBS with complete media also containing 200 μM leupeptin.
11. The cells were then fixed and subject to immunological characterization (Subheading 3.2) (2, 6).
12. Using fresh cells, the unknown material was incubated with the marker (WGA-TxR or BSA-TxR) under experimental conditions that result in a known distribution of the WGA-TxR/BSA-TxR.
13. Following the pulse and chase phases, the cells are fixed with aldehyde, and the permeabilization step is omitted completely. The cells are then mounted as previously described and imaged (2, 4, 6).

3.4 Live-Cell Imaging

Sterile 6-well TC treated plates (or individual sterile 35 mm TC treated dishes) were ideal for this application. Live-cell imaging was performed using an inverted microscope with a heated stage, preheated to 37°C. This temperature was maintained throughout the experiment. To minimize photobleaching and oxidative cell damage, the exposure of the cells to all light sources was kept to a minimum. The lid was removed from the dish containing the cells in order to remove condensation. If necessary, the media was replaced when evaporation was deemed to be a problem (2, 4).

1. At an appropriate time, complete media was removed and the cells washed three times with sterile 1× PBS (pH 7.2).
2. Prewarmed (37°C) RP media (~2 ml) containing 200 μM leupeptin (where appropriate) was then applied to the cells.
3. Once the region of interest was selected using phase contrast microscopy (and the appropriate fluorescent channels), images were taken every 10 min over 4 h.
4. After completing the imaging process, the pictures from both channels were merged using the software provided by the microscope manufacturer (i.e., NIS-Elements Advanced Research software by Nikon). The data was exported as an avi file and edited in Final Cut Studio (Apple Computing Ltd. Cupertino, USA).

3.5 Pre-Fixation Extraction

This methodology was deployed in order to remove a cytosolic pool of immunoreactive protein. This may impact upon the signal to noise ratio of protein pools that may be hard to image (6) (see Note 27).

1. Cells were prepared as described (Subheading 3.1, steps 1–5)
2. Following the first set of three PBS washes, *saponin extraction buffer* (~1 ml) was added to the cells prior to fixation.
3. The *saponin extraction buffer* was left on the cells at room temperature for 60 s before the cells were then washed three times with PBS.
4. Cells were then fixed either using solvent or aldehyde.
5. If the cells have been fixed using aldehyde, do not incubate in *Triton permeabilization buffer* for more than 5 min (see Note 20).
6. Cells can be subject to immunolabeling or mounted (Subheading 3.2) as previously described.

4 Notes

1. This is an isosmotic reagent that is buffered to pH 7.4. It is very useful for washing cells and removing serum components. It is also occasionally used (when its pH is dropped below its buffering capacity) to remove non-covalently bound material from the outside of cells. It can be autoclaved or used non-sterile depending upon the application.
2. This is a proprietary cell culture media for live-cell imaging and, as it uses a phosphate buffering system as opposed to a carbonate buffering system, does not require an atmosphere of 5% v/v CO_2 to maintain its pH (2). RP media also contains no phenol red indicator. This media has been used for 2–4 h at a time successfully, though much longer time frames are probably ill-advised. This is because once the cells are on the microscope stage, the lid of the dish is often removed to prevent condensation building up, or to allow reagents to be added. Consequently, after the lid is removed, the media is no longer sterile. Also, if possible it is a good idea to change the media regularly to allow for evaporation, which will raise the salt concentration. Osmotic stress is known to modulate endocytosis (7) potentially introducing artifacts into an experimental system.
3. Formaldehyde, under normal atmospheric pressure and at ambient temperature, is a gas, and when dissolved in water, the resultant solution is formalin. One solid precursor of formaldehyde is paraformaldehyde (PAF). Fresh isosmotic, buffered

formalin can be prepared from PAF. For best results, make this solution fresh, less than an hour before using. When preparing buffered formalin, do not add the PAF to the liquid before heating, as this will cause the release of a considerable volume of formaldehyde gas. Occasionally it may be desirable to decrease the fixation time, and if this is the case, then a 4% w/v PAF solution may be prepared. This will reduce the fixation time from 20 min at room temperature to 4 min at room temperature.

4. This buffer is applied to the cells prior to fixation and may be used to deplete an immunoreactive cytosolic pool of material prior to fixation and immunolabeling. Make fresh every time (6).

5. This buffer extracts lipid from cellular membranes, postfixation, and allows antibodies to assimilate their intracellular targets without the impediment of biological barriers. The removal of intracellular barriers will also allow the free movement of non-fixed material. It is necessary to make this solution fresh every time. Do not store Triton-X-100 as dilute stock solution. If this buffer is being used in conjunction with saponin extraction buffer (Subheading 3.5), do not incubate the cells for more than 5 min with this buffer at room temperature.

6. Store serum as multiple 1ml aliquots at –20°C in sterile 1.5 ml Eppendorf tubes. *Blocking buffer* can be kept at 4°C if sterile. Serum from any species that will not react with the primary or secondary antibody will provide an adequate block, though serum from the same species, the secondary antibody was raised in is ideal. Fetal bovine serum or goat serum offers adequate noise suppression when used in conjunction with goat anti-mouse or goat anti-rabbit secondary antibodies.

7. Using products such as Vectashield® (Vector Laboratories, Peterborough, UK) for mounting works well if a high volume of fluorescence microscopy is being undertaken. However, the short shelf life of this relatively expensive product can be prohibitive. After a couple of months at 4°C, there is a tendency to see unacceptably high levels of autofluorescence in the red channel. Consequently, an inexpensive and perfectly acceptable alternative can be made (*mounting media*).

8. Warm the preparation gently (do not boil) in a water bath or microwave oven. Typically multiple bursts of energy (<10 s at full power) will be sufficient when using a commercial 800 W microwave oven. Store this at room temperature in the dark for up to 1 week.

9. These methodologies relate to the use of squamous epithelial cells, which lend themselves to microscopic examination, especially by epifluorescence microscopy by virtue of being intrinsically "flat" (looking a little like a small "fried egg" (see Notes

2 and 3)). Consequently the cell densities quoted here have been optimized for Vero and other squamous epithelial cells. More cuboidal cell morphologies may be better suited to imaging using a confocal aperture.

10. Momentarily release your grip to allow the alcohol to cover the entire coverslip. A gentle but firm grip was maintained upon the coverslip during manipulations, as if the coverslip was gripped too firmly, or if there was excess alcohol upon ignition, the coverslip was prone to cracking or shattering. This was not dangerous, though having a sharp disposal container to hand was convenient.
11. For live imaging experiments, the coverslip may be omitted.
12. Cell density and the timing of experiments in relation to seeding cell density are particularly important as it is much harder to image cells at confluence than it is at 80% confluence (on account of an increased volume in the *Z*-axis at confluence). If noise is a problem associated with signal above and below the focal plane, a confocal aperture may be used.
13. For an eGFP fusion protein-encoding plasmid, utilizing either an SV40 or CMV promoter.
14. Transgene expression is strongly evident after 24 h and optimal after 48 h. It is still normally evident 72 h after transfection. Expression levels varied greatly from transfected cell to transfected cell, with up to 20% of the living cells expressing the transgene. At 24 h post lipofection, many LAMP positive structures appeared vacuolated as a consequence of the transfection process. Some toxicity was also seen.
15. It was vitally important to know which side of the coverslip the majority of the cells were attached (i.e., the side facing up during culture) and to keep track of this during the experiment (i.e., when manipulating the coverslips).
16. With practice, this can be poured directly into a sink rather than aspirated without losing the coverslip.
17. The PBS, unless otherwise stated, was at room temperature. If required, PBS may be prechilled to 0°C or acidified to pH 5. If acidified PBS has been used, the acid wash was followed (rapidly) with 3× washes using PBS at pH 7.4. Using a 500 ml wash bottle filled with PBS, run down the wall of the plate, was least likely to disrupt the monolayer of cells.
18. Either aldehyde or solvent fixation was used depending upon the antibodies being deployed. Solvent fixation will remove lipids, which spatially constrain macromolecular material within the intracellular vesicle. Consequently solvent fixation should not be used with "hard to fix" material (Subheading 3.2, step 4).

19. In the instance of solvent extraction, the first PBS wash was added directly to the methanol. The cells were then washed another two times with PBS. In order to prevent the cells drying out, (methanol is volatile) the next PBS wash was added immediately after removing the previous one. Each time, the cells were left in wash for 5 min before the PBS was removed. After the final wash, the cells were incubated with *blocking buffer* (Subheading 3.2, step 2). Do not extract the cells with *Triton permeabilization buffer* as the cells have already had their membrane lipids extracted by the solvent (4).
20. If the cells have not been subject to transfection or saponin pre-fixation extraction, an incubation time of 20 min has worked well. If the cells were transfected or subject to saponin pre-fixation extraction, then a maximum incubation time of 5 min should be used. Longer incubation times resulted in very few intact cells remaining on the coverslip. Following incubation with the *Triton permeabilization buffer*, the cells were washed 3× with PBS as before (4).
21. A strip of Parafilm "M"™ was placed on top of two paper towels moistened with PBS, in a sealable container. The primary antibody was diluted using 1:1 blocking buffer (Subheading 2.2, item 6). The coverslip was placed face (cells side) down onto the antibody, and the preparation was left at room temperature for 60 min. Antibody final dilution should be about 10× less than is used for immunoblotting. After the incubation period, the coverslip was transferred back into the 6-well plate (cells side up) and washed 3× in PBS pH 7.4.
22. It was critical to avoid any sheer force coming from lateral movement or rotation of the coverslip once all the excess mounting media was removed.
23. The glycerol in the mounting media prevented sample damage from crystal formation at low temperature. Once the excess mounting media was removed, repositioning the coverslip was avoided as this also could sheer cells. Trapping air bubbles between the coverslip and the microscope slide was avoided through the use of excess mounting media.
24. Such as albumin (i.e., BSA) (8), conjugated to Texas red® *N*-hydroxy-succinimidyl-ester (TxR-NHS) (2, 4).
25. This can be done the day before.
26. This preparation should contain minimal free (unconjugated) TxR. The exact quantity of unconjugated TxR can be determined by further fractionation of the purified sample over a PD10 column, and the Mol% loading can be determined by following the manufacturer's instructions. Unused TxR-NHS can be stored at −20°C. This reagent is best used at a final concentration of BSA of 5–10 mg/ml. The high concentration of

BSA-TxR (relative to WGA-TxR) was necessary as endocytic capture was inefficient (i.e., fluid phase capture).

27. This extraction resulted in vesicular structures adopting a more expanded "donut" like appearance. This procedure can be used to remove an immunoreactive cytosolic pool of a protein that may mask the decoration of the target protein on other intracellular structures (4). Other labs have successfully used streptolysin O (9) in place of saponin.

References

1. Galush W, Nye J, Groves T (2008) Quantitative fluorescence microscopy using supported lipid bilayer standards. Biophys J 95:2512–2519
2. Richardson SC, Wallom KL, Ferguson EL, Deacon SP, Davies MW, Powell AJ, Piper RC, Duncan R (2008) The use of fluorescence microscopy to define polymer conjugate localisation to late endocytic compartments in fixed and live target cells. J Control Release 127:1–11
3. Richardson SC, Pattrick NG, Lavignac N, Ferruti P, Duncan R (2010) Intracellular fate of bioresponsive poly(amidoamine)s in vitro and in vivo. J Control Release 142:78–88
4. Richardson SC (2010) Tracking intracellular polymer localisation via fluorescence microscopy. In: Weissig V, D'Souza G (eds) Organelle-specific pharmaceutical nanotechnology. Wiley, New Jersey, pp 177–192. ISBN 978-0-470-63165-2
5. Dyer PDR, Richardson SC (2011) Delivery of biologics to select organelles—the role of biologically active polymers. Expert Opin Drug Deliv 8(4):403–407
6. Richardson SC, Winistorfer SC, Poupon V, Luzio JP, Piper RC (2004) Mammalian late Vps orthologues participate in early endosomal fusion and interact with the cytoskeleton. Mol Biol Cell 15:1197–1210
7. Dove SK, Cooke FT, Douglas MR, Sayers LG, Parker PJ, Michell RH (1997) Osmotic stress activates phosphatidylinositol-3, 5-bipgosphate synthesis. Nature 390:187–192
8. Mullock B, Bright N, Fearon C, Gray S, Luzio JP (1998) Fusion of lysosomes with late endosomes produces a hybrid organelle of intermediate density and is NSF dependent. J Cell Biol 140:592–601
9. Mallard F, Antony F, Tenza D, Salamero J, Goud B, Johannes L (1998) Direct pathway from early/recycling endosomes to the Golgi apparatus revealed through the study of shiga toxin B-fragment transport. J Cell Biol 143(4): 973–990

Chapter 20

Real-Time Particle Tracking for Studying Intracellular Trafficking of Pharmaceutical Nanocarriers

Feiran Huang, Erin Watson, Christopher Dempsey, and Junghae Suh

Abstract

Real-time particle tracking is a technique that combines fluorescence microscopy with object tracking and computing and can be used to extract quantitative transport parameters for small particles inside cells. Since the success of a nanocarrier can often be determined by how effectively it delivers cargo to the target organelle, understanding the complex intracellular transport of pharmaceutical nanocarriers is critical. Real-time particle tracking provides insight into the dynamics of the intracellular behavior of nanoparticles, which may lead to significant improvements in the design and development of novel delivery systems. Unfortunately, this technique is not often fully understood, limiting its implementation by researchers in the field of nanomedicine. In this chapter, one of the most complicated aspects of particle tracking, the mean square displacement (MSD) calculation, is explained in a simple manner designed for the novice particle tracker. Pseudo code for performing the MSD calculation in MATLAB is also provided. This chapter contains clear and comprehensive instructions for a series of basic procedures in the technique of particle tracking. Instructions for performing confocal microscopy of nanoparticle samples are provided, and two methods of determining particle trajectories that do not require commercial particle-tracking software are provided. Trajectory analysis and determination of the tracking resolution are also explained. By providing comprehensive instructions needed to perform particle-tracking experiments, this chapter will enable researchers to gain new insight into the intracellular dynamics of nanocarriers, potentially leading to the development of more effective and intelligent therapeutic delivery vectors.

Key words Real-time, Particle tracking, Nanocarrier, Confocal microscopy, Mean square displacement, ImageJ, MATLAB

1 Introduction

1.1 Real-Time Particle Tracking

Nanocarriers for drug or gene delivery have a number of potential advantages over systemic drug delivery, including more specific targeting and higher efficiency. However, the success of these therapeutics can often be limited by how effectively they reach target organelles—a process potentially made difficult by the number of intracellular biological barriers to nanocarrier delivery. For example, nanocarriers can be hindered by the highly crowded cytoplasm, which includes a dense cytoskeletal network. Even if gene delivery

Volkmar Weissig et al. (eds.), *Cellular and Subcellular Nanotechnology: Methods and Protocols*, Methods in Molecular Biology, vol. 991, DOI 10.1007/978-1-62703-336-7_20, © Springer Science+Business Media New York 2013

vectors or other nucleus-targeted nanocarriers are able to navigate the cytoplasm and reach the perinuclear region, they still may be blocked from entering the nucleus. Understanding the dynamics and transport mechanisms of nanoparticles inside cells will allow for the design of more intelligent and efficient nanocarriers. Many fluorescence techniques, such as fluorescence recovery after photobleaching (FRAP), only reveal ensemble information about nanocarrier transport dynamics. Real-time particle tracking, in contrast, can reveal both ensemble and individual particle transport information. This technique can be used to extract detailed quantitative nanoparticle transport information, such as diffusivity, velocity, transport mode, and directionality—all of which can be used to characterize the dynamics of intracellular nanocarrier transport (1, 2). Such insight can be harnessed to improve the design and development of novel delivery systems for pharmaceuticals, such as drug and genes.

Despite these potential advantages, real-time particle tracking has not been widely used in the field of nanotherapeutics, partially due to the difficulty in understanding the calculations behind the technique. In a typical particle-tracking experiment (Fig. 1), fluorescently labeled nanocarriers are added to live cells and imaged using a fluorescence confocal microscope. Captured high-resolution time-series images are processed to extract x, y coordinate data over time, which are then used to reconstruct the trajectories of the particles. The data are subsequently used to calculate mean square displacement (MSD) and other transport parameters, such as diffusion coefficient, or velocity. This chapter begins by describing the MSD calculation, the core of the data analysis and often a confusing concept, in an approachable way designed for researchers inexperienced with this technique. Additionally, programmatic language is provided in Subheading 3 that explains how to perform the MSD calculation using MATLAB.

Comprehensive instructions are given that detail how to successfully perform the basic procedures in a typical particle-tracking experiment. Specifically, real-time particle tracking of nanoparticles in live cells and in glue samples (for determining tracking resolution) will be described in parallel in Subheading 3. The following procedures will be explained: preparing the glue samples to determine tracking resolution, preparing the live cell imaging samples, imaging samples with a confocal microscope, tracking particles (to determine trajectories) with ImageJ and MATLAB programs, computing MSD, and determining tracking resolution with glued nanoparticles.

1.2 MSD Calculation

MSD is a measure of a particle's motion which is typically calculated in the analysis stage of a particle-tracking experiment. It is calculated after particle trajectories have been determined, and once calculated, it can be used to derive other transport parameters, such as

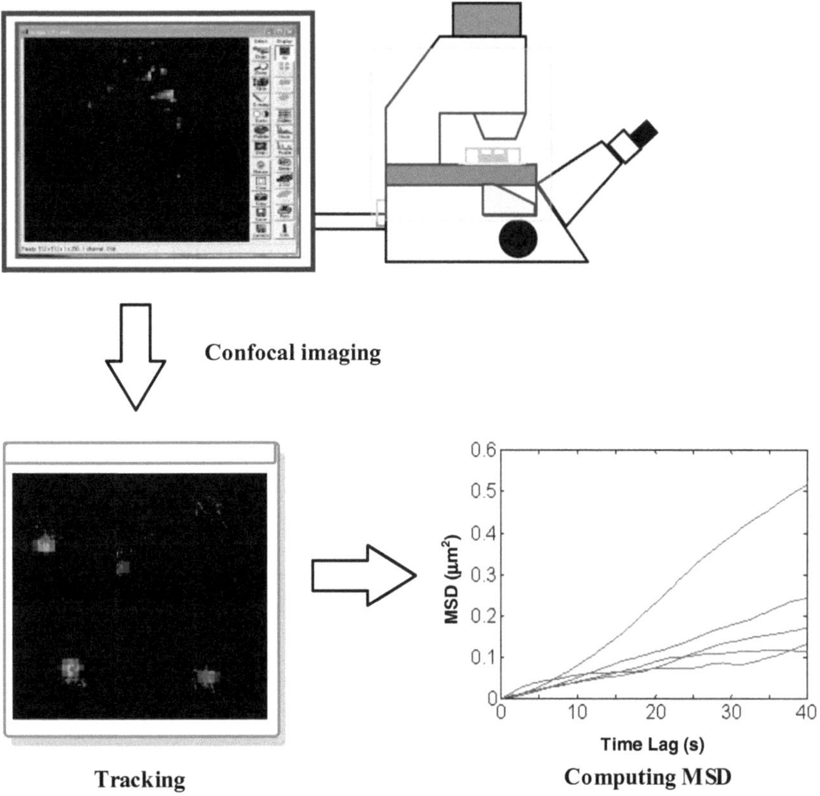

Fig. 1 Steps in a typical real-time particle-tracking experiment. Images are taken of fluorescently labeled nanoparticles inside live cells using confocal microscopy. Next, image analysis is performed to find the particles' positions in each frame and then to determine the particle trajectories. Finally, mean square displacement (MSD) and other transport parameters are calculated using the trajectory data

diffusivity. Two-dimensional MSD is most commonly used for intracellular tracking and therefore will be discussed in this chapter. MSD, $<\Delta r^2(\tau)>$, is determined as a function of time lag, $\tau_n = n\Delta t$, where Δt is the time between frames and n is the number of time intervals. MSD is calculated by finding the average particle square displacement for all possible time lags for a particle trajectory, where square displacements between frames can easily be calculated using the x and y coordinate data from the particle trajectories,

$$\Delta r^2 = \left(x_{i+n} - x_i\right)^2 + \left(y_{i+n} - y_i\right)^2$$

For the smallest τ ($n = 1$), the particle square displacements are simply the square displacements between each frame. For time lags longer than one frame interval, the particle displacements are calculated between all frames that are of that particular time lag apart. For example, if the time lag is 3-frame intervals ($n = 3$), then particle displacement would be calculated between frames 1 and 4, frames 2 and 5, frames 3 and 6, etc. Note that each τ_n for all integer values of n from 1 to $N-1$ should be used, where N is the total number of frames for that trajectory. As an example, Fig. 2 demonstrates the mean square displacement calculation for the case of a five-frame trajectory. The number of displacements, m, that can be computed for a particular time lag is $m = N - n$. For example, for the smallest time lag $\tau = \Delta t$, there are $N-1$ displacement values; for the largest time lag, between frame 1 and frame N, $\tau = (N-1)\Delta t$, there is only 1 displacement value.

Once all of the displacements have been found for a particular time lag, the MSD for that time lag can be calculated by finding the arithmetic mean of the square of the displacements $<\Delta r^2(\tau)> = <\Delta r_1^{\ 2}, \Delta r_2^{\ 2}, \ \ldots, \ \Delta r_{N-n}^{\ \ 2}>$, where <...> denotes mean. Mathematically, the following equation summarizes the calculation of two-dimensional MSD of a given trajectory of x,y coordinates at a particular τ_n:

$$\langle \Delta r^2(\tau) \rangle = \langle \Delta r^2(n\Delta t) \rangle = \frac{1}{N-n} \sum_{i}^{N-n} [(x_{i+n} - x_i)^2 + (y_{i+n} - y_i)^2]$$

Next, ensemble MSD can be determined across all particles. To calculate ensemble MSD, a simple arithmetic mean of the MSD's of all particle trajectories in an experiment is determined. Note that longer trajectories and smaller time scales both result in more displacement values, and thus more statistically meaningful MSD calculations. Therefore, MSD values calculated using the upper range of the time lag or using very short trajectories are considered unreliable since few displacement values were used in these calculations. More discussions of MSD calculation and various transport parameters that can be derived from it are found in references (3–10).

2 Materials

2.1 Cell Culture

1. HeLa cells (ATCC).
2. Complete medium: Dulbecco's Modified Eagle's Medium (Life Technologies, Grand Island, NY, USA) supplemented with 10% fetal bovine serum and 1% penicillin/streptomycin.
3. Phosphate buffered saline (PBS): 137 mM NaCl, 2.7 mM KCl, 10 mM Na_2HPO_4, 2 mM KH_2PO_4, pH 7.4.
4. 0.25% trypsin–EDTA for cell passing.

Fig. 2 (continued) calculated between particles in every second frame ($P_1 \rightarrow P_3$, $P_3 \rightarrow P_5$). Rather, displacements are calculated between particles in all frames that are 2 time intervals apart ($P_1 \rightarrow P_3$, $P_2 \rightarrow P_4$, $P_3 \rightarrow P_5$), as shown by the arrow. Once all the square displacements have been found for a particular time lag, the mean of those displacements is calculated, $\Delta r^2 = <\Delta r_1^{\ 2}, \ldots, \Delta r_m^{\ 2}>$. Finally, MSD is typically plotted as a function of time lag, as shown in (**c**), to allow for further analysis of transport parameters

For a 5 frame trajectory, N = 5:

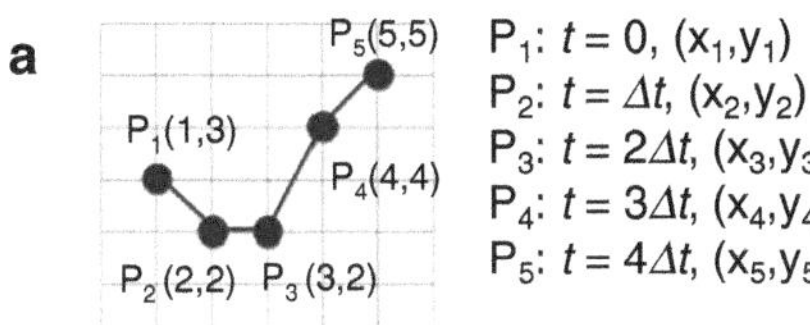

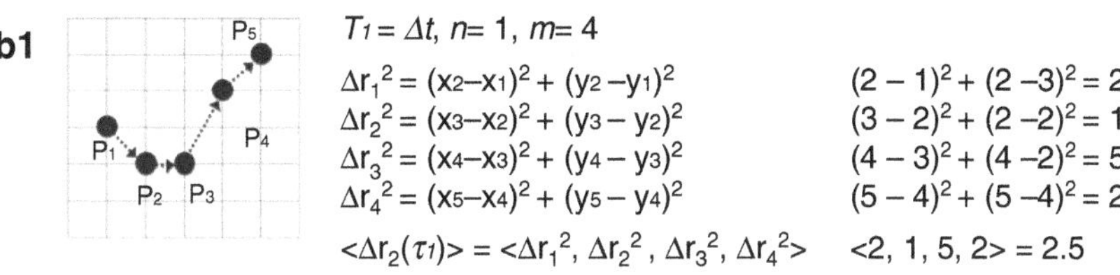

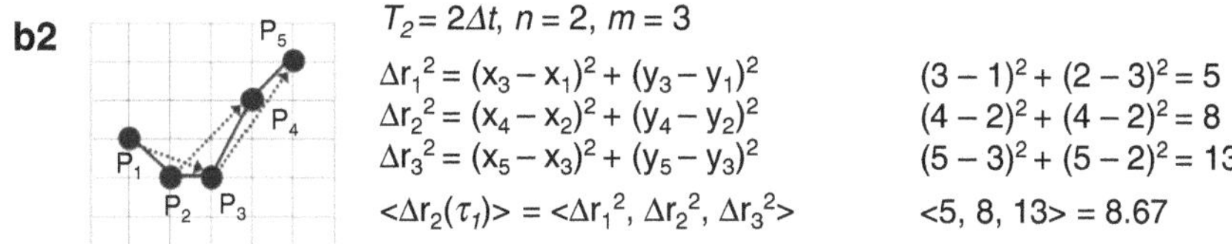

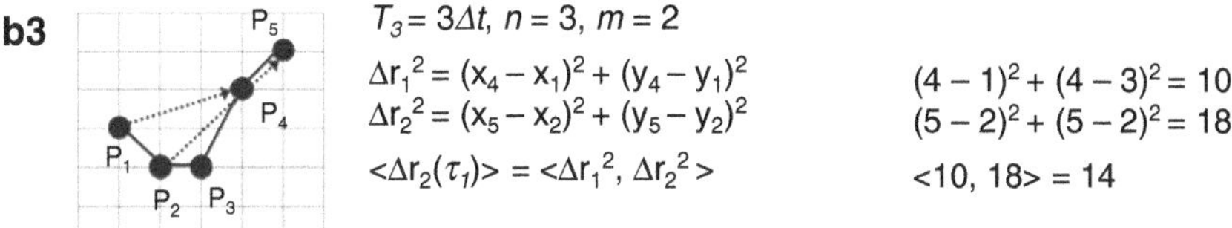

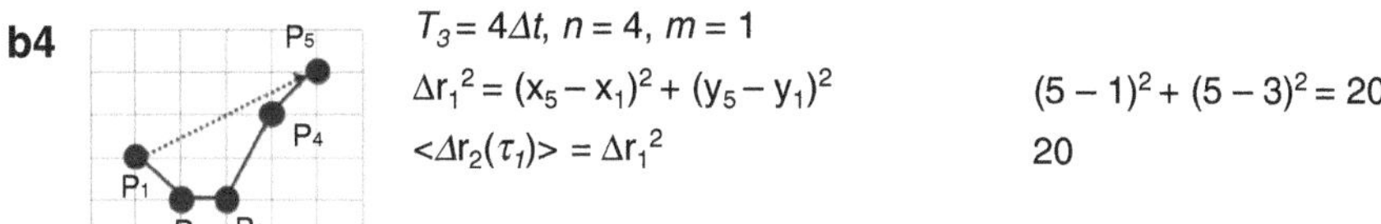

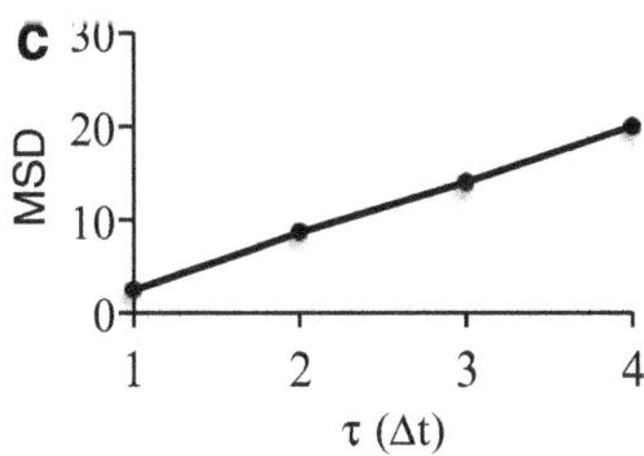

Fig. 2 Example MSD calculation for a five-frame trajectory. (**a**) Particle positions in each frame, where P_i denotes the and position for each frame, *i*. The displacement squared, $\Delta r_i^2 = (x_{i+n} - x_i)^2 + (y_{i+n} - y_i)^2$, is calculated between the *x,y* particle positions in frames that are time intervals apart, with n increasing by 1 for each successive time lag. (**b1**–**b4**) Displacements used for a time lag of $\tau_n = n\Delta t$, where = 1, 2, 3, and 4, and the corresponding calculations to find MSD for each time lag. $m = N - n$ is the number of displacements, . The displacements used are denoted by the arrow. For example, (**b2**) depicts the displacements used for $\tau = 2\Delta t$, so the square displacements are calculated between particle positions in frames that are 2 time intervals apart. Note that displacements are not simply

2.2 Imaging

1. Carboxylate-Modified FluoSpheres polystyrene nanoparticles (Life Technologies, Grand Island, NY, USA), red (580/605 nm), 100 nm, 2% in weight.
2. Hoechst 34580 if nucleus identification is desired.
3. Microscope slides and coverslips.
4. Loctite 495 superglue.
5. Lab-Tek™ 8-well chambered coverglass (culture area = 0.8 cm^2/well) (Thermo Fisher Scientific, Waltham, MA, USA).
6. LSM 5 LIVE confocal microscope (Zeiss, Thornwood, NY, USA) with a 63×/1.4 oil objective, the appropriate filters for the nanoparticles being studied, and an environmental chamber maintained at 37°C and 5% CO_2.

2.3 Image Processing and Data Analysis

1. For tracking particles with ImageJ, a "spot tracker" plug-in needs to be downloaded from http://bigwww.epfl.ch/sage/soft/spottracker/ and installed. Documentation of the plug-in is in Ref. 11.
2. MATLAB 7.0 (MathWorks) or newer needs to be installed for MSD calculation and particle tracking with MATLAB. For tracking particles with MATLAB, download publicly available subroutines and instruction document "Instructions_feature_track_pretrack.pdf" from the Maria Kilfoil research group Web site, http://people.umass.edu/kilfoil/downloads.html in "Particle pretracking and tracking and 2D feature finding" section. Store all M-files in the current directory. MATLAB's Image Processing Toolbox is also required.

3 Methods

3.1 Cell Culture

1. Maintain cells in complete medium at 37°C in 5% CO_2.
2. Pass cells at 1:10 ratio every 3–4 days.

3.2 Preparing Glue Sample to Determine Tracking Resolution

1. Dilute PSNP 1:100 in ultrapure water.
2. Spot 10 μL of the PSNP solution onto a microscope slide, and spot a small amount of superglue (about 50–100 μL) on top of the PSNP.
3. Press down a coverslip over the glue and wait until dry.

3.3 Preparing Live Cell Imaging Sample

1. Plate cells onto a Lab-Tek 8-well chambered coverglass at density of 2×10^4 cells/well (20% confluency), and incubate in complete medium for 24 h.
2. Replace the medium with 0.4 mL fresh complete medium containing 0.1 mg/mL PSNP (see Note 1).

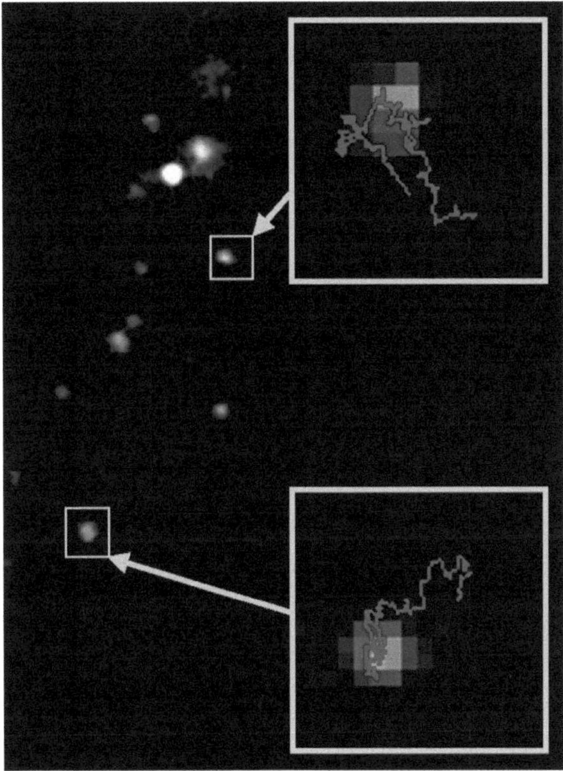

Fig. 3 Example trajectories (*insets*) obtained with ImageJ spot tracker plug-in for PSNPs

3. Incubate samples for desired time (usually 15 min to 24 h), at 37°C and 5% CO_2.
4. If nucleus identification is desired, add medium containing 1 µg/mL Hoechst to cells and incubate at 37°C for 30 min before imaging (see Note 2 for an alternative nuclear labeling method).
5. Wash cells with PBS and replenish with medium.

3.4 Imaging Sample with Confocal Microscope

Using a high-magnification objective (63× oil), perform time-series imaging for 200 frames at 50 ms per frame with a 2× zoom (see Note 3). For imaging glued particles, find a population of beads near the glue/particle interface where the particles are immobilized by the glue. For imaging particles in living cells, find the best focal plane with reference to the Hoechst-stained cell nucleus. A single image with both Hoechst channel and nanoparticle channel can be taken before and after the time-series images to ensure that the images stay on the same optical plane.

3.5 Tracking Particles with Image J

The ImageJ plug-in explained below can only track one particle at a time in a time-series image stack that was obtained from the confocal imaging. See Fig. 3 for PSNP trajectories obtained with ImageJ.

1. Open the time-series file with ImageJ, and crop the stack of images to the particle to be tracked by drawing a rectangle around it and selecting image > crop, making sure that the particle stays within the field of view during the frames to be tracked.
2. Select plug-ins > spot tracker > spot tracker 2D.
3. Draw a small rectangle around the particle and select Add to specify the initial location of the particle.
4. Modify the parameters of the plug-in to fit the time-series properties. Maximum displacement refers to the maximum movement in pixels below which the particle can be successfully tracked. The other parameters (intensity factor, intensity variation, movement constraint, and center constraint) specify the different weights for the cost function of the algorithm. More information can be obtained by clicking Help.
5. Click Track to obtain a table of the tracking data, which includes x position, y position, confidence interval, pixel intensity at the center, velocity from one frame to the next, and a mean intensity over a 1×1 pixel window. At this point, Display Results can be chosen to visualize the trajectory of the particle.
6. After performing the tracking, the plug-in will output the data in a table that includes frame number, x position, y position, and various other quantities that may be desired. This table can be copied and further analyzed to calculate other useful parameters such as MSD. Select File > Save As… to save table in .xls format.

3.6 Tracking with MATLAB Programs

The MATLAB programs from Maria Kilfoil research group can track multiple particles simultaneously. The following procedure is based on a complete tutorial written by Vincent Pelletier and Maria Kilfoil (Instructions_feature_track_pretrack.pdf). Small changes must be made to the program files to run on Macintosh computers (see Note 4).

1. The following explains the default arrangement of the files (for custom arrangement, see Note 5). Organize images from each experiment into one root folder containing subfolders named "fov#" with a different # for each field of view. Each subfolder should contain time-series images named "fov#_####.tif," with the first # referring to the subfolder and the second #### referring to the index of the frame in four-digit format. The root folder should contain time information vectors called "time" and saved as "fov#times.mat" with the # referring to the corresponding subfolder. All downloaded M-files should also be stored in the root folder. In MATLAB, make the root folder the current directory.
2. Run *mpretrack_intit.m* to determine the correct parameters for accepting real particles and rejecting false features. This function requires inputting *basepath*, *featsize* (size of the feature),

barint (minimum intensity), *barrg* (maximum Rg squared), *barcc* (maximum eccentricity), *IdivRg* (minimum ratio of Intensity/pixel), *fovn* (ID# for the series of images), *frame* (frame number of a representative image), and three more optional parameters. The inputs for all functions are detailed in the comments found in the function files and the tutorial.

3. The output of *mpretrack_init.m* will be a matrix containing accepted features (MT) and a matrix containing rejected features (M2). These matrices contain the following columns from left to right (all units in pixels or pixels^2): *x* positions, *y* positions, integrated intensity, square radius of gyration, and eccentricity. A figure with accepted features surrounded by green circles and rejected features in red will be displayed. It takes several runs to optimize the input parameters until the correct features are accepted.

4. Run *mpretrack.m* using as inputs *basepath*, *fovn*, *numframes* (number of frames in the series), the parameters optimized in step 3, and three other optional parameters. This function creates a matrix named *MT* containing accepted features for all frames in that field of view. MT has the same first five columns as the output of *mpretrack.init.m*, as well as a sixth column for frame number and a seventh column for the image time (taken from *fov#_times.mat*). This matrix is stored as the following file:

   (*basepath 'Feature_finding MT_#_Feat_Size_#.mat'*).

5. Run *fancytrack.m* to determine trajectories from particle data determined by *mpretrack.m*. The following are inputs for *fancytrack.m*: *basepath*, *FOVnum* (field of view number), *featsize* (feature size for accessing the right MT file), and optional trajectory parameters, *maxdisp* (maximum displacement the particle may make between successive frames), *goodenough* (minimum length requirement, in number of frames, for a trajectory to be retained), and *memory* (how many consecutive frames a feature is allowed to skip). If unsatisfactory trajectories are determined, an empty matrix is output, or errors occur, try optimizing these optional trajectory parameters. For example, the default of optional parameter *goodenough* is 100 and may cause errors if the total number of frames is near or less than 100.

6. The output of *fancytrack.m* is a file named *res_fov#.mat* (# refers to the field of view number) in the folder *(basepath 'Bead_tracking\res_files\')*. This file contains a matrix, *res*, with the same first seven columns as the output from step 4 (*MT*) and an additional eighth column containing the trajectory ID number. The matrix is automatically saved in .mat form. If desired, MATLAB's built-in function *xlswrite* can be used to save the matrix as an excel spreadsheet; however, this is not necessary if MSD is calculated using MATLAB as described in Subheading 3.7.

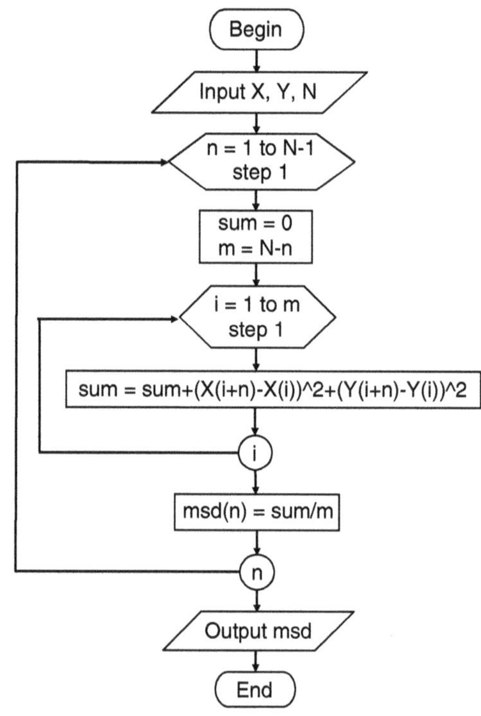

Fig. 4 Flowchart of a subroutine for computing MSD of a single trajectory. *X* and *Y* are the particle position vectors, *N* is the total number of frames, *n* is the number of time intervals for a particular time lag, *m* is the number of displacements for a particular *n*, and *i* is the index of *m*

3.7 Computing MSD with MATLAB

1. Format data outputted from ImageJ or other tracking software to X, Y vectors. If the trajectories are different in length, then the data need to be trimmed to the same length. Usually a trajectory that has more than 200 frames is sufficient for statistically significant calculations (see Note 6).
2. Create a subroutine for computing MSD of one trajectory. A main program needs to be created to calculate multiple trajectories by calling this subroutine in a loop. The following pseudo-code of the subroutine can be translated easily to MATLAB code (see Fig. 4 for flowchart). *X*, *Y* vectors are the position data, *N* is number of frames, and msd is the calculated MSD vector. Note that the frame interval and the time lag vectors are not needed in the computation.

   MSD computing subroutine:

```
Begin
   input X, Y, N
   for n = 1 to N-1
      sum = 0
      m = N-n
      for i = 1 to m
```

```
            sum = sum + (X(i + n)-X(i))^2 + (Y(i + n)-Y(i))^2
        msd(n) = sum/m;
    output msd
End
```

3. Ensemble MSD can be computed by averaging MSD of all particles. For an MSD with N time lags, the first $N - 100$ points are considered significant (see Note 6).

3.8 Determining Tracking Resolution from Glued Nanoparticles

The glued particles are immobile and will reflect the vibrational noise of the experimental setup. Thus, by calculating the MSD of glued PSNP, the tracking resolution of the system is determined. Calculate the arithmetic mean (M) of the first 50 points of the ensemble MSD. Then, use $\sigma = (M/2)^{1/2}$ to obtain the tracking resolution σ (see Note 7 and Fig. 5).

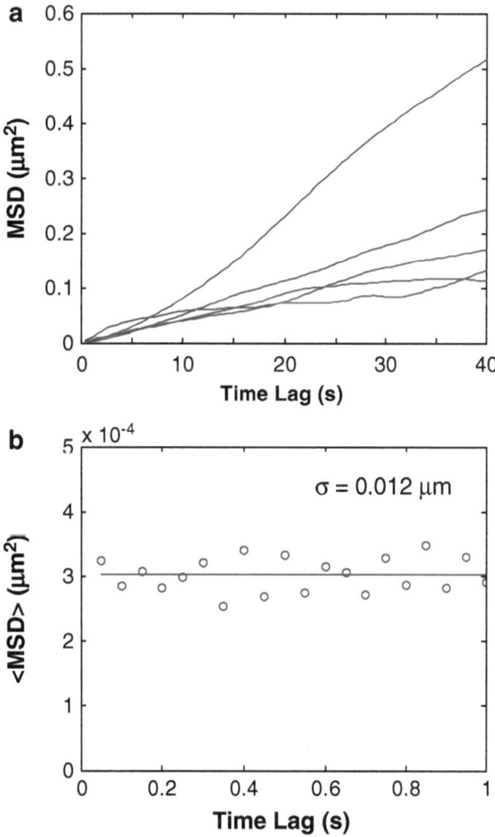

Fig. 5 Example MSD plots of PSNPs. (**a**) Five PSNPs were tracked in live HeLa cells. (**b**) PSNPs immobilized with superglue were tracked to determine tracking resolution. Tracking resolution was calculated by taking the mean of the ensemble MSD of the first 20 time lags

4 Notes

1. Too high of a concentration of nanoparticles may complicate particle-tracking image analysis.
2. Labeling the nucleus helps to locate the plane crossing the cell cytoplasm and nucleus and to assure the nanoparticles are inside the cell. An alternative way to outline the cell is by transfecting it with green fluorescent protein (GFP) plasmid DNA. 24–48 h after transfection, the cytoplasm and nucleus will be labeled with expressed GFP and ready for nanoparticle addition.
3. Optimal time interval depends on fluorescence properties of particles and how fast the particles are moving. For PSNP endocytosed by mammalian cells, a time interval of 50–400 ms works well. The duration of the movie also depends on whether the cell moves during the course of the movie. To maximize the temporal resolution, excitation laser power needs to be optimized so that the fluorescence signal is strong enough for particle tracking but will not cause sample photobleaching within the recording time.
4. Mac users need to make minor changes to the program files to account for differences in file path naming. Backward slashes (\) should be changed to forward slashes (/) in the following locations: line 72 of *mpretrack_init.m*, lines 70 and 77 of *mpretrack.m*, and lines 51 and 59 of *fancytrack.m*.
5. The default arrangement of the files and folders, *basepath* can be set to null. Readers can customize the arrangement and name format of the folders and image files by changing lines 71 and 72 in *mpretrack_init.m* and lines 74 and 77 in *mpretrack.m*, accordingly.
6. The error of an MSD data point is $m^{-1/2}$, where m is the number of displacements for a particular time lag. If 10% is an acceptable error, then $m = 100$ and $n = N - 100$. Only the MSD of first n time lags are acceptable. If $N = 200$, then $n = 100$, i.e., the first 100 points of the MSD are significant.
7. In practice, $<\Delta r^2(0)> = 2\sigma^2 \neq 0$, due to the errors introduced by experimental conditions and particle-tracking algorithm. Glued PSNPs are assumed to be fixed on the slide during the tracking, thus any displacements are considered to be errors. Therefore, the system tracking resolution is determined by the $<\Delta r^2>$ of the glued PSNPs.

Acknowledgment

This work was supported by the National Institutes of Health (RC2GM092599 and T32EB009379).

References

1. Suh J et al (2003) Efficient active transport of gene nanocarriers to the cell nucleus. Proc Natl Acad Sci USA 100:3878–3882
2. Huang F et al (2011) Quantitative nanoparticle tracking: applications to nanomedicine. Nanomedicine 6(4):693–700
3. Qian H et al (1991) Single particle tracking. Analysis of diffusion and flow in two-dimensional systems. Biophys J 60:910–921
4. Kusumi A et al (1993) Confined lateral diffusion of membrane receptors as studied by single particle tracking (nanovid microscopy). Effects of calcium-induced differentiation in cultured epithelial cells. Biophys J 65:2021–2040
5. Saxton MJ, Jacobson K (1997) Single-particle tracking: applications to membrane dynamics. Annu Rev Biophys Biomol Struct 26:373–399
6. Suh J et al (2005) Real-time multiple-particle tracking: applications to drug and gene delivery. Adv Drug Deliv Rev 57:63–78
7. Lai SK, Hanes J (2008) Real-time multiple particle tracking of gene nanocarriers in complex biological environments. Methods Mol Biol 434:81–97
8. Kawai M et al (2009) Dynamics of different-sized solid-state nanocrystals as tracers for a drug-delivery system in the interstitium of a human tumor xenograft. Breast Cancer Res 11:R43
9. Chen C, Suh J (2010) Real-time particle tracking for studying intracellular transport of nanotherapeutics. In: Weissig V, D'Souza GGM (eds) Organelle-specific pharmaceutical nanotechnology. Wiley, Hoboken
10. Wirtz D (2009) Particle-tracking microrheology of living cells: principles and applications. Annu Rev Biophys 38:301–326
11. Sage D et al (2005) Automatic tracking of individual fluorescence particles: application to the study of chromosome dynamics. IEEE Trans Image Process 14:1372–1383

Chapter 21

Interactions of Nanoparticles with Proteins: Determination of Equilibrium Constants

Lennart Treuel and Marcelina Malissek

Abstract

The behavior of nanoparticles towards proteins is an important aspect across wide areas of nanotoxicology and nanomedicine. In this chapter, we describe a procedure to study the adsorption of proteins onto nanoparticle surfaces. Circular dichroism (CD) spectroscopy is utilized to quantify the amount of free protein in a solution, and the experimental information is evaluated to derive equilibrium constants for the protein adsorption/desorption equilibrium. These equilibrium constants are comparable parameters in describing the interactions between proteins and nanoparticles.

Key words Nanoparticles, Protein structure, Protein adsorption, Protein corona, Equilibrium constant, Dissociation constant, Circular dichroism spectroscopy

1 Introduction

The behavior of nanoparticles (NPs) towards proteins is a key aspect in understanding the fate of NPs in biological systems. The efficiency of this interaction can be a decisive factor for the effect of a nanoparticle within a biological system (1–5). Most interactions of proteins with nanoparticles lead to structural changes in the protein (6–10), and an effect of ligand chemistry on the quantity of adsorbed protein and the extent of denaturation has also been reported (11). These structural changes upon adsorption to a nanoparticle surface can result in the loss of biological activity and also in an activation of immune response (12, 13).

In this chapter, we describe how circular dichroism (CD) spectroscopy can be used for quantitative studies of the protein corona formation around colloidal nanoparticles. This technique is utilized to monitor the loss of protein structure as a function of nanoparticle surface area present in an observed volume. From the spectral information, the amount of free protein can be determined and this experimental information can be evaluated to derive

Volkmar Weissig et al. (eds.), *Cellular and Subcellular Nanotechnology: Methods and Protocols*, Methods in Molecular Biology, vol. 991, DOI 10.1007/978-1-62703-336-7_21, © Springer Science+Business Media New York 2013

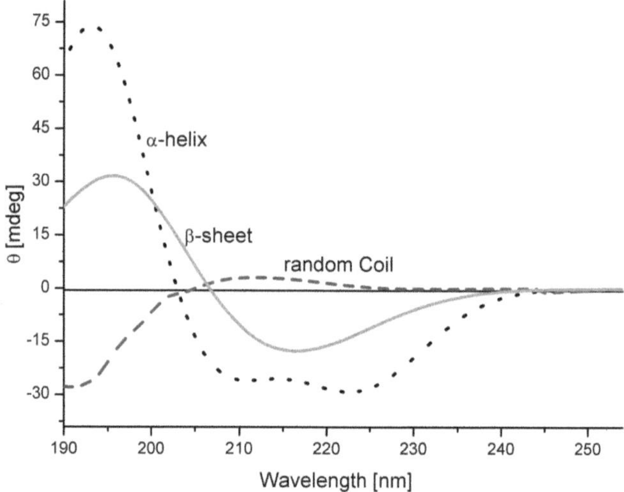

Fig. 1 CD structure shows typical spectra for pure secondary structures occurring in proteins

equilibrium constants for the protein adsorption / desorption equilibrium. These equilibrium constants are comparable parameters in describing the interactions between proteins and nanoparticles, and we describe how they can be determined from CD spectroscopic measurements largely independent of the nanoparticle properties (5).

1.1 Principle of Circular Dichroism Spectroscopy

Circular dichroism (CD) spectroscopy essentially measures differences in the absorption of left-handed and right-handed polarized light. These differences arise from structural asymmetries. In the UV spectral region (260–180 nm), characteristic signals arising from secondary protein structures can be detected. These signals are related to peptide bonds located in differently folded environments. For a more detailed discussion of the physical basics of circular dichroism spectroscopy, the reader is referred to the many excellent textbooks that have been published on this topic among which those by G.D. Fasman (14) and by N. Berova, K. Nakanishi, and R.W. Woody (15) can be especially recommended by the authors. Figure 1 shows typical spectra for pure secondary structures occurring in proteins. Spectral deconvolution methods (16–20) can be used to extract the relative contributions of the individual structural elements from protein spectra.

1.2 Derivation of Equilibrium Constants

The following descriptions regarding the derivation of equilibrium constants, the determination of free protein content, and the determination of surface sites have been published by L. Treuel et al. (5), and they are in parts reproduced from this source with kind permission from the copyright holder, Wiley-VCH Verlag GmbH & Co. KGaA.

For the derivation of the necessary equation we assume that free proteins [P] and surface adsorption sites [S] are in equilibrium with proteins adsorbed on the surface sites [PS]:

$$\mathrm{P} + \mathrm{S} \rightleftharpoons \mathrm{PS}. \quad (1)$$

Thus, the following equilibrium constant K_D can be derived for the adsorption/desorption process:

$$K_D = \frac{[\mathrm{P}]\cdot[\mathrm{S}]}{[\mathrm{PS}]}. \quad (2)$$

The total surface available for protein adsorption is quantitatively expressed by the number of free surface sites per unit volume. Provided, a monodispersed distribution of particles is assumed, the number of surface sites is determined by the sum over all particles n_i times the maximum number of protein molecules that can fit on the surface one nanoparticle N_{max} divided by the Avogadro constant N_A.

$$[\mathrm{S}] = \frac{\sum_i n_i \cdot N_{max}}{N_A}. \quad (3)$$

This number can be calculated using a spherical geometry and literature size values for the protein. The amount of protein adsorbed [PS] onto the surface at equilibrium is given by the initial amount of protein present in the mixture $[P_0]$ minus the amount of free protein in solution $[P]$.

$$[P_0] - [P] = [\mathrm{PS}] \quad (4)$$

After substitution of this expression into Eq. 2 and simple rearrangements the following expression is obtained:

$$\frac{[P_0]}{[P]} - 1 = [\mathrm{S}]\frac{1}{K_D} \quad (5)$$

To determine K_D, $(P_0/P) - 1$ is plotted against the surface site concentration in mol/L, yielding a linear relation with a gradient of $1/K_D$.

2 Materials

Pipettes and pipette tips in the range of 10–1,000 μL are needed for the preparation of the samples. For handling proteins at these low concentrations, we advise to use specially functionalized test tubes that have a low affinity to bind proteins (e.g., LoBind® tubes, Eppendorf). All solutions should be prepared using ultrapure deionized water (18 MΩ × cm at 25°C). Glassware must be carefully cleaned (aqua regia, ultrapure deionized water).

2.1 Circular Dichroism Spectrometer

The sensitivity and spectral range (for these studies wavelengths between 260 and 180 nm should be accessible) might require modifications of the spectrometer (see Note 1).

2.2 Preparation of the Cuvettes

In the UV range used here (from 260 to 180 nm) special quartz cuvettes are necessary to ensure a sufficient UV transparency (e.g., 1 mm Suprasil® quartz cuvette, Helma). Moreover, these cuvettes are temperature resistant and resistant to most chemicals, for example, acids which can therefore be used as cleaning agents. The cuvettes must be protein free and free of any nanoparticle residues and must be cleaned between two measurements. In the following, we describe a general cleaning procedure, and the reader is referred to Note 2 for further considerations.

"To remove protein residues and ions, clean your cuvettes carefully between the individual measurements with a cuvette cleaning agent (e.g., Helmanex®, Helma).

1. Preparing cleaning agent: Mix the stock solution of the cuvette cleaning agent (e.g., Helmanex® Fa. Helma) with ultrapure water as indicated by the supplier.
2. Place the cuvette into a beaker containing the cleaning solution. Heat the solution to between 50 and 60°C and keep it at this temperature for about 20 min. Thereafter, wash the cuvette several times with ultrapure water.
3. Preparation of aqua regia: To remove metallic nanoparticles from the cuvette, clean with aqua regia (see Notes 2 and 3). To produce aqua regia 3 aliquots of concentrated hydrochloric acid (HCl_{aq}) are mixed with 1 aliquot concentrated nitric acid ($HNO_{3\,(aq)}$). These are all very strong acids and also gaseous Cl_2 is produced in the reaction (see Note 3). Pay attention to the relevant safety instructions for all components! Mixing the two acids produces aqua regia, which can be identified by a golden yellow color.
4. Boil your cuvette with a little (as little as possible to fully submerge your cuvette) aqua regia for about 5 min.
5. After cleaning with aqua regia, the cuvettes have to be carefully rinsed with ultrapure water 3–5 times. To check if all aqua regia is rinsed off, test the pH of the rinsing water and wash until a neutral pH is reached. Avoid traces of fingerprints.

2.3 Preparation of Protein Solution

Prepare a protein stock solution in a suitable buffer depending on the type of protein and the type of nanoparticle used in the study (see Note 4). Typical protein concentrations for protein–nanoparticles interaction studies are 0.05–0.5 mg/mL depending on the affinity. The stock solution should have a concentration between 10 and 100 mg/mL. Test spectra should be acquired at the diluted concentrations with protein contents between 0.5

and 0.05 mg/mL. The final concentrations need to yield a sufficient CD signal. Attention must also be paid to the dependence on the nanoparticle concentration and size as described in Subheading 2.4.

1. Weigh in your protein into a low bind test tube and add buffer solution to prepare the stock solution with the desired concentration.

2.4 Preparation of NP Solution

1. Calculating the concentration range of the sample: To choose the concentration of the nanoparticles relative to the protein concentration, calculate the number of proteins needed to form a monolayer on the nanoparticles surface. Select your nanoparticle concentration assuring that the total nanoparticle surface in the mixture will be sufficient to theoretically accommodate all proteins present.
2. Test the stability of the nanoparticle suspension in the buffer used for the preparation of the protein solution (see Note 4).

3 Methods

3.1 Preparation of Protein–Nanoparticles Samples

1. Prepare low binding tubes.
2. Pipette the pre-calculated amount of protein stock solution into each of the tubes
3. Fill the tubes with the calculated amount of water (see Note 5).
4. Shake your protein solutions and incubate them for about 5 min at room temperature (20°C).
5. Add the different pre-calculated volumes of your nanoparticle suspension to the protein solution. This procedure produces samples with a range of nanoparticle concentrations and a constant protein concentration.
6. Incubate your prepared samples for 3–5 h at room temperature (20°C) until an adsorption equilibrium is established (see Note 6).
7. Perform a final test of NP stability after the equilibration. The sample should be stable for at least 12 h.

3.2 Measuring of the Protein–Nanoparticle Samples

1. Record a CD spectrum of the buffer used for the sample preparation. The measuring parameters (slid width, scanning step, integration time/scanning speed) have to be the same as those that will be used for measuring the samples. This background spectrum is necessary, because some buffers can also produce a circular dichroism signal in this wavelength region.
2. Record CD spectra of the residual chemicals that may be present in your nanoparticle solution from the synthesis procedure or

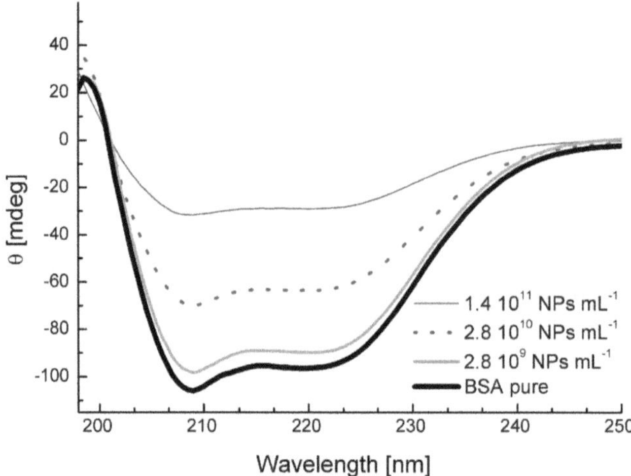

Fig. 2 Circular dichroism spectra of pure bovine serum albumin (*black line*) and bovine serum albumin at three different nanoparticle concentrations (citrate stabilized gold nanoparticles, diameter = 20 nm)

as surface ligands. Some of these chemicals (e.g., TPPTS) may by themselves influence the protein structure (see Note 7).

3. Test the UV absorption of the nanoparticles used in the study (see Note 8).
4. Carry out CD measurements with the incubated protein–nanoparticle solutions. Test the reproducibility of the CD measurements (see Note 9) (Fig. 2).
5. Subtract the background.

3.3 Determination of Adsorption Equilibrium Constant (K_D)

The following descriptions and the corresponding notes regarding the derivation of equilibrium constants, the determination of free protein content, and the determination of surface sites have been published by L. Treuel et al. (5) and are in parts reproduced from this source with kind permission from the copyright holder, Wiley-VCH Verlag GmbH & Co. KGaA:

1. *Conversion of the measured data from mdeg to the molar ellipticity*: First, the acquired CD signal has to be converted to the mean residue ellipticity, MRE (symbol [Θ]) using the following equation:

$$[\Theta] = \frac{\Theta_{208\,\mathrm{nm}} \cdot M}{n \cdot l \cdot c \cdot 10}. \qquad (6)$$

   With $\Theta_{208\,\mathrm{nm}}$ being the observed CD in mdeg, M the molar mass of the protein, n the number of amino acid residues in the protein, l the path length in cm, and c the protein concentration in g/L.

2. *Calculation of structural content*: Determine the relative structural contribution of α-helix structural elements in the protein

using spectral deconvolution (see Note 9). For alternative procedures: (see Note 10). Commercial as well as open source software exists for this procedure (e.g., CDNN (21, 22)).

3. *Calculation of free protein content*: This helical content is used to determine the amount of free protein in solution assigning the loss of CD spectroscopic signal intensity to the loss of protein structure in the interaction process. This allows the determination of the amount of free protein present in the solution $[P]$. The α-helix content in the pure protein solution (without nanoparticles present) is set to be 100% of intact protein. This is referring to $[P_0]$ as described in the introduction of this chapter. Any loss in the α-helix content is attributed to loss of free protein and hence adsorption of proteins on NP surface sites. Subsequently the α-helix content of the protein is determined from the CD signal at various nanoparticle concentrations as described. The amount of free protein $[P]$ can then be determined in relation to $[P_0]$ according to Eq. 7.

$$\frac{\text{Helix}_{\text{protein+NP}}}{\text{Helix}_{\text{pure protein}}} \cdot [P_0] = [P]. \tag{7}$$

4. *Calculation of* $(P_0/P)-1$: For the determination of K_D values, a plot of $(P_0/P)-1$ against the surface site concentration is required. Having determined $[P]$, and knowing $[P_0]$, the initial protein concentration in mol/L, $[(P_0/P)-1]$, can now be calculated as described in the introduction.

5. For the determination of K_D values, the surface site concentration is required. This is calculated according to the following procedure (the determination of surface sites on the nanoparticle follows the assumption of a spherical particle geometry). The total number of protein molecules that can adsorb onto these surface sites is obtained by dividing the surface area of the nanoparticle by the interacting area of the protein (information about the size, shape, and structure of many proteins can be found in the literature or in protein data bases, e.g., RCSB protein data bank (23) or swissprot (24)). With this information, it is possible to calculate the number of proteins fitting on one nanoparticle assuming a monolayer coverage by dividing the surface area of a nanoparticle by the surface area of the protein according to the following equation:

$$\text{Number of surface sites / NP} = \frac{\text{NP}_{\text{surface area}}\left[\text{nm}^2\right]}{\text{protein}_{\text{surface area}}\left[\text{nm}^2\right]}. \tag{8}$$

Now, the surface site concentration can be determined by simply multiplying this value with the number of nanoparticles in 1 L and dividing by the Avogadro constant ($N_A = 6.022 \times 10^{23}\ \text{mol}^{-1}$).

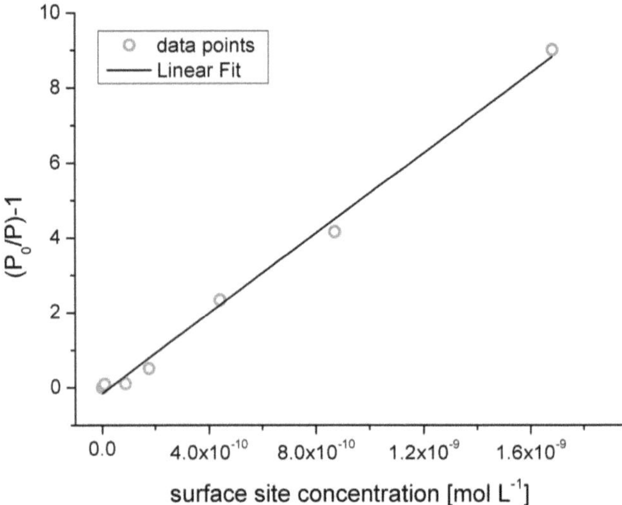

Fig. 3 KD plot—a linear fit of this plot yields a gradient of $1/K_D$ and thus K_D

$$C_{\text{surface site concentration}}[\text{mol} / \text{L}] = \frac{c_{\text{surface sites}}\left[\text{L}^{-1}\right]}{N_{\text{A}}\left[\text{mol}^{-1}\right]}. \tag{9}$$

6. *Determination of the K_D value with a linear plot*: To now determine K_D, $(P_0/P) - 1$ is plotted against the surface site concentration. A linear fit of this plot yields a gradient of $1/K_D$ and thus K_D. An example of such a plot is shown in Fig. 3.

4 Notes

1. A UV lamp emitting between around 260 nm and 180 nm is needed for the measurements described here. Typically, Xenon lamps will be used for this purpose. At these wavelengths oxygen will be photolyzed and subsequently ozone will be formed. This is a health hazard and also the ozone will damage optical parts in the CD spectrometer. Therefore, the spectrometer has to be oxygen free which is usually achieved by flowing nitrogen through the housing. If in doubt, please refer to the literature (14, 25) or contact the manufacturer of your spectrometer for further advice.
2. Most of the metallic nanoparticle residues can be removed with aqua regia. However, the procedure will not work for some other nanoparticles. In this case, more suitable solvents (depending on the individual type of nanoparticle used in the experiment) are advised.
3. Aqua regia is produced by mixing hydrochloric acid and nitric acid according to the following reaction:

$$HNO_{3(aq)} + 3HCl_{(aq)} \rightarrow NOCl_{(aq)} + 2Cl^{nasc}(g) + 2H_2O_{(l)}. \quad (10)$$

This and the gaseous chlorine formed in the reaction are very hazardous chemicals and close attention must be paid to the relevant safety instructions. The synthesis must be carried out in a fume hood and should only be carried out with a sufficient chemical background and relevant practical experience.

4. The buffer usually has to be chosen according to the requirements of the protein under consideration. However, the presence of the ionic components from this buffer might compromise the stability of the colloidal nanoparticle system. This is an important point since agglomeration of the nanoparticles will falsify the calculation of the surface sites and lead to complex light scattering effects during the measurements (26–28). It is therefore necessary to test the stability of the nanoparticle suspension in the buffer at the relevant concentrations. This can be done using dynamic light scattering (DLS) (29–31) or Brownian motion nanoparticle sizing (BMNS) (5, 32). The suspensions have to be stable in buffer for at least 12 h. Some proteins might be stable at this pH without a buffer making it possible to refrain from using a buffer system.
5. To keep the protein concentration constant during the measurements, the total volume of the samples is kept constant throughout the measurement series. This is achieved by always using the same amount of protein stock solution. Different relative quantities of water / buffer and nanoparticle suspension are then added to vary the nanoparticle concentration. In this procedure, the sum of the solvent and the nanoparticle suspension volume is kept constant (only the ratio is varied) and hence a constant total volume is achieved (usually 1–2 mL).
6. The described values of 3–5 h are equilibration times for typical nanoparticle and protein concentrations. However, these times will strongly depend on the nanoparticle and protein concentrations (absolute and relative concentrations) used in the measurements. Therefore, it is strongly advisable to test the equilibrium time by performing time dependent CD studies at a constant relative NP/protein ratio. Equilibrium is reached when the acquired CD spectra show no further change.
7. It has to be tested, if any residual chemicals present in the nanoparticle suspension affect the protein structure. These can be residues from the nanoparticle synthesis but also some ligands, e.g., TPPTS (3,3′,3″-Phosphinidynetris(benzenesulph onic acid) trisodium salt). To exclude the possibility of such an influence, CD spectra of the protein under consideration together with these chemicals have to be recorded and compared to the spectra of the pure protein. The concentrations of the

individual chemical substances should be chosen to represent the maximum possible concentration in the final solution.

8. Test the UV absorption of the nanoparticles by recording a UV/Vis spectrum of your nanoparticles. If the nanoparticles absorb too strongly in the wavelength region between 260 and 180 nm, the particles might not be suitable for this type of experiment. It has also to be tested if the particles themselves produce a CD spectrum. In this case, they might also be unsuitable for the experiment.
9. To test the reproducibility of the CD measurements, repeat the experiment for a sufficient number of times. If the data shows a bad reproducibility, this can point to an insufficient equilibration time or to a destabilized colloid.
10. The deconvolution of CD spectra can be carried out with different methods many of which are used by commercial software solutions (e.g., CDNN (17, 21, 22), SELCON3 (33), CONTIL/LL (34), CDSSTR (34)). Alternatively, simplified procedures to determine structural content such as that described by Lu et al. (18, 35–37) can be used and yield comparable results in many cases. In the procedure the α-helix content of the protein is determined based on the different spectral intensities at 220 and 208 nm, respectively. This method is commonly used in the contemporary interpretation of CD spectroscopic measurements (5, 35–38).

References

1. Watari F, Takashi N, Yokoyama A, Uo M, Akasaka M, Sato Y, Abe S, Totsuka Y, Tohji K (2009) Material nanosizing effect on living organism: non-specific, biointeractive, physical size effects. J R Soc Interface 6(S3):S371–S388
2. Fillafer C, Friedl DS, Ilyes AK, Wirth M, Gabor F (2009) Bionanoprobes to study particle-cell interactions. J Nanosci Nanotechnol 9(5):3239–3245
3. Jiang X, Weise S, Hafner M, Röcker C, Zhang F, Parak WJ, Nienhaus GU (2010) Quantitative analysis of the protein corona on FePt nanoparticles formed by transferrin binding. J R Soc Interface 7(Suppl 1):S5–S13
4. Röcker C, Pötzl M, Zhang F, Parak WJ, Nienhaus GU (2009) A quantitative fluoresence study of protein monolayer formation on colloidal nanoparticles. Nat Nanotechnol 4(9):577–580
5. Treuel L, Malissek M, Gebauer JS, Zellner R (2010) The influence of surface composition of nanoparticles on their interactions with serum albumin. ChemPhysChem 11(14): 3093–3099
6. Zhou HS, Aoki S, Honma I, Hirasawa M, Nagamune T, Komiyama H (1997) Conformational change of protein cytochrome b-562 adsorbed on colloidal gold particles; absorption band shift. Chem Commun:605–606.
7. Jiang X, Jiang J, Jin Y, Wang E, Dong S (2005) Effect of colloidal gold size on the conformational changes of adsorbed cytochrome c: probing by circular dichroism, UV-visible, and infrared spectroscopy. Biomacromolecules 6(1):46–53
8. Aubin-Tam ME, Hamad-Schifferli K (2005) Gold nanoparticle-cytochrome c complexes: the effect of nanoparticle ligand charge on protein structure. Langmuir 21(26): 12080–12084
9. Roach P, Farrar D, Perry CC (2006) Surface tailoring for controlled protein adsorption: effect of topography at the nanometer scale and chemistry. J Am Chem Soc 128(12):3939–3945
10. Medintz IL, Konnert JH, Clapp AR, Stanish I, Twing ME, Mattoussi H, Mauro JM, Deschamps JR (2004) A fluorescence

resonance energy transfer-derived structure of a quantum dot-protein bioconjugate nanoassembly. Proc Natl Acad Sci U S A 101(26): 9612–9617

11. Verma A, Rotello VM (2005) Surface recognition of biomacromolecules using nanoparticle receptors. Chem Commun:303–312
12. Baron MH, Revault M, Servagent-Noinville S, Abadie J, Qui-Quampoix HJ (1999) Chymotrypsin adsorption on montmorillonite: enzymatic activity and kinetic FTIR structural analysis. J Colloid Interface Sci 214(2): 319–332
13. Brandes N, Welzel PB, Werner C, Kroh LW (2006) Adsorption-induced conformational changes of proteins onto ceramic particles: differential scanning calorimetry and FTIR analysis. J Colloid Interface Sci 299(1):56–69
14. Fasman GD (ed) (1996) Circular dichroism and the conformational analysis of biomolecules. Springer, New York
15. Berova N, Nakanishi K, Woody RW (eds) (2000) Circular dichroism—principles and applications, 2nd edn. Wiley, New York
16. Sreerama N, Woody RW (2004) On the analysis of membrane protein circular dichroism spectra. Protein Sci 13(1):100–112
17. Perczel A, Fasman GD (1992) Convex constraint analysis: a natural deconvolution of circular dirchroism curves of proteins. Protein Eng 4(6):669–679
18. Lu ZX, Cui T, Shi QL (1987) Application of circular dichroism and optical rotatory dispersion in molecular biology, 1st edn. Science Press, Beijing
19. Alder AJ, Greenfield NJ, Fasman GD (1973) Dichroism and optical rotary dispersion of proteins and polypeptides. Methods Enzymol 27:675–735
20. Greenfield N, Fasman GD (1969) Computed circular dichroism spectra for the evaluation of protein conformation. Biochemistry 8(10):4108
21. Boehm G (1996) New approaches in molecular structure prediction. Biophys Chem 59(1–2):1–32
22. Boehm G, Muhr R, Jaenicke R (1992) Quantitative analysis of protein far UV circular dichroism spectra by neural networks. Protein Eng 5:191–195
23. (RCSB protein data bank) www.pdb.org.
24. (swissprot) www.expasy.org.
25. Wallance BA, Janes RW (eds) (2009) Modern techniques for circular dichroism and synchrotron radiation circular dichroism spectroscopy, vol 1. Ios, Amsterdam
26. van de Hulst H (1981) Light scattering by small particles. Dover, New York
27. Bohren C, Huffman D (1983) Absorption and scattering of light by small particles. Wiley, New York
28. Kerker M (1969) The scattering of light and other electromagnetic radiation. Academic, New York
29. Scherer C, Utech S, Scholz S, Noshov S, Kindvater P, Graf R, Thünemann AF, Maskos M (2010) Synthesis, characterization and fine-tuning of bimodal poly(organosiloxane) nanoparticles. Polymer 51:5432–5439
30. Scherer C, Noskov S, Utech S, Bantz C, Mueller W, Krohne K, Maskos M (2010) Characterization of polymer nanoparticles by asymmetrical flow field flow fraction (AF-FFF). J Nanosci Nanotechnol 10(10):6834–6839
31. Panacek A, Kvitek L, Prucek R, Kolar M, Vecerova R, Pizurova N, Sharma VK, Nevecna T, Zboril R (2006) Silver colloid nanoparticles: synthesis, characterization, and their antibacterial activity. J Phys Chem B 110(33): 16248–16253
32. Gebauer JS, Treuel L (2011) Influence of individual ionic components on the agglomeration kinetics of silver nanoparticles. J Colloid Interface Sci 354(2):546–554
33. Johnson WC (1999) Analyzing protein circular dichroism spectra for accurate secondary structures. Proteins 35(3):307–312
34. Sreerama N, Woody RW (2000) Estimation of protein secondary structure from circular dichroism spectra: comparison of CONTIN, SELCON, and CDSSTR methods with an expanded reference set. Anal Biochem 287:252–260
35. Shang L, Wang Y, Jiang J, Dong S (2007) pH-dependent protein conformational changes in albumin: gold nanoparticle bioconjugates: a spectroscopic study. Langmuir 23(5): 2714–2721
36. Xiao Q, Huang S, Liu Y, Tian F, Zhu J (2009) Thermodynamics, conformation and active sites of the binding of Zn–Nd hetero-bimetallic schiff base to bovine serum albumin. J Fluoresc 19(2):317–326
37. Ying L, WenYing H, Jianniao T, Jianghong T, Zhide H, Xingguo C (2005) The effect of berberine on the secondary structure of human serum albumin. J Mol Struct 743(1–3):79–84
38. Wang Y, Sun H, Wang H, Liu Y (2001) In vitro interaction of nicotine and hemoglobin under liver cell metabolizing condition. Chinese Chem Lett 12(5):449–452

Chapter 22

Tracing the Endocytic Pathways and Trafficking Kinetics of Cell Signaling Receptors Using Single QD Nanoparticles

Katye M. Fichter and Tania Q. Vu

Abstract

Cellular signaling is the fundamental process through which cells communicate with each other and respond to their environment. Regulation of this cellular signaling is crucial for healthy cellular function. Malfunctions in signaling are the cause for many diseases and disorders and therefore are under heavy investigation. The molecular mechanisms that underlie cellular signaling rely upon complex and dynamic processes of receptor intracellular trafficking. The specific endosomal pathways and kinetics through which receptors are intracellularly transported regulate the strength and duration of cellular signaling. In even more subtle and complex aspects, the cell orchestrates the individual motions of many receptors, through multiple different pathways, simultaneously. Despite the fundamental role of endosomal trafficking in signal regulation, it has been technically challenging to study since intracellular trafficking is complex and dynamic, with millions of individual receptors simultaneously undergoing trafficking in different endocytic stages. Here, we describe the use of single nanoparticle quantum dot (QD) probes to quantitatively investigate the endocytic trafficking pathways that receptors undergo following ligand activation. This new capability to directly visualize and quantitate cellular signaling at the level of individual receptors inside the cell has broad and important value for understanding fundamental cell signaling processes and the action and effect of therapeutics upon signaling.

Key words Quantum dots, Receptors, Cellular signaling, Intracellular trafficking, Endocytosis, Receptor recycling, Serotonin, GPCRs

1 Introduction

Cellular signaling is the fundamental process through which cells communicate with each other and their environment and as such represents a large field of study with many biomedical applications (1–8). Once thought to be distinct from each other, endocytosis and cell signaling have found significant functional overlap (9). In recent years a significant role for endocytosis has been implicated in the regulation of cellular signaling strength (10). For example, a delicate balance of internalization and intracellular traffic controls the number of receptors that are available at the surface of the

Volkmar Weissig et al. (eds.), *Cellular and Subcellular Nanotechnology: Methods and Protocols*, Methods in Molecular Biology, vol. 991, DOI 10.1007/978-1-62703-336-7_22, © Springer Science+Business Media New York 2013

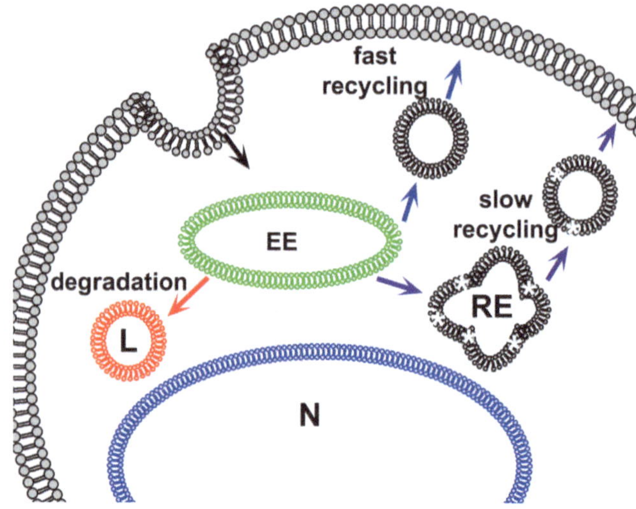

Fig. 1 Some of the typical intracellular pathways involved in the trafficking of GPCRs. Vesicles bud from the plasma membrane and are directed toward early endosomes (EE) where the cargo is sorted by the cell. The receptors can be directed toward lysosomes (L) for degradation or recycled either quickly, through Rab 4-associated vesicles, or slowly through recycling endosomes (RE) studded with rab coupling proteins (*asterisks*). *N* denotes the nucleus

plasma membrane, and thus the number of receptors capable of activation (11). It is the intracellular trafficking through membrane-bound endosomes in which the cell may direct receptors either back to the surface of the cell for subsequent rounds of activation, or toward lysosomes for degradation and signal downregulation (11–14).

The pathways and kinetics of intracellular trafficking are very responsive to subtle nuances such as the duration of agonist exposure, concentration of agonist, presence of drugs, and other biochemicals. (15–18). Investigating these processes is very difficult since intracellular trafficking is a very complex process that employs multiple different intracellular pathways that individual receptors may use simultaneously with different trafficking kinetics (Fig. 1) (13). Additionally, recent studies are uncovering the existence of signaling endosomes from which activated receptors may continue to initiate signaling cascades while traveling intracellularly (19, 20). Unfortunately, classical techniques for investigating cellular internalization and intracellular trafficking lack the degree of resolution necessary to observe how subpopulations of receptors are regulated through this process. Surface biotinylation and radiological assays must be performed on a large number of cells, which blinds investigators from the contributions of individual receptors (21, 22). While many organic fluorophores have been used to observe single cells, it is very difficult to visualize single or small groups of receptors with these labels because of the relative dimness of organic fluorophores and their susceptibility to photobleaching.

Quantum dots are extremely bright and photostable nanoparticles that have proved valuable for studying the molecular tracking of receptors on the surface of cells because of their ability to label single or small groups of receptors (23–26). Here, we present new straightforward and robust methods to identify the intracellular trafficking pathways of receptors and the kinetics through which they proceed through those pathways. Our lab has expertise in using single QD nanoparticles for superior resolution in labeling single or small groups of receptor (27–32). Here, our labeling scheme involves genetically encoding three copies of the hemagglutinin (HA) epitope (33) onto the extracellular N terminus of the human serotonin subtype 1A (5-HT1A) receptor. A single QD probe, consisting of a biotinylated anti-HA antibody combined with streptavidin-coated QDs at 1:1 stoichiometry, is then used to label the HA-tagged receptors with high affinity (see Note 1). However, this general scheme is widely applicable to just about any receptor system with a corresponding high-affinity epitope tag. Recently, we have shown that the presence of the QD, as bound to the receptor through this method, does not appear to alter the trafficking behavior (pathway or kinetics) of the 5-HT1A receptor in live cells (27).

Here, we describe methods to investigate the intracellular pathways and trafficking kinetics of 5-HT1A receptors with our QD labeling system. Cellular internalization is initiated with exposure to an agonist (i.e., serotonin). After activation, the receptors are allowed to traffic for a set of predetermined time points (0–60 min). The cells are then fixed and immunolabeled with markers for various endocytic compartments. The cells are imaged and then image analysis is used to verify the presence of single QDs and to co-localize the QD-labeled receptors with the various endocytic compartments. This data is then plotted with time to give a picture of the dynamics of the receptors through each endosomal compartment. While we have used this method to investigate the trafficking of 5-HT1A (27), it is generally applicable most other receptor systems. Furthermore, this method presents an ideal assay platform to determine the effect of therapeutics on the trafficking/signaling behavior of receptors, such as the effect of selective serotonin reuptake inhibitors (SSRIs) on the dynamics of serotonin receptor signaling.

2 Materials

2.1 General Reagents/Materials

1. N2a (neuroblastoma) cells, transfected with pDNA encoding the human 5-HT1A gene tagged with the hemagglutinin epitope (HA-5-HT1A) (see Note 1). Culture N2a cells on six glass coverslips (one coverslip for each duration of 5-HT activation) in each of six separate, small Petrie dishes to facilitate the receptor activation protocol (Subheading 3.2 below).
2. Dulbecco's Modified Eagle Medium (D-MEM).

3. Serotonin HCl (chemically known as 5-hydroxytrytamine, or 5-HT).
4. Phosphate-buffered saline (PBS), pH 7.4.
5. 0.1 M borate buffer, pH 8.5.
6. Streptavidin-conjugated QD_{655} (Strep-QD_{655}) (Life Technologies, Grand Island, NY, USA).
7. NHS-PEO_4-biotin.
8. 10,000 MWCO slide-a-lyzer dialysis unit (Thermo Scientific).
9. 4% paraformaldehyde (PFA) in PBS.
10. 0.2% Triton X-100 in PBS.
11. 10% normal goat serum (NGS) in PBS and/or 10% bovine serum albumin (BSA) in PBS.

2.2 Immunoreagents

1. Monocolonal mouse anti-HA.
2. Polyclonal rabbit anti-Rab 4.
3. Polyclonal goat anti-Rab11-F1P1(N-15).
4. Monocolonal rat anti-LAMP1.
5. Alexa-conjugated secondary antibodies: anti-rabbit, anti-goat, and anti-rat (Life Technologies, Grand Island, NY, USA).

2.3 Imaging Equipment/Analysis Software

1. Inverted Zeiss Axiovert 200M (or other comparable epifluorescent or confocal microscope) equipped with 63× (or higher), 1.4 NA oil objective, and digital CCD camera. A digital CCD camera capable of capturing time-lapse movies at a frame rate of at least 30 frames per second is necessary to verify the presence of single QDs (Subheading 3.4).
2. An imaging chamber, compatible with the coverslips of choice.
3. Appropriate filters for QDs and organic dyes (e.g., Chroma, QD_{655} filter set # 32012).
4. Appropriate acquisition software for your digital camera.
5. ImageJ (34) with the JACoP plug-in (35) (both are available for free download at http://rsbweb.nih.gov/ij).
6. Microsoft Excel or similar spreadsheet program.
7. Autoquant X or similar deconvolution software (if not using a confocal microscope).

3 Methods

3.1 Biotinylation of Anti-HA Antibody and Preparation of QD Stock Solution

Store all QD solutions at 4°C. Never freeze QD solutions. Because of potential QD aggregation, it is not recommended to store QD solutions at a concentration less than 100 nM.

1. Biotinylate the anti-HA antibody by reacting 10 μg of antibody with a threefold molar excess of NHS-PEO_4-biotin in PBS.

Allow the reaction to incubate 30 min at room temperature (RT). Purify the biotinylated antibody from excess NHS-PEO_4-biotin via dialysis against 1 L PBS using a 10,000 MWCO slide-a-lyzer unit. Dialysis typically consists of 4–5 changes of dialysate over the course of 4–5 h. The biotinylated antibody should then be aliquoted into 10-μL aliquots and unused aliquots should be stored at −80°C.

2. Prepare separate, 50-μL dilutions of both the biotinylated anti-HA and Strep-QD_{655} at a concentration of 100 nM in 10% NGS. Store solutions at 4°C. Allowing the QDs to incubate in 10% NGS for 24–48 h before forming the QD probes has been found to help decrease the nonspecific binding of QDs to cells.

3.2 Agonist-Induced Activation of 5-HT Receptors

Ensure all reagents to be used with live cells are warmed to 37°C prior to cell incubation (including 4% PFA).

1. Aspirate media from cells and replace with D-MEM (or media containing dialyzed serum) (see Note 2). Allow cells to incubate at 37°C and 5% CO_2 for at least 4 h before beginning the activation.
2. Set a timer to facilitate the time-dependent 5-HT activation step. Create one alarm at 15 min to alert you to end 5-HT activation. Create separate alarms to alert you to fix cells at 5, 10, 20, 30, and 60 min after 5-HT activation. Label each dish with the appropriate incubation time (5, 10, 20 min, etc.). Leave one dish to the side as a control (no activation).
3. Prepare QD-anti-HA conjugates. This step will prepare 6 mL of QD-Ab conjugates (combined at a 1:1 stoichiometry) at a final concentration of 250 pM. Add 15 μL 100 nM biotinylated anti-HA to 5,970 μL 10% NGS and mix. To this mixture add 15-μL Strep-QD655 and mix again. Allow the solution to incubate for 2 min, and then use immediately.
4. Label cells with QD-anti-HA conjugates. Aspirate media from each dish. Add 1 mL of the above QD/Ab conjugate solution to each dish. Incubate for 5 min RT. Wash the cells six times with warm PBS rapidly, with no incubation in between. Leave the cells in PBS temporarily until completing the next step.
5. Prepare a solution of 10-μM 5-HT in warmed D-MEM (or media containing dialyzed serum). Use this solution immediately after preparing.
6. Aspirate PBS from cells, replace with 5-HT solution, and activate the timers. Incubate cells for 5, 10, 20, 30, and 60 min at 37°C and 5% CO_2.

7. At the time of the alarm to fix cells, (5 and 10 min) remove the appropriate coverslip from the incubator and fix in 4% PFA for 30 min. Store cells in 0.1 M borate buffer, pH 8.5 (see Note 3) at 4°C until immunolabeling.
8. At the time of the alarm to end 5-HT activation (15 min), cells activated for 5 and 10 min should already be fixed. Remove the remaining cells from the incubator, aspirate the 5-HT solution and replace with D-MEM (or media containing dialyzed serum). Return cells to the incubator and continue to fix the rest of the cells at the appropriate time (20, 30, and 60 min). Remember to also fix the control dish (no activation). Store all cells in 0.1 M borate buffer, pH 8.5 (see Note 3) at 4°C until immunolabeling.

3.3 Immunolabeling Cells for Endocytic Compartments

For extended storing time (<12 h), store cells in 0.1 M borate buffer, instead of PBS, to maximize QD fluorescence.

1. Permeabilize cells by incubating them with 0.2% Triton X-100 for 20 min and wash three times with PBS, with a 15 min incubation in between each wash.
2. Block cells in 10% NGS for at least 1 h.
3. Incubate cells in the first primary antibody (e.g., polyclonal rabbit anti-Rab 4 at a dilution of 1:500 in 10% NGS for 4 h at room temperature). Afterwards, wash the cells three times with PBS, with a 15-min incubation between each wash.
4. Block cells again in 10% NGS for at least an hour (if using an anti-goat antibody, substitute with 10% BSA).
5. Incubate cells in the first secondary antibody (e.g., Alexa488 donkey anti-rabbit at a dilution of 5 μg/mL in 10% NGS or BSA for 1 h at room temperature). Wash the cells three times with PBS, with a 15-min incubation between each wash.
6. Repeat steps 2–5 for each additional primary and secondary antibody you wish to use (see Note 4).

3.4 Verify the Presence of Single QDs

To ensure the data collected from these experiments reveal information on the level of single or small groups of receptors, it is necessary to verify that the QDs in the cell sample are single QDs.

1. Load a coverslip containing QD-labeled cells into an imaging chamber and fill the chamber with 0.1 M borate buffer pH 8.5 (see Note 5).
2. Observe the fluorescence from the QD channel. All QDs should be blinking rapidly and should exhibit approximately the same fluorescence intensity. Larger, brighter puncta that are blinking slowly, or not at all, are evidence of QD aggregates.
3. Record a time-lapse fluorescence movie in the QD channel at a rate of 30 frames per second for 500 frames.

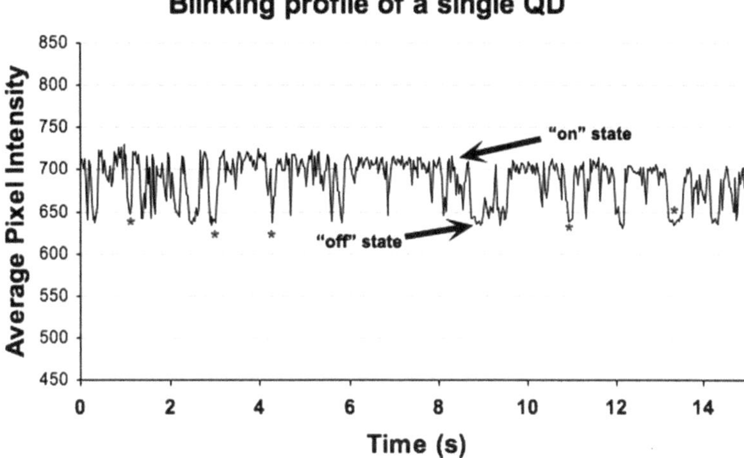

Fig. 2 Exemplary intensity trace of a single QD displaying multiple square on/off wave pulses (some examples marked with an *asterisk*) and constant intensity of "on" and "off" states

4. Use ImageJ to open the movie file. Zoom in on an individual QD puncta as far as possible. Use the rectangle marquee to select the brightest pixels of the QD puncta (i.e., 4–6 pixels for an image with a resolution of ~0.13 μm/pixel).
5. Generate a plot of the fluorescence intensity rate of the QD puncta by choosing Plugins → T-functions → Intensity vs. Time Monitor. Copy the frame number and average pixel intensity output to a worksheet in Excel. Repeat steps 4 and 5 several times to obtain a representative population of QD puncta.
6. For each QD puncta, create a fluorescence intensity trace. Create an *xy* scatter plot using the frame number (or convert to time) as *x* input and the average pixel intensity as *y* input. Inspect the plot carefully. The trace should demonstrate discrete square on/off wave pulses. The intensity of the "on" and "off" states should remain constant throughout the plot (Fig. 2).

3.5 Image Acquisition

1. Place a coverslip in an imaging chamber and fill the chamber with 0.1 M borate buffer pH 8.5.
2. Find a cell expressing HA-5-HT1A by looking for high QD binding.
3. Capture a z-stack of each fluorescent channel of the cell. Ensure that the stack includes the entire cell from top to bottom. Use a reasonable distance between slices (z-slice) to ensure sampling of the entire vertical height of the cell (e.g., 0.275 μm between slices for a 100× 1.4 NA objective).
4. Capture one bright-field (DIC or phase) image for each cell.
5. Repeat steps 2–4 to capture image data for at least 5–10 cells.
6. Repeat steps 2–5 for each coverslip at each time point.

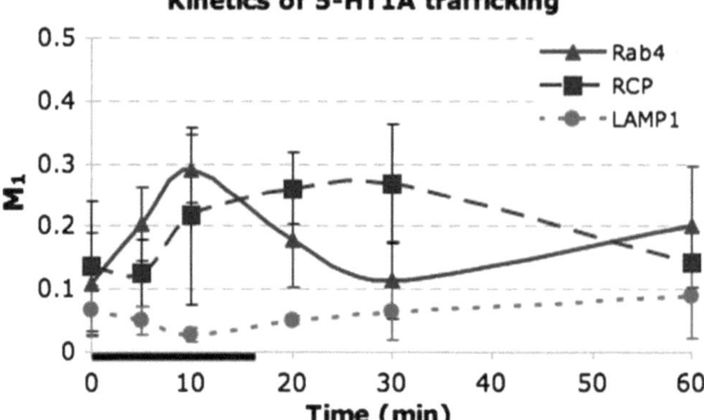

Fig. 3 The intracellular trafficking kinetics of the 5-HT1A receptor. Receptor trafficking through the fast recycling pathway (Rab 4) maximized at 10 min, while recycling through the slow pathway (rab coupling protein, RCP) maximized around 30 min. 5-HT1A receptors do not appear to significantly accumulate in lysosomes (LAMP1) after short-term exposure to 5-HT. Bar depicts duration of exposure to 5-HT

3.6 Image Analysis

1. If images were not acquired using a confocal microscope, deconvolve each z-stack using Autoquant X or other similar deconvolution software.
2. Open a deconvolved QD stack and corresponding deconvolved endosome stack (e.g., Rab 4) from one cell in ImageJ. If necessary, perform a background subtraction of each stack and adjust the brightness and contrast.
3. Run the JACoP plug-in by selecting Plugins→JACoP. Ensure the QD stack is listed as Image A and the Rab 4 stack is listed as Image B. Click the threshold button and use the slider to choose a threshold for each image. Slide though the entire stack to ensure choosing an appropriate threshold. Push the analyze button.
4. Copy the M_1 value to an excel spreadsheet under the duration of agonist exposure for that particular coverslip. Repeat steps 2–3 for each cell images. Calculate the average M_1 and the standard deviation of M_1 for each coverslip.
5. Plot the average M_1 against time for each coverslip. This plot will allow you to visualize the time course through with receptors traffic through Rab 4 endosomes (Fig. 3).

4 Notes

1. The hemagglutinin (HA) epitope is a peptide sequence (YPYDVPDYA) (33) for which very high-affinity antibodies have been created. We prefer to use HA-tagged receptors for targets since antibodies targeting native epitopes of many

GPCRs have been found to have less-than-optimal affinity. This HA peptide can be fused to your receptor sequence via basic molecular cloning techniques. In this protocol, we have fused three copies of the HA epitope to the N-terminal (extracellular) end of the receptor. By fusing the HA tag to the extracellular end of the receptor, we aim to avoid interferences with the signaling and trafficking machinery that interacts with the C-terminal (intracellular) end of the receptor.

2. Because serum typically contains serotonin, cells cultured in full serum may downregulate the activation of serotonin receptors to mitigate this exposure. We recommend removing the serum from cells before ligand activation to minimize the impact of this downregulation on receptor trafficking. Alternatively, cells can be cultured in dialyzed serum for a few days before activation to alleviate cellular downregulation before receptor activation.
3. QDs may quench when stored in PBS for longer durations. When necessary to store cells labeled with QDs for longer than 12 h, 0.1 M borate buffer, pH 8.5 is recommended as the storage buffer to maintain QD fluorescence.
4. To avoid complications due to cross-species reactivity with secondary antibodies, separate experiments may be designed to investigate receptor co-localization with multiple different endocytic compartments.
5. Many mounting mediums may cause QDs to quench or stop blinking. At the time of this writing, most mounting mediums compatible with QDs require dehydration in ethanol and washing with toluene, which may destroy the integrity of cellular structures. For this reason, we recommend imaging coverslips in an imaging chamber.

References

1. Bunnett NW, Cottrell GS (2010) Trafficking and signaling of G protein coupled receptors in the nervous system: implications for disease and therapy. CNS Neurol Disord Drug Targets 9:539–556
2. Williams C, Hill SJ (2009) GPCR signaling: understanding the pathway to successful drug discovery. Methods Mol Biol 552: 39–50
3. Tressel SL, Koukos G, Tchernychev B, Jacques SL, Covic L, Kuliopulos A (2011) Pharmacology, biodistribution, and efficacy of GPCR-based pepducins in disease models. Methods Mol Biol 683:259–275
4. Wei CJ, Li W, Chen JF (2011) Normal and abnormal functions of adenosine receptors in the central nervous system revealed by genetic knockout studies. Biochim Biophys Acta 1808:1358–1379
5. Thathiah A, De Strooper B (2011) The role of G protein-coupled receptors in the pathology of Alzheimer's disease. Nat Rev Neurosci 12:73–87
6. Callahan R, Hurvitz S (2011) Human epidermal growth factor receptor-2-positive breast cancer: current management of early, advanced, and recurrent disease. Curr Opin Obstet Gynecol 23:37–43
7. Wagner M, Zollner G, Trauner M (2011) Nuclear receptors in liver disease. Hepatology 53:1023–1034
8. Yurchenko M, Sidorenko SP (2010) Hodgkin's lymphoma: the role of cell surface receptors in regulation of tumor cell fate. Exp Oncol 32:214–223
9. Sorkin A, von Zastrow M (2009) Endocytosis and signalling: intertwining molecular networks. Nat Rev Mol Cell Biol 10:609–622

10. Wiley HS, Burke PM (2001) Regulation of receptor tyrosine kinase signaling by endocytic trafficking. Traffic 2:12–18
11. Tsao P, Cao T, von Zastrow M (2001) Role of endocytosis in mediating downregulation of G-protein-coupled receptors. Trends Pharmacol Sci 22:91–96
12. Innamorati G, Le Gouill C, Balamotis M, Birnbaumer M (2001) The long and the short cycle. Alternative intracellular routes for trafficking of G-protein-coupled receptors. J Biol Chem 276:13096–13103
13. Sheff DR, Daro EA, Hull M, Mellman I (1999) The receptor recycling pathway contains two distinct populations of early endosomes with different sorting functions. J Cell Biol 145:123–139
14. Moser E, Kargl J, Whistler JL, Waldhoer M, Tschische P (2010) G protein-coupled receptor-associated sorting protein 1 regulates the postendocytic sorting of seven-transmembrane-spanning G protein-coupled receptors. Pharmacology 86:22–29
15. French AR, Tadaki DK, Niyogi SK, Lauffenburger DA (1995) Intracellular trafficking of epidermal growth factor family ligands is directly influenced by the pH sensitivity of the receptor/ligand interaction. J Biol Chem 270:4334–4340
16. Arden JR, Segredo V, Wang Z, Lameh J, Sadee W (1995) Phosphorylation and agonist-specific intracellular trafficking of an epitope-tagged mu-opioid receptor expressed in HEK 293 cells. J Neurochem 65:1636–1645
17. Coutts AA, Anavi-Goffer S, Ross RA, MacEwan DJ, Mackie K, Pertwee RG, Irving AJ (2001) Agonist-induced internalization and trafficking of cannabinoid CB1 receptors in hippocampal neurons. J Neurosci 21:2425–2433
18. Bernard V, Levey AI, Bloch B (1999) Regulation of the subcellular distribution of m4 muscarinic acetylcholine receptors in striatal neurons in vivo by the cholinergic environment: evidence for regulation of cell surface receptors by endogenous and exogenous stimulation. J Neurosci 19:10237–10249
19. Moises T, Dreier A, Flohr S, Esser M, Brauers E, Reiss K, Merken D, Weis J, Kruttgen A (2007) Tracking TrkA's trafficking: NGF receptor trafficking controls NGF receptor signaling. Mol Neurobiol 35:151–159
20. Murphy JE, Padilla BE, Hasdemir B, Cottrell GS, Bunnett NW (2009) Endosomes: a legitimate platform for the signaling train. Proc Natl Acad Sci USA 106:17615–17622
21. Lin JW, Ju W, Foster K, Lee SH, Ahmadian G, Wyszynski M, Wang YT, Sheng M (2000) Distinct molecular mechanisms and divergent endocytotic pathways of AMPA receptor internalization. Nat Neurosci 3:1282–1290
22. Pals-Rylaarsdam R, Gurevich VV, Lee KB, Ptasienski JA, Benovic JL, Hosey MM (1997) Internalization of the m2 muscarinic acetylcholine receptor. Arrestin-independent and -dependent pathways. J Biol Chem 272: 23682–23689
23. Dahan M, Levi S, Luccardini C, Rostaing P, Riveau B, Triller A (2003) Diffusion dynamics of glycine receptors revealed by single-quantum dot tracking. Science 302:442–445
24. Lidke DS, Nagy P, Heintzmann R, Arndt-Jovin DJ, Post JN, Grecco HE, Jares-Erijman EA, Jovin TM (2004) Quantum dot ligands provide new insights into erbB/HER receptor-mediated signal transduction. Nat Biotechnol 22:198–203
25. Haggie PM, Kim JK, Lukacs GL, Verkman AS (2006) Tracking of quantum dot-labeled CFTR shows near immobilization by C-terminal PDZ interactions. Mol Biol Cell 17:4937–4945
26. Chung I, Akita R, Vandlen R, Toomre D, Schlessinger J, Mellman I (2010) Spatial control of EGF receptor activation by reversible dimerization on living cells. Nature 464:783–787
27. Fichter KM, Flajolet M, Greengard P, Vu TQ (2010) Kinetics of G-protein-coupled receptor endosomal trafficking pathways revealed by single quantum dots. Proc Natl Acad Sci USA 107:18658–18663
28. Vu TQ, Maddipati R, Blute TA, Nehilla BJ, Nusblat L, Desai TA (2005) Peptide-conjugated quantum dots activate neuronal receptors and initiate downstream signaling of neurite growth. Nano Lett 5:603–607
29. Rajan SS, Vu TQ (2006) Quantum dots monitor TrkA receptor dynamics in the interior of neural PC12 cells. Nano Lett 6:2049–2059
30. Rajan SS, Liu HY, Vu TQ (2008) Ligand-bound quantum dot probes for studying the molecular scale dynamics of receptor endocytic trafficking in live cells. ACS Nano 2: 1153–1166
31. Fichter KM, Ardeshiri A, Vu TQ (2009) Tracking single biomolecules in live cells using quantum dots. In: Rege K, Medintz IL (eds) Methods in bioengineering: nanoscale bioengineering and nanomedicine. Artech House, Boston, pp 75–84
32. Long BR, Vu TQ (2010) Spatial structure and diffusive dynamics from single-particle trajectories using spline analysis. Biophys J 98:1712–1721

33. Field J, Nikawa J, Broek D, MacDonald B, Rodgers L, Wilson IA, Lerner RA, Wigler M (1988) Purification of a RAS-responsive adenylyl cyclase complex from *Saccharomyces cerevisiae* by use of an epitope addition method. Mol Cell Biol 8:2159–2165
34. Abramoff MD, Magelhaes PJ, Ram SJ (2004) Image processing with imageJ. Biophotonics Int 11:36–42
35. Bolte S, Cordelières FP (2006) A guided tour into subcellular colocalization analysis in light microscopy. J Microsc 224:213–232

Chapter 23

Cellular Internalization of Quantum Dots

Yue-Wern Huang, Han-Jung Lee, Betty Revon Liu, Huey-Jenn Chiang, and Chi-Heng Wu

Abstract

Cell-penetrating peptides (CPPs) can facilitate uptake of quantum dots (QDs) for a variety of basic and applied sciences. Here we describe a method that utilizes simple noncovalent interactions to complex QDs and CPPs. We further describe methods to study uptake mechanisms of the QD/CPP complex. The inhibitor study coupled with the RNA interference (RNAi) technique provides a comprehensive approach to elucidate cellular entry of the QD/CPP complex.

Key words Quantum dots, Cell-penetrating peptides, Inhibitors, siRNA, Clathrin, Caveolin, Lipid raft, Macropinocytosis

1 Introduction

Nanomaterials have found numerous biomedical applications in recent years. The applications include, but not limited to, drug delivery, disease staging and therapeutic planning, sentinel lymph node mapping and removal, cellular receptor trafficking monitoring, and nanoelectronic biosensing. Among nanomaterials, fluorescent semiconductor quantum dots (QDs) have been widely used to deliver and monitor biologically active molecules into cells (1–3). QDs have unique physical and chemical properties such as photostability, high quantum yield, narrow emission peak, resistance to degradation, and broad size-dependent photoluminescence (4). These properties enable QDs for long-term multiplexing imaging. QDs are extremely slow in entering the cell. One solution to overcome the slow uptake is to conjugate with cell-penetrating peptides (CPPs) (5–7). We have developed protocols to (1) enhance efficiency of cellular uptake of QDs mediated by cell-penetrating peptides (8, 9) and (2) study molecular mechanisms of action of QD internalization.

Volkmar Weissig et al. (eds.), *Cellular and Subcellular Nanotechnology: Methods and Protocols*, Methods in Molecular Biology, vol. 991, DOI 10.1007/978-1-62703-336-7_23, © Springer Science+Business Media New York 2013

2 Materials

2.1 General Components

1. Quantum dots (QDs): Carboxyl-functionalized QDs with a CdSe/ZnS core-shell have emission and excitation peak wavelengths at 520 and 505 nm, respectively. Store at 4°C.
2. Synthetic nona-arginines (SR9): SR9 are synthesized by solid-phase peptide synthesis and then purified by high performance liquid chromatography (HPLC) using a reverse phase column. The desired purity of SR9 should be ≥90%. Store at −20°C.
3. Cell culture medium: Ham's F-12 modified medium (Cellgro, VA, USA) supplemented with 10% fetal bovine serum (FBS) (Hyclone, Utah, USA) and 1% penicillin–streptomycin. Add 50 mL FBS and 5 mL penicillin–streptomycin into 445 mL Ham's F-12 modified medium. Store at 4°C.
4. Phosphate buffered saline (PBS): Dissolve 8 g NaCl, 0.2 g KCl, 1.44 g Na_2HPO_4, and 0.24 g KH_2PO_4 in 800 mL distilled H_2O. Adjust pH to 7.4 and make up to 1 L with distilled H_2O. Sterilize by autoclaving. Store at 4°C.
5. Trypsin–EDTA: 0.25% (w/v) trypsin supplemented with 0.53 mM EDTA in PBS. Mix 5 mL trypsin (5×) and 50 μL of 53 mM EDTA together. Make up to 50 mL with PBS. Store at −20°C.
6. Tris-acetate–EDTA (50× TAE) buffer: Weigh 242 g Tris and then transfer into a graduated cylinder containing 500 mL double distilled water (ddH_2O). Add 57.1 mL glacial acetic acid and 100 mL of 0.5 M EDTA. Make up to 1 L with ddH_2O. Store at 4°C.

2.2 Components of Gel Retardation

1. 0.6% agarose gel: Dissolve 3 g agarose in a 100-mL graduated cylinder containing 50 mL TAE buffer (1×). Heat agarose solution in a microwave until agarose is completely dissolved. Cool down the agarose solution to 60°C. Insert the comb onto the glass plate and then pour the agarose liquid into glass plate. Remove the air bubbles under or between the teeth of the comb. Wait the agarose gel to become solid before use.
2. UV transilluminator.

2.3 Components of Fluorescence Image Studies

1. Medium for fluorescence imaging: Ham's F-12 modified medium supplemented with 1% FBS. Add 5 mL FBS into 495 mL Ham's F-12 modified medium. Store at 4°C.
2. Dish for fluorescence imaging: 35-mm glass-bottom tissue culture plates (MatTek, Massachusetts, USA).
3. Phenol red-free medium (Invitrogen, California, USA).
4. Epifluorescent microscopy or confocal microscopy.

2.4 Components of Energy-Dependent Cellular Uptake

1. Temperature treatment: Cells are treated at either 37°C or 4°C.
2. A combination of metabolic inhibitors: 0.15% sodium azide, 15 mM sodium fluoride, and 2 μg/mL antimycin A.

2.5 Components of Pathway-Specific Inhibitors

1. Inhibitors of clathrin-dependent pathway: Chlorpromazine stabilizes the nascent clathrin-coated vesicles while monodansylcadaverine (MDC) inhibits relocation of clathrin and adaptor protein complex-2 (AP-2) from the plasma membrane to the endosomal membrane. The mechanism of hypertonic sucrose (0.45 M) involves the dispersion of clathrin lattices on the plasma membrane. Mix chlorpromazine or MDC stock solution in Ham's F-12 modified medium supplemented with 10% FBS and 1% penicillin–streptomycin at final concentrations of 10 μM and 25 μg/mL, respectively. For hypertonic studies, dissolve 0.154 mg sucrose in 1 mL Ham's F-12 modified medium supplied with 10% FBS and 1% penicillin–streptomycin.
2. Inhibitors of caveolin-dependent pathway: Filipin and nystatin are used to deplete cholesterol. Mix the filipin or nystatin stock solution in Ham's F-12 modified medium supplemented with 10% FBS and 1% penicillin–streptomycin at final concentrations of 3 and 20 μg/mL, respectively.
3. Inhibitors of macropinocytosis: 5-(*N*-ethyl-*N*-isopropyl) amiloride (EIPA) and cytochalasin D (CytD) are used to inhibit Na+/H+ exchange or cap the barbed, fast-growing ends of actin filaments, respectively. Mix the EIPA or CytD stock solution in Ham's F-12 modified medium supplemented with 10% FBS and 1% penicillin–streptomycin at final concentrations of 30 μM or 1 μg/mL, respectively.

2.6 Components of siRNA Experiments

1. Small interfering RNA (siRNA): The sequences of siRNAs for clathrin heavy chain and caveolin-1 are as follows:

   Clathrin heavy chain:

   5′-CCCUAAACACCUCAACGAU[dT][dT]-3′ (sense)

   5′-AUCGUUGAGGUGUUUAGGG[dT][dT]-3′ (antisense)

   Caveolin-1:

   5′-CAUUAUGACCGGGCUCAUA[dT][dT]-3′ (sense)

   5′-UAUGAGCCCGGUCAUAAUG[dT][dT]-3′ (antisense)
2. Lipofectamine 2000: Dilute lipofectamine stock solution in the OPTI-MEM I-reduced serum medium according to the manufacturer's instructions (Invitrogen).
3. Treatment medium: Desired siRNA is premixed with OPTI-MEM I-reduced serum medium (Invitrogen) before complexing with lipofectamine.

4. Lysis buffer: 150 mM NaCl, 1% NP-40, and 50 mM Tris (pH 8.0) supplemented freshly with 1% protease inhibitor cocktail. Weigh 0.876 g NaCl and then transfer to a 100-mL cylinder containing 50 mL ddH_2O. Add 1 mL NP-40 and 5 mL of 1 M Tris solution (pH 8.0). Make up to 100 mL with ddH_2O. Store at −20°C. Add 1% protease inhibitor cocktail before use.
5. Bio-Rad protein assay (1×): Dilute 1 mL of Bio-Rad protein assay (5×) (Bio-Rad, California, USA) with 4 mL ddH_2O.
6. Standard solutions: Prepare bovine serum albumin in a series of five dilutions. The range depends on predicted concentrations in samples.
7. Resolving buffer: 1.5 M Tris–HCl (pH 8.8). Add 100 mL ddH_2O to a graduated cylinder. Weigh 181.7 g Tris and 4 g sodium dodecyl sulfate (SDS), and then transfer them into the cylinder. Make up to 900 mL with ddH_2O. Mix and adjust pH to 8.8 with HCl. Make up to 1 L with ddH_2O. Store at 4°C.
8. Stacking gel buffer: 0.5 M Tris–HCl (pH 6.8). Weigh 60.6 g Tris and 4 g SDS. Prepare a 1-L solution as in the previous step. Store at 4°C.
9. 30% acrylamide/bisacrylamide solution: Weigh 29.2 g acrylamide monomer and 0.8 g *N*,*N*′-methylenebisacrylamide. Transfer them to a graduated cylinder containing 80 mL ddH_2O. Make up to 100 mL ddH_2O and filter through a 0.45-μm filter (Corning, Massachusetts, USA). Store at 4°C in the darkness.
10. 10% ammonium persulfate solution: Dissolve 0.1 g ammonium persulfate in 1 mL ddH_2O. Store at 4°C in the darkness.
11. *N*,*N*,*N*′,*N*′-tetramethyl-ethylenediamine (TEMED): Store at 4°C.
12. SDS-PAGE buffer (5×): 0.125 M Tris–HCl (pH 8.3), 0.96 M glycine, 0.5% SDS. Weigh 15.1 g Tris, 72 g glycine, and 5 g SDS. Transfer all of them into a 1-L cylinder containing 700 mL ddH_2O. Adjust pH to 8.3 with HCl. Make up to 1 L with ddH_2O. Store at 4°C.
13. SDS loading buffer (5×): 300 mM Tris–HCl (pH 6.8), 10% SDS, 0.05% bromophenol blue, and 40% glycerol. Prepare 1 mL of 3 M Tris–HCl (pH 6.8), 1 g SDS, and 0.5 g bromophenol blue. Transfer them to 5 mL ddH_2O. Make up to 9 mL with ddH_2O. Store at 4°C. Mix 100 μL mercaptoethanol and 900 μL SDS loading buffer before use.
14. Nitrocellulose membranes (Bio-Rad).
15. Western blot transfer buffer: 0.025 M Tris, 0.192 M glycine, and 20% methanol. Weigh 14.4 g glycine and 3 g Tris, and then transfer them to 200 mL methanol. Make up to 1 L with ddH_2O.

16. Phosphate buffered saline-Tween 20 (PBS-T): Dissolve 8 g NaCl, 0.2 g KCl , 1.44 g Na_2HPO_4, 0.24 g KH_2PO_4, and 1 mL Tween 20 in 800 mL distilled H_2O. Adjust pH to 7.4 and make up to 1 L with distilled H_2O. Sterilize by autoclaving. Store at 4°C.
17. Blocking solution: 5% milk in PBS. Dissolve 2.5 mg nonfat dry milk in PBS and then make up to 50 mL with PBS. Store at 4°C.
18. Diluent solution: 5% milk in PBS-T. Dissolve 2.5 mg nonfat dry milk in PBS and then make up to 50 mL with PBS-T. Store at 4°C.
19. Mini PROTEAN® 3 System glass plates: Bio-Rad (catalogue #1653311).
20. Pierce ECL western blotting substrate (catalogue #32106).

3 Methods

3.1 Formation of QDs and SR9 Noncovalent Binding

To test whether SR9 peptides complex with QDs, QDs are mixed with SR9 at various molar ratios (1:10, 1:20, 1:30, and 1:60) followed by separation with a 0.6% agarose gel. That the complexes mobility decrease as the amount of SR9 increases indicates the formation of noncovalent QD/SR9 complexes.

3.1.1 QDs/SR9 Noncovalent Binding

1. The concentrations of QDs and SR9 stock solutions are 10 and 625 μM, respectively.
2. Mix 5 μL of QDs stock solution with SR9 stock solution to reach various molecular ratios; make up to a final volume of 40 μL with PBS.
3. Incubate the QDs/SR9 mixture at room temperature for 20 min before use.

3.1.2 Gel Retardation Assay

1. Prepare the QDs/SR9 mixture of various molecular ratios.
2. Prepare 0.6% agarose gel. After gel becomes solid, fill the electrophoresis tank with TAE buffer (1×).
3. Load QDs/SR9 mixture into wells.
4. Turn on the power supply. Set the voltage at 130 V and then perform the electrophoresis for 60 min.
5. After 60 min, stop the electrophoresis and then capture the image by a UV transilluminator.

3.2 Dependent Uptake of Qds/SR9

We select the final concentration of QDs at 150 nM as this concentration is below the cytotoxic level. To determine the optimal molar ratio of cellular uptake, cells are incubated with QDs and SR9 at different ratios. The ratio is determined by two criteria: the kinetics of uptake and the imaging quality. For instance, QD uptake increases as the molar ratio of SR9 increases from 1:10 to 1:30; the uptake reaches the highest level at 1:30. However, good imaging quality can be obtained at 1:20. Thus this ratio is chosen. The following procedure describes time-dependent uptake:

1. Seed 1.2×10^5 cells into 35-mm glass-bottom dishes and then allow attaching for 48 h.
2. To achieve a molar ratio of 1:20 (QD:SR9). Prepare stock solutions of QD (10 μM) and SR9 (625 μM). Mix 15 μL QD and 4.8 μL SR9 in 980.2 μL Ham's F-12 modified medium supplemented with 1% FBS.
3. Incubate the QDs/SR9 mixture at room temperature for 20 min before use.
4. Discard the old medium.
5. Wash cells with 1 mL PBS three times.
6. Add 1 mL phenol red-free medium into the dish and then detect fluorescence intensity using fluorescent microscopy.

3.3 Energy-Dependent Endocystosis

To determine whether uptake of QDs/SR9 is energy dependent, cells are incubated with QDs/SR9 under varying metabolic conditions. In temperature studies, cells are treated at either 37°C or 4°C. In metabolic inhibition experiments, cells are incubated in the absence or presence of a mixture of metabolic inhibitors.

3.3.1 Temperature Treatment

1. Seed 1.2×10^5 cells into 35-mm glass-bottom dishes and then allow cells to attach for 48 h.
2. Preincubate cells at 4°C or 37°C for 1 h.
3. To achieve a molar ratio of 1:20 (QD:SR9). Prepare stock solutions of QD (10 μM) and SR9 (625 μM). Mix 15 μL QD and 4.8 μL SR9 in 980.2 μL Ham's F-12 modified medium supplemented with 1% FBS.
4. Incubate the QDs/SR9 mixture at room temperature for 20 min before use.
5. Discard the old medium.
6. Add 1 mL of the QDs/SR9 mixture into the dish and place the dish at either 4°C or 37°C for another hour.
7. Wash cells with 1 mL PBS three times.
8. Add 1 mL phenol red-free medium into the dish and then detect fluorescence intensity using fluorescent microscopy.

3.3.2 Metabolic Inhibition Studies

1. Seed 1.2×10^5 cells into 35-mm glass-bottom dishes and then allow cells to attach for 48 h.
2. Preincubate cells with a combination of metabolic inhibitors (0.15% sodium azide, 15 mM sodium fluoride, and 2 μg/mL antimycin A.) for 1 h at 37°C.

3. Discard the old medium.
4. To achieve a molar ratio of 1:20 (QD:SR9). Prepare stock solutions of QD (10 μM) and SR9 (625 μM). Mix 15 μL QD and 4.8 μL SR9 in 980.2 μL Ham's F-12 modified medium supplemented with 1% FBS.
5. Incubate the QDs/SR9 mixture at room temperature for 20 min before use.
6. Add 1 mL QDs/SR9 mixture into the 35-mm glass-bottom dish and place the dish at 37°C for another hour.
7. Wash cells with 1 mL PBS three times.
8. Add 1 mL phenol red-free medium into the dish and then detect fluorescence intensity using fluorescent microscopy.

3.4 Pathway-Specific Inhibitor Experiments

To investigate the uptake mechanism of QDs/SR9, inhibitors are used to disrupt three major pathways of endocytosis: clathrin-dependent, caveolin-dependent, and macropinocytosis.

1. Seed 1.2×10^5 cells into 35-mm glass-bottom dishes and then allow cells to attach for 48 h.
2. To achieve a molar ratio of 1:20 (QD:SR9). Prepare stock solutions of QD (10 μM) and SR9 (625 μM). Mix 15 μL QD and 4.8 μL SR9 in 980.2 μL Ham's F-12 modified medium supplemented with 1% FBS.
3. Incubate the QDs/SR9 mixture at room temperature for 20 min before use.
4. Mix a desired inhibitor in Ham's F-12 medium supplemented with 10% FBS.
5. Discard old culture medium. Incubate cells with a specific inhibitor for 30 min at 37°C. See Subheading 2.5 for final concentrations of inhibitors. Add QDs/SR9 mixture into the cells to incubate for another hour.
6. Wash cells with 1 mL PBS three times.
7. Add 1 mL phenol red-free medium into the dish and then detect fluorescence intensity using fluorescent microscopy.

3.5 siRNA Experiments

RNAi technique is applied to complement the use of pharmacological inhibitors. Clathrin heavy chain and caveolin-1 siRNAs depress expression of critical components of the clathrin and caveolar pathways.

3.5.1 siRNA Treatment and Imaging Experiment

1. Seed 1.0×10^5 cells into 35-mm glass-bottom dishes and then allow cells to attach for 48 h.
2. Desired siRNA is complexed with Lipofectamine 2000 reagent in OPTI-MEM-reduced serum medium for 30 min according to the manufacturer's instruction.

3. Discard old culture medium.
4. Add 1 mL siRNA-lipofectamine mixture into the dish, and then incubate for 4 h at 37°C.
5. After 4 h, add FBS containing Ham's F-12 modified medium to cells to achieve a final concentration of 5%.
6. After another 68 h, mix 15 μL QDs stock solution (10 μM) and 4.8 μL SR9 stock solution (625 μM) in 980.2 μL Ham's F-12 modified medium supplemented with 1% FBS.
7. Incubate the QDs/SR9 mixture at room temperature for 20 min before use.
8. Add 1 mL QDs/SR9 mixture into the dish for 1 h.
9. Wash cells with 1 mL PBS three times.
10. Add 1 mL phenol red-free medium into the dish and then detect fluorescence intensity using fluorescent microscopy.

3.5.2 siRNA-Mediated Protein Downregulation Experiment

1. Seed the 3.0×10^5 cells into 60-mm dishes and then allow cells to attach for 48 h.
2. Desired siRNA is complexed with Lipofectamine 2000 reagent in OPTI-MEM I-reduced serum medium for 30 min according to the manufacturer's instruction. A nonspecific-targeting siRNA is employed as a negative control.
3. Discard old culture medium.
4. Add 5 mL siRNA-lipofectamine mixture into the 60-mm dish for 4 h at 37°C.
5. After 4 h, add FBS containing Ham's F-12 modified medium to cells to achieve a final concentration of 5%.
6. After another 68 h, follow the instruction in Subheading 3.6.3.

3.5.3 Western Blot Analysis

1. The analysis is conducted 3 days after transfection with siRNA.
2. Discard old medium.
3. Wash the cells with 2 mL PBS three times and then place the dish on ice.
4. Add ice-cold 200 μL lysis buffer into the dish.
5. Scrape cells off the dish using an ice-cold plastic cell scraper, and then transfer the cell lysate into 1.5 mL Eppendorf tubes.
6. Set the 1.5 mL Eppendorf tubes on ice for 30 min.
7. After 30 min, centrifuge 1.5 mL Eppendorf tubes at 4°C, 12,000 rpm for 30 min.
8. Transfer the supernatant to another 1.5 mL Eppendorf tubes. Store at 4°C for fresh use. Store the rest of samples at −20°C for future use.
9. Measure the protein concentration of the sample according to the Bio-Rad protein assay.

10. Mix the 10 μL SDS loading buffer (5×) with 40 μL samples. Boil the mixture at 95°C for 5 min. Then set the samples on ice for at least 10 min.
11. Make 10% ammonium persulfate in ddH_2O.
12. Assemble the gel casting apparatus, making sure that the sandwich of glass plates and spacers seal properly.
13. Prepare the separating gel solution according to the acrylamide concentration needed. Vortex before use.
14. Various percentages of separating gels

Final acrylamide conc	5%	6%	7%	8%	9%	10%	12%	13%	15%
30% acryl/0.8% bisacryl (mL)	2.5	3.0	3.5	4.0	4.5	5.0	6.0	6.5	7.5
H_2O (mL)	8.8	8.3	7.8	7.3	6.8	6.3	5.3	4.8	3.8
4× Tris–HCl/SDS pH 8.8 (mL)	3.7	3.7	3.7	3.7	3.7	3.7	3.7	3.7	3.7
10% ammonium persulfate (μL)	200	200	200	200	200	200	200	200	200
TEMED (μL)	10	10	10	10	10	10	10	10	10

15. Load the apparatus with 4.5 mL of the separating gel solution.
16. Top with 100 μL of isoamyl alcohol.
17. After polymerization, pour off the isoamyl alcohol, and rinse with distilled water.
18. Remove any water droplets from the inside of the casting apparatus with a Whatman paper or a paper towel. Insert the comb for the stacking gel.
19. Prepare the stacking gel solution. Add 3 mL ddH_2O, 1.3 mL Tris–HCl/SDS (pH 6.8, 4×), 0.9 mL of 30% acrylamide/0.8% bisacrylamide, 80 μL of 10% ammonium persulfate, and 5 μL TEMED. Vortex mixtures.
20. Load the stacking gel solution; do not introduce air bubbles underneath or between the teeth of comb. Remove air bubbles by pipetting up and down. Allow the stacking gel to polymerize completely (~45 min) before removing comb.
21. Remove the glass and gel sandwich from the casting apparatus.
22. Clip the sandwich to the electrophoresis apparatus. Carefully remove the comb from the gel and fill the top of the apparatus with SDS-PAGE electrophoresis buffer (1×).
23. Flush the wells with buffer. Carefully load samples and 5 μL pre-stained protein molecular weight markers into the bottom

of the wells using a flat-tipped pipette tip. The sample protein loaded into each wells is 30 mg.

24. Fill the bottom of the electrophoresis apparatus with SDS-PAGE electrophoresis buffer (1×) and connect the apparatus to the power supply.
25. Perform the gel at 80 V. After the protein markers pass through the stacking gel, increase the voltage to 120 V.
26. When the dye reaches the bottom of the separating gel, turn off the power supply. Remove the gel sandwich.
27. Remove the stacking gel.
28. Use the Bio-Rad gel blotting sandwiches kit to transfer protein from the gel to nitrocellulose membrane.
29. Set the transfer system and filled with transfer buffer. The ampere is 200 mA and transferring time is 2 h.
30. After 2 h, block the nitrocellulose with 5 mL blocking solution for 1 h.
31. Wash with 5 mL PBS-T for 5 min three times.
32. Prepare 1st antibody with diluent solution. Clathrin heavy chain is detected using a mouse monoclonal antibody at a dilution of 1:200. Mix 10 μL antibody with 2 mL diluent solution and keep on ice. Caveolin-1 protein is detected using a rabbit monoclonal antibody at a dilution of 1:1,000. Mix 2 μL antibody with 2 mL diluent solution and keep on ice.
33. Incubate the nitrocellulose in 1st antibody-containing diluent solution. Keep shaking for 12 h at 4°C.
34. After 12 h, wash with 5 mL PBS-T for 5 min three times.
35. Prepare 2nd antibody with diluent solution. Goat anti-mouse IgG-HRP secondary antibody is used for clathrin 1st antibody at dilution of 1:1,000. Goat anti-rabbit IgG-HRP secondary antibody is used for caveolin 1st antibody at dilution of 1:1,000. Mix 2 μL antibody with 2 mL diluent solution and keep on ice. Incubate the nitrocellulose in 2nd antibody-containing diluent solution. Keep shaking for 1 h at room temperature.
36. After 1 h, wash with 5 mL PBS-T for 5 min three times.
37. The blots are probed with the ECL Western blot detection system.

4 Notes

1. It is important that both inhibitors and siRNAs are used, as they are complementary methods to investigate uptake mechanisms. In general, inhibitors are relatively nonspecific. Thereby, they can inhibit multiple pathways of cellular uptake, which mislead false results.

2. Imaging of cellular uptake can be tricky. An image from a single plane sometimes can be difficult to determine whether QD/CPPs are on the membrane surface or inside the cell. A 3D image would be the most appropriate venue to resolve this issue.
3. The uptake efficiency and kinetics are QDs-, CPPs-, and cell type specific.

Acknowledgments

The authors thank Robert S. Aronstam for technical editing. This work was supported by Award No. R15EB009530 from the National Institute of Biomedical Imaging and Bioengineering (to Y.-W. Huang) and the National Science Council (NSC 97-2621-B-259-003-MY3 to H.-J. Lee), Taiwan.

References

1. Michalet X, Pinaud FF, Bentolila LA, Tsay JM, Doose S, Li JJ, Sundaresan G, Wu AM, Gambhir SS, Weiss S (2005) Quantum dots for live cells, in vivo imaging, and diagnostics. Science 307:538–544
2. Bharali DJ, Lucey DW, Jayakumar H, Pudavar HE, Prasad PN (2005) Folate-receptor-mediated delivery of InP quantum dots for bio-imaging using confocal and two-photon microscopy. J Am Chem Soc 127: 11364–11371
3. Hoshino A, Fujioka K, Oku T, Nakamura S, Suga M, Yamaguchi Y, Suzuki K, Yasuhara M, Yamamoto K (2004) Quantum dots targeted to the assigned organelle in living cells. Microbiol Immunol 48:985–994
4. Gerion D, Cheng F (2004) Fluorescent CdSe/ZnS nanocrystal - peptide conjugates for long-term, nontoxic imaging and nuclear targeting in living cells. Nano Lett 4: 1827–1832
5. Dietz GP, Bahr M (2004) Delivery of bioactive molecules into the cell: the Trojan horse approach. Mol Cell Neurosci 27:85–131
6. Wang YH, Hou YW, Lee HJ (2007) An intracellular delivery method for siRNA by an arginine-rich peptide. J Biochem Biophys Methods 70:579–586
7. Tunnemann G, Ter-Avetisyan G, Martin RM, Stockl M, Herrmann A, Cardoso MC (2008) Live-cell analysis of cell penetration ability and toxicity of oligo-arginines. J Pept Sci 14:469–476
8. Xu Y, Liu BR, Lee HJ, Shannon KB, Winiarz JG, Wang TC, Chiang HJ, Huang YW (2010) Nona-arginine facilitates delivery of quantum dots into cells via multiple pathways. J Biomed Biotechnol 2010:1–11
9. Liu BR, Li JF, Lu SW, Leel HJ, Huang YW, Shannon KB, Aronstam RS (2010) Cellular internalization of quantum dots noncovalently conju gated with arginine-rich cell-penetrating peptides. J Nanosci Nanotechnol 10:6534–6543

Chapter 24

Electrochemical Scanning Tunneling Microscopy and Spectroscopy for Single-Molecule Investigation

Andrea Alessandrini and Paolo Facci

Abstract

The technique of electrochemical scanning tunneling microscopy (ECSTM) and spectroscopy (ECSTS) for studying electron transport through single redox molecules is here described. Redox molecules of both biological and organic nature have been studied by this technique with the aim of understanding the transport mechanisms ruling the flow of electrons via a single molecule placed in a nanometer-sized gap between two electrodes while elucidating the role of the redox density of states brought about by the molecule. The obtained results provide unique clues to single-molecule transport behavior and support the concept of single-molecule electrochemical gating.

Key words ECSTM, ECSTS, Redox molecules, Redox metalloproteins, Blue copper proteins, Cyclic voltammetry

1 Introduction

Scanning tunneling microscopy and spectroscopy with full control of substrate and tip potential in an electrochemical cell represent powerful tools for investigating the phenomenology and mechanisms of interfacial electron transport in redox molecules down to single-molecule level (1). These techniques have been exploited for studying transition metal complexes (2–4), molecular wires containing viologen moieties (5, 6), redox metalloproteins (e.g., the blue copper protein azurin (7–9), cytochrome c (10)), and derivatives of redox cofactors (e.g., quinones with thio-alkyl chains) (11) chemisorbed at noble metal surfaces (e.g., Pt and Au). Both ECSTM and ECSTS allow the retrieval of information on the role of the molecular redox moieties in mediating the electron transport process in the tunneling gap between two electrodes. In comparison to UHV-performed measurements, these techniques allow operation in buffered aqueous media, especially important for metalloproteins, while enabling control of the potential (i.e., Fermi level) of

Volkmar Weissig et al. (eds.), *Cellular and Subcellular Nanotechnology: Methods and Protocols*, Methods in Molecular Biology, vol. 991, DOI 10.1007/978-1-62703-336-7_24, © Springer Science+Business Media New York 2013

both substrate and tip with respect to a reference electrode. Such a goal is reached by assembling a four-electrode electrochemical cell where substrate and tip play the role of two working electrodes (12); the tip has to be insulated in order to minimize capacitance and Faradaic contributions to the overall current, and a bipotentiostat is needed to drive the potential of the two working electrodes (13). The result is an imaging/electron tunneling spectroscopy in physiologic-like conditions and the possibility of scanning the energy scale in search for electronic levels brought about by the molecular adsorbate, possibly involved in electronic transport. These possibilities make the ECSTM and ECSTS techniques very promising for the characterization of single molecules in the context of molecular electronics. The exploitation of the potential of the electrolyte solution allows assembling a transistor-like device in which the positioning of a third electrode near to the tip and the substrate (drain and source) is not required. In fact, the role of the gate electrode is assured by the electrolyte solution potential which is able to affect the levels of the involved redox molecule. Many experimental studies have demonstrated the possibility of obtaining tunneling current enhancement at constant bias voltage by means of electrolyte gating for a redox molecule sandwiched between tip and substrate. The enhancement is obtained by an alignment of the redox energy levels of the molecule with the Fermi levels of both tip and substrate. Typically, in an ECSTM investigation the focus of the measurements is the apparent height of the molecules with respect to an internal reference feature in the image. The same area of the sample is imaged at constant bias voltage between the tip and the substrate for different values of the substrate potential. The images are acquired in the constant-current mode of the STM technique. As a consequence, an enhancement of the tunneling current is reflected in an increased height of the imaged structure. The apparent height of the redox molecules is measured with respect to that of a structure whose apparent height is not affected by changes in the substrate potential. The substrate potential is chosen in a range which includes the redox equilibrium potential of the molecule investigated as determined by electrochemical methods such as cyclic voltammetry. The above technique is usually named “spectroscopy-like imaging.” In ECSTS experiments the tip is brought into close proximity to the sample until a specified tunneling current set point is reached. At that point the z-feedback is switched off and the potential of the working electrode on which the molecule is adsorbed is swept while tunneling current is measured. By this approach the correlation of tunneling current and potential is directly obtained and the results are easier to compare even to theoretical models (14). The results of the investigation are reported in a plot of the tunneling current as a function of the potential of the electrode on which the redox molecule is immobilized.

Here, the methods used to study the interfacial electron transport properties of the metalloprotein azurin will be discussed in detail.

Both the strategies for ECSTM spectroscopy-like imaging and for the ECSTS technique with the molecule immobilized on the tip will be discussed in the case of asymmetric junction conditions. The methods can be easily transferred to the study of other metalloproteins once a suitable technique to immobilize them on a conductive substrate has been developed. Similarly, the methods to study the electron transport features of a molecule carrying the hydroquinone/benzoquinone redox couple will be discussed.

2 Materials

Prepare all solutions using ultrapure water (prepared by purifying deionized water to attain a resistivity of 18.2 MΩ cm at 25°C) and analytical grade reagents. Prepare and store all reagents at room temperature (unless indicated otherwise). Follow waste disposal regulation when disposing waste materials. Follow strictly regulations on personal lab safety when dealing with hazardous reagents, wearing personal protection equipment, and always operating under aspiration hood.

2.1 Solutions and Reagents

1. Imaging buffers: 50 mM NH_4Ac pH 4.6 or 7.6. Store at 4°C.
2. Azurin: commercial azurin has been used throughout without any further purification at typical concentration of 1 mg/ml. Add 910 μL of pure water to the lyophilized protein directly in the commercial vial to reconstitute solution. Store at 4°C.
3. 2-(6-Mercaptoalkyl)hydroquinone: 10^{-5} M in EtOH (99.8%).
4. 6-mercapto-1-hexanol (MCH) 10^{-5} M in EtOH (99.8%).
5. Gold for evaporation, purity 99.99+.
6. Freshly cleaved muscovite mica substrates (see Subheading 3).
7. Imaging substrates: thermally evaporated and recrystallized 200-nm-thick gold on mica (see Note 1).
8. Tips: electrochemically etched Pt/Ir (80:20) and Au 0.25-mm-thick wires (see Note 2) with UHV wax (see Subheading 3) and/or electropolymerizable varnish (ClearClad Co.) insulation (see Note 3).
9. Electrodes: 0.5-mm-thick Pt wire as counter electrode; 0.5-mm-thick Ag wire as quasi-reference electrode (see Note 4).
10. Substrate electrical-contact clip: 1-mm-thick stainless steel wire.
11. "Piranha" solution: H_2SO_4/H_2O_2 (70/30 V/V) freshly prepared in a 50-ml beaker (see Note 5).
12. Imaging cell: The bottom of the cell is formed by the imaging substrate; walls consist of a 10-mm-high and 2-mm-thick PTFE ring (Molecular Imaging. Co.).
13. Plastic pipettes.

3 Methods

Here the methods to assemble the sample, to prepare the tip, and to perform the measurements both for azurin and 2-(6-Mercaptoalkyl)hydroquinone will be presented.

Carry out all the procedures at room temperature unless otherwise specified.

3.1 Substrate Preparation

1. Cut a piece of muscovite mica in shape of a square, 20 mm in side. Cleave it with a sharp blade by inserting the blade laterally on a corner of the slide and lifting some sheets in order to expose a freshly cleaved surface. Pay attention handling the blade.
2. Evaporate 200-nm-thick Au film on the prepared mica sheet (see Note 1).
3. Flame anneal the evaporated substrate and check surface quality by STM in air (see Note 6). Repeat the procedure until flat triangular terraces, some hundreds of nanometers in size with surface reconstruction of Au(111), are evident (Fig. 1).

3.2 Tip Preparation

3.2.1 Pt/Ir Tip Preparation

1. Cut a 30-mm-long, 0.25-mm-thick Pt/Ir wire and install it in the etching cell (see Note 2).
2. Insulate the tip by mounting it in a device similar to that reported in Fig. 2. Put a small piece of Apiezon wax (about 1 mm^3) across the slit cut in a piece of bulk copper (cubic shape, 8 mm in side) mounted at the end of a soldering iron (Fig. 2) and melt it. Then, with the help of a micrometer screw, pass the tip in vertical direction from below through the melted wax film spanning the slit. The very tip dewets (phenomenon

Fig. 1 STM images of Au (111) surfaces used as substrates for ECSTM experiments. (**a**) Image showing the typical triangular features of Au(111) surfaces. (**b**) Au(111) surface showing the characteristic herringbone $23 \times \sqrt{3}$ reconstruction of Au(111)

Fig. 2 Set-up for the coating of ECSTM tips. The electrochemically etched tip is passed through a film of melted wax. The tip will be coated by the wax but in the very apex, due to dewetting. The inset shows the soldering iron details

neither visible by eyes nor under optical microscope) leaving a small (some tens of square microns) metal area uncoated.

3. Test the tip for proper insulation by:
   (a) Installing the electrochemical cell (see Subheading 3.3 below).
   (b) Installing the tip in the microscope measuring chamber filled with the imaging buffer (the tip should be far from the substrate).
   (c) Applying a series of tip potentials and checking that the leakage current be below 5 pA at each potential value. The potential range in which the tip current response will remain under the aforementioned threshold will define the operation window for the tip.
   (d) Save the tip for the experiment.

3.2.2 Au Tip Preparation

1. Cut a 30-mm-long, 0.25-mm-thick Au wire and install it in the etching cell (see Note 2).
2. Etch it in HCl 37% applying a bias of 5 V tip positive with respect to a Pt counter electrode.
3. Go through points b and c in step 3 of Subheading 3.2.1 for insulation and tip leakage test.

3.3 Installing the Electrochemical Cell (Both ECSTM and ECSTS)

1. At the beginning and at the end of each experimental session, check the potential of the silver wire against a standard reference electrode (here SCE) and refer all the data to the selected standard reference electrode.

2. Install the substrate on the sample holder plate by pressing over it the PTFE ring and fixing it with two clips. Connect the stainless steel wire to act as a substrate electrical contact so that it touches the substrate outside the cell. With the help of a multimeter check for electrical continuity between the substrate and the wire.
3. Connect counter electrode and quasi reference, paying attention they neither touch the metal substrate nor each other.
4. Fill the cell with the imaging buffer (typically 500 μL) while controlling the substrate potential by the potentiostat. Check carefully for buffer leaks; if any, disassemble the cell, dry carefully the substrate and the ring with N_2 flow, and assemble it again adjusting the clip pressure to fix the problem.

3.4 Sample Preparation

3.4.1 Azurin for ECSTM

1. Proceed as for installing the electrochemical cell (step 1 of Subheading 3.3).
2. Fill the cell with 200 μL protein solution (1 mg/ml).
3. Keep it incubating for about 30 min at room T.
4. Remove protein solution with a micropipette paying attention to leave enough residual solution on the surface to avoid surface dewetting (see Note 7).
5. Refill the measuring chamber with the imaging buffer to rinse the chamber.
6. Remove the rinsing buffer almost completely, avoiding to dewet completely the sample surface.
7. Repeat the last two steps three times.
8. Fill the measuring chamber with 500 μL imaging buffer.

3.4.2 Azurin for ECSTS

1. Incubate insulated gold tips with azurin solution by fixing tips on the bottom of a Petri dish and dropping 100 μL of azurin solution over the tip apex. Place the Petri dish at 4°C and let incubate overnight.
2. Rinse the tip in 50 mM NH_4Ac, pH 4.6.
3. Install the tip into the tip holder.

3.4.3 2-(6-Mercaptoalkyl) hydroquinone for ECSTM

1. Flame anneal a Au(111) substrate and immediately immerse it in a 10^{-5} M 6-mercapto-1-hexanol (MCH) solution in ethanol.
2. Incubate the substrate overnight in a sealed container to avoid ethanol evaporation.
3. Rinse copiously the substrate with ethanol and immerse it in a 10^{-5} M 2-(6-mercaptoalkyl)hydroquinone solution in ethanol (see Note 8).
4. Incubate the substrate for 30 min.
5. Rinse the substrate copiously with ethanol.

6. Mount the substrate in the ECSTM and assemble the electrochemical cell, filling it with 500 μl of 50 mM NH_4Ac pH 4.6 or 7.6 aqueous solution while controlling the substrate potential by the potentiostat.
7. Install the tip into the tip holder.

3.4.4 2-(6-Mercaptoalkyl) hydroquinone for ECSTS

1. Flame anneal a Au(111) substrate and immediately immerse it in a 10^{-5} M 2-(6-Mercaptoalkyl)hydroquinone solution in ethanol.
2. Incubate the substrate overnight in a sealed container to avoid ethanol evaporation.
3. Rinse the substrate copiously with ethanol.
4. Mount the substrate in the ECSTM and assemble the electrochemical cell, filling it with 500 μl of 50 mM NH_4Ac pH 4.6 or 7.6 aqueous solution while controlling the substrate potential by the potentiostat.
5. Install the tip into the tip holder.

3.5 Performing ECSTM Measurements on Azurin

1. Install the insulated tip into the tip holder.
2. Using a dummy substrate approximately of the same thickness than the real one, regulate tip–substrate distance at less than about 0.5 mm by optical inspection.
3. Assemble the electrochemical cell as described.
4. Set the substrate potential at a value within the ideal capacitance window of the sample (see Note 9).
5. Set the V_{bias} at 400 mV (tip positive) and the current set point at 1 nA.
6. Start approach.
7. Once set point is reached, start scanning at scanning frequency not exceeding 4 Hz (number of lines per second. The exact value depends upon the microscope used and on the scan size: larger scan sizes require lower scan frequency).
8. Check for thermal drift by controlling that subsequent frames are not "shifted" (see Note 10).
9. Change the substrate potential stepwise. Wait for capacitance currents to be over by checking the substrate current on the bipotentiostat. Once current is stabilized, start a new scan on the same area (see Note 11).

An example of the above procedure is shown in Fig. 3. After obtaining the sequence of images, the apparent height of azurin molecules with respect to an internal reference for each substrate potential should be measured (in the specific case of Fig. 3, the reference is constituted by non-electroactive Zn-azurin molecules) and reported in a plot as a function of the substrate potential (it is

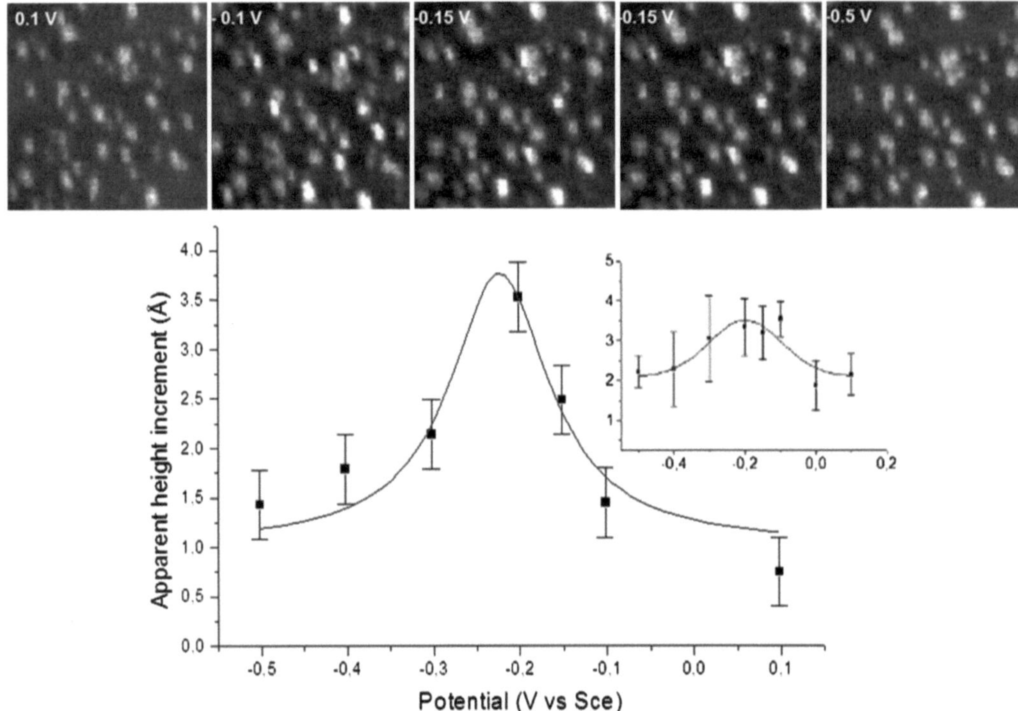

Fig. 3 Sequence of ECSTM images of a sample containing a mixture redox-active and redox-inactive azurin molecules for different values of the substrate potential (reported in each image). Scan size: 130 nm. The redox-active molecules change their apparent height with respect to the redox-inactive ones. The height of one molecule with respect to the non-changing molecules is reported in the plot as a function of the substrate potential. In the inset the average of the same measurement over six molecules is shown

also possible to refer the apparent height to the overpotential, which represents the difference between the substrate potential and the equilibrium redox potential of the azurin molecules).

3.6 Performing ECSTS Measurements on Azurin

1. Follow the same steps as in the case of ECSTM (steps 1–6, Subheading 3.5) but using the azurin-coated gold tip.
2. Position the tip potential in a potential range in which no tunneling current enhancement is expected, according to data from ECSTM.
3. Approach the tip towards a Au(111) substrate.
4. Once the set point current is reached, wait for thermal drift to be negligible, by checking the tip piezo voltage to be stable.
5. Disable the feedback on tunneling current by lowering the servo gain parameters to about 0.1.
6. Start scanning the tip potential in the potential range (−0.2 ÷ 0.4 V vs. SCE) while recording the tip current.
7. Check for the stability of the tunneling gap by controlling the value of the tunneling current at the start and at the end of the

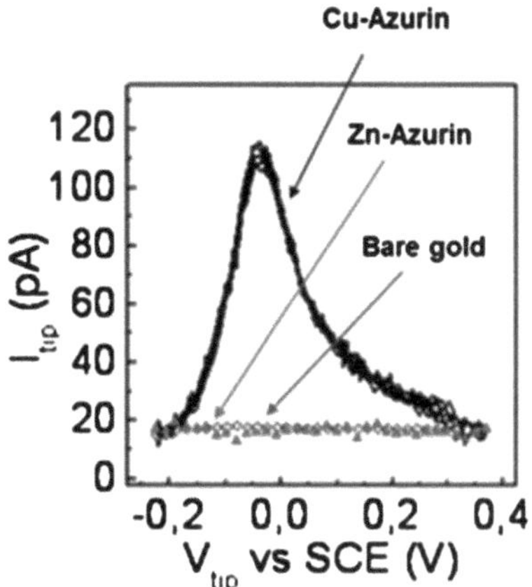

Fig. 4 ECSTS curves obtained with an azurin-coated tip. The continuous curves (*black Curves*) represent curve obtained with the azurin-coated tip. The other curves have been performed as a check: (*triangles*) represent data obtained with a tip coated with non-electroactive azurin; (*circles*) represent data obtained with a tip not modified with azurin (bare gold)

cyclic potential scan (the positions should be in regions of non-enhancement in tunneling current).

8. Discard the curves if the initial and final values of the tunneling current differ too much.
9. Repeat the measurement for different values of both the tunneling current initial set point and the bias voltage between the tip and the substrate.
10. Compare the obtained results with the case in which there are no proteins on the tip surface or with another molecule which is not electroactive.

An example of the results of this kind of measurement is reported in Fig. 4.

3.7 Performing ECSTM Measurements on 2-(6-Mercaptoalkyl) hydroquinone/MCH Adlayer

1. Perform the same steps (1–10) as in Subheading 3.5.
2. The presence of protruding bright spots is a signature of the presence of 2-(6-mercaptoalkyl)hydroquinone molecules inserted in the MCH layer.
3. Once a sequence of ECSTM images has been obtained, measure the apparent height of the bright spots which are present in the images and report them in a plot as a function of the substrate potential. In this case the apparent height is measured with respect to the non-electroactive MCH layer as shown in Fig. 5.

3.8 Performing ECSTS Measurements on 2-(6-Mercaptoalkyl) hydroquinone Adlayer

1. Mount the 2-(6-Mercaptoalkyl)hydroquinone modified Au(111) substrate in the electrochemical cell while controlling the substrate potential by the potentiostat.
2. Mount an insulated tip into the tip holder.

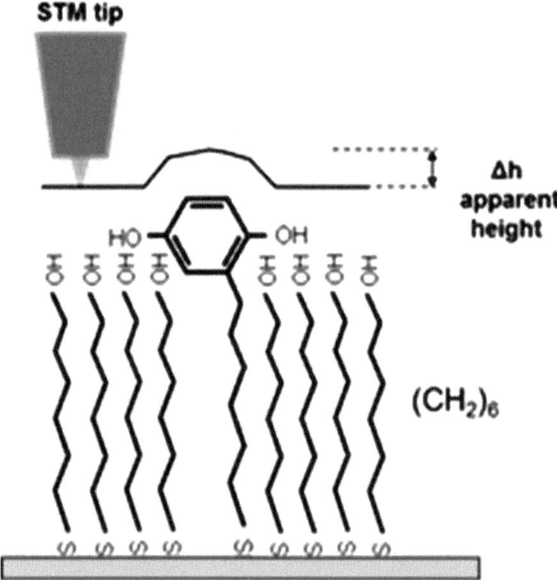

Fig. 5 Scheme of the molecular layer in the case of a mixed self-assembled monolayer and the apparent height Δh which has to be measured in each image

3. Select a substrate potential which does not induce a tunneling current enhancement, according to ECSTM data on the same molecule.
4. Select a tip/substrate bias voltage (typically 100 mV) and a set point tunneling current (typically 0.1 nA).
5. Approach the tip to the sample until the set point tunneling current is reached.
6. Wait for stabilization of the tunneling junction with the *z*-feedback on the current on.
7. Switch off the feedback system.
8. Start scanning the substrate potential measuring the corresponding tunneling current.
9. Check for the stability of the set-up by comparing the initial tunneling current with the final value (both initial and final potential should be outside the region of tunneling current enhancement).
10. Switch on the feedback system to reestablish control on the tunneling gap.
11. Repeat the same measurement for different values of the bias voltage and for different values of current set point.

4 Notes

1. Use a metal thermal evaporator. Insert freshly cleaved muscovite mica sheet and fix it in the sample holder. Load a boot with gold (pellets or wire). Start the evaporator and operate it to 10^{-7} mbar. Once the target pressure has been reached, heat the substrate at 450°C for 4 h in order to degas it and to remove all residual water and organics. After that, start evaporation at a rate of $0.1 \div 0.2$ nm/s up to a thickness of 200 nm. After that, anneal the sample for 6 h at 450°C. Subsequently, switch the pumps off along with the sample heater and wait for room temperature thermalization (typically 4 h). Open the vacuum chamber and extract the sample.
2. Pt/Ir (80:20) tips are made by electrochemical etching in saturated $CaCl_2$ using a two-electrode electrochemical cell. One electrode is the Pt/Ir wire, whereas the counter electrode is a piece of nuclear graphite. An ac voltage of 24 V (p–p) and a current of 0.27 A are used. The wire has to be immersed 3 mm below the solution meniscus. Full erosion of the immersed part of the Pt/Ir wire takes typically 10 min, as it is revealed by the zeroing of the current flowing between the two electrodes. Pay attention to avoid the formation of waves at the air–solution interface (e.g., due to mechanical vibration) since that will result in a blurred etching plane and in blunt tips. As long as the solution tends to age, etching time and performances will worsen. If you notice that etching times increase significantly above 10 min, change the saturated solution for a fresh one.
3. ClearClad™ electropolymerizable varnish (HSR 401 from LVH Coatings Ltd., UK) can be used to insulate the tip before or in place of wax coating. Insulation is accomplished by filling a 50-mL beaker with the varnish diluted in water (1:3 V:V). In the beaker insert your tip and around it a Pt-coiled wire (diameter of the coil 3 mm) acting as a counter electrode. Apply 8 V dc between the tip (negative) and the counter. Monitor the current flow and continue till current is zeroed. Afterwards, lift the tip, rinse it in pure H_2O, and cure it in an oven at 90°C for 30 min.
4. Silver wire is not exactly a reference electrode, but it is used more conveniently in ECSTM and ECSTS experiments due to its reduced size. Its potential in the working solution has to be calibrated against that of a real reference (e.g., Ag/AgCl, Hg/HgCl). It is strongly recommended to test Ag wire potential in the working buffer before and after any experiment in order to check for potential shifts. If any, the average between the two values (before and after the experiment) is usually taken for quoting any further potential value. To measure the

Ag wire potential against a reference, connect a voltmeter between the Ag wire and a reference electrode and immerse both of them in the working solution, recording the readout value.

5. Pay extreme attention to "piranha" solution. It is extremely oxidizing and reacts violently with organics. Always wear antiacid gloves and protective goggles and never leave the solution unattended. When you prepare it, it is extremely important to add acid to H_2O_2 and not the other way around, since it can result in splashes of acid. Dispose it by first neutralizing it with KOH.
6. Flame annealing can be performed with different types of flames. The best is hydrogen flame but is very dangerous and not everywhere allowed. We use to work with an ethanol flame. The aim is to heat up the gold film in order to get atomically flat terraces suitable for high-resolution imaging in STM. Switch on the flame; pick the substrate up with stainless steel tweezers; facing the flame tip with the gold surface, touch the flame tip with the substrate inclined at about 45° with respect to the flame vertical axis. Do not leave the sample in the flame still for more than 1–2 s; otherwise you run the risk of overheating it. Pay attention to avoid burning the underlying mica. Check with STM imaging in air the formation of large, flat Au(111) terraces (Fig. 1) and repeat the procedure till you get the optimal quality of your sample. Pay attention that overheating will destroy the film, enhancing its roughness and making it useless for STM/STS experiments.
7. It is of paramount importance to avoid that the protein sample crosses the liquid meniscus. In fact such an action can easily induce protein denaturation due to the action of water-based solution surface tension. Therefore, once wet, protein sample should always remain immersed in liquid.
8. The method described allows, by a substitution mechanism, to assemble a mixed monolayer in which the molecules in the second incubation step are able to replace some of the previously immobilized molecules or to insert themselves in defects present in the previous self-assembled monolayer. The concentration of the second incubation solution should be adjusted in order to have a significant number of substituted molecules in the monolayer.
9. Although there are no strict rules on this aspect, it is usually desirable to avoid any redox reaction at surface during tip–sample approach in order to avoid disturbance to the feedback system. A good strategy is that of presetting the substrate potential at a value where both the substrate and the molecular adsorbates thereby present are known to be redox inactive. One can check in advance for such a condition by performing

cyclic voltammety measurements in the imaging buffer and selecting regions where no Faradaic contributions are present.

10. Small thermal drifts or failure in tip repositioning can be corrected either relocating the tip starting point before each frame scan or by post-processing image realignment if some internal markers are present in the images (e.g., Au monoatomic steps, a surface defect, some adsorbed features other than the molecules under investigation).
11. This aspect is important, since the presence of residual capacitance currents will superimpose to the tunneling one hampering molecular behavior identification.

Acknowledgment

The authors acknowledge partial financial support by Italian MIUR FIRB "Italnanonet."

References

1. Alessandrini A, Corni S, Facci P (2006) Unraveling single metalloprotein electron transfer by scanning probe techniques. Phys Chem Chem Phys 8:4383–4397
2. Albrect T, Guckian A, Ulstrup J et al (2005) Transistor-like behavior of transition metal complexes. Nano Lett 5:451–1455
3. Albrecht T, Moth-Poulsen K, Christensen JB et al (2006) In situ scanning tunnelling spectroscopy of inorganic transition metal complexes. Faraday Discuss 131:265–279
4. Albrecht T, Guckian A, Ulstrup J et al (2005) Transistor effects and in situ STM of redox molecules at room temperature. IEEE Trans Nanotechnol 4:430–434
5. Li Z, Han B, Meszaros G et al (2006) Two-dimensional assembly and local redox-activity of molecular hybrid structures in an electrochemical environment. Faraday Discuss 131:121–143
6. Pobelov IV, Li Z, Wandlowski T (2008) Electrolyte gating in redox-active tunneling junctions—an electrochemical STM approach. J Am Chem Soc 130:16045–16054
7. Facci P, Alliata D, Cannistraro S (2001) Potential-induced resonant tunneling through a redox metalloprotein probed by electrochemical scanning probe microscopy. Ultramicroscopy 89:291–298
8. Alessandrini A, Gerunda M, Canters G et al (2003) Electron tunneling through azurin is mediated by the active site Cu ion. Chem Phys Lett 376:625–630
9. Alessandrini A, Salerno M, Frabboni S et al (2005) Single-metalloprotein wet biotransistor. Appl Phys Lett 86:133902
10. Chi Q, Zhang J, Arslan T et al (2010) Approach to interfacial and intramolecular electron transfer of the diheme protein cytochrome c4 assembled on Au(111) surfaces. J Phys Chem B 114:5617–5624
11. Petrangolini P, Alessandrini A, Berti L et al (2010) An electrochemical scanning tunneling microscopy study of 2-(6-mercaptoalkyl)hydroquinone molecules on Au (111). J Am Chem Soc 132:7445–7453
12. Alessandrini A, Facci P (2007) Electrochemically assisted scanning probe microscopy: a powerful tool in nano(bio)science. In: Erokhin V, Ram MK, Yavuz O (eds) Biophysical aspects of nanotechnology. Elsevier, Amsterdam
13. Siegenthaler H (1995) STM in electrochemistry. In: Wiesendanger R, Güntherodt H-J (eds) Scanning tunneling microscopy II, 2nd edn. Springer, Berlin
14. Albrecht T, Guckian A, Kuznetsov AM et al (2006) Mechanism of electrochemical charge transport in individual transition metal complexes. J Am Chem Soc 128: 17132–17138

Chapter 25

Intracellular Delivery of Biologically Active Proteins with Peptide-Based Carriers

Seong Loong Lo and Shu Wang

Abstract

The medical applications of protein-based therapeutics are hampered by low bioavailability associated with inefficient intracellular delivery. Various delivery materials have been developed and tested to interact with protein cargos in a manner of stabilizing proteins extracellularly and facilitating cellular uptake of proteins, thus enhancing delivery efficiency. Peptides that can form stable complexes with proteins through non-covalent interaction appear to be a promising tool to improve intracellular delivery of proteins. Here we describe the preparation of complexes formed between β-galactosidase and peptide-based carrier, protein transfer of the complexes, and the methods to evaluate delivery efficiency qualitatively and quantitatively.

Key words Protein delivery, Protein transduction, Peptide, Self-assembling complex, Non-covalent

1 Introduction

Rapid advances in functional genomics and proteomics have revealed numerous attractive diseased-related intracellular targets and hence significantly accelerated the discovery and development of protein-based therapeutics. However, delivery of biologically active proteins into cells is always challenging due to various extra- and intracellular barriers that affect protein stability and delivery efficiency. Over past few decades, there has been significant effort put into this field. Physical methods (1–9), covalent modification of protein (10–12), and lipid-based carriers (13, 14) have been applied for protein delivery. However, the use of these methods usually suffers from different inherent limitations including high rate of cell mortality (15, 16), loss of protein function (17, 18), and allergic-type immune reactions (19, 20).

Peptide-based carriers have emerged as one of the protein delivery vectors with great potential (21) because of lower immunogenicity, large-scale synthesis with well-established chemical methods, and incorporation of natural or synthetic sequences with

Volkmar Weissig et al. (eds.), *Cellular and Subcellular Nanotechnology: Methods and Protocols*, Methods in Molecular Biology, vol. 991, DOI 10.1007/978-1-62703-336-7_25, © Springer Science+Business Media New York 2013

biological activities such as cell targeting domain and cell penetrating domain. This chapter describes the basic methods for (1) non-covalent complex formation between protein cargo and peptide-based carrier, (2) protein delivery using these complexes, and (3) evaluation of delivery efficiency by quantitative and qualitative methods.

2 Materials

2.1 Reagents

1. Mammalian cells (exponentially growing).
2. Complete cell culture medium: Dulbecco's modified Eagle's medium, 10% fetal bovine serum (FBS), 1% penicillin–streptomycin, 1% L-glutamine.
3. Opti-MEM® reduced serum media (Life Technologies, Grand Island, NY, USA).
4. Ultrapure laboratory grade water (deionized water with sensitivity of 18 MΩ cm at 25°C).
5. Phosphate-buffered saline (PBS): 0.1 M sodium phosphate (pH 7.4), 0.15 M NaCl.
6. Lyophilized peptides.
7. β-galactosidase 119 kDa subunit (Active Motif, Carlsbad, CA, USA).
8. β-galactosidase staining kit (Active Motif, Carlsbad, CA, USA).

   For each well in a 96-well plate.

   1× fixing solution: 90 μl of PBS, 10 μl of 10× fixing solution.

   Complete staining solution: 138 μl of PBS, 7.5 μl of X-gal, 1.5 μl of staining solution 1, 2, and 3 respectively.
9. Galacto-Light Plus™ beta-Galactosidase Reporter Gene Assay System (Life Technologies, Grand Island, NY, USA).

   For each well in a 96-well plate.

   Lysis solution: 30 μl.

   Reaction buffer: 69.3 μl of reaction buffer diluent, 0.7 μl of 100× Galacto-Plus substrate.

   Accelerator-II: 100 μl.
10. Bovine serum albumin (BSA).
11. D_c protein assay kit (Bio-rad, Hercules, CA, USA).

    For each 5 μl of sample.

    Reagent A': 24.5 μl of reagent A, 0.5 μl of reagent S.

    Reagent B: 200 μl.

2.2 Equipment

1. Humidified CO_2 incubator for cell culture
2. Tissue culture vessels
3. Cell culture well plate
4. Inverted microscope
5. Single-tube luminometer
6. Microplate spectrophotometer (capable of absorbance measurement at wavelength of 280 nm and 650–750 nm).

3 Methods

Prepare all solutions using ultrapure laboratory grade water or PBS at room temperature (unless indicated otherwise).

3.1 Preparation of Peptide/β-Galactosidase Complex

1. Prepare peptide stock solution by reconstituting lyophilized peptide in appropriate volume of ultrapure water (see Note 1).
2. Prepare the working solution by further diluting peptide stock solution to a concentration of 50 μM with ultrapure water (see Note 2).
3. Prepare β-galactosidase solution of 0.25 mg/ml by reconstituting lyophilized protein in PBS (see Note 2).
4. Dilute 0.5 μg of β-galactosidase (for delivery into cells seeded in a well of a 96-well plate) in 8 μl of PBS.
5. Based on the required peptide–β-galactosidase molar ratio (typically ranging from 5 to 150), dilute appropriate quantity of peptide working solution in PBS to 10 μl.
6. Add the peptide solution from step 5 into β-galactosidase solution from step 4. Mix immediately by vortexing for 10 s.
7. Incubate the mixture for 30 min at room temperature.

3.2 Protein Delivery into Mammalian Cells

1. The day before protein delivery, harvest mammalian cells, grown in complete cell culture medium, by trypsinization and replate them into 96-well plate at density of 10,000–12,000 cells/well. Incubate the cells for 24 h at 37°C in a humidified incubator with 5% CO_2.
2. Remove the medium from the wells and wash the cells with 100 μl of PBS (prewarmed to 37°C) (see Note 3).
3. Add 50 μl/well of Opti-MEM reduced serum media to the cells, followed by 20 μl/well of the peptide/β-galactosidase complex.
4. Incubate the cells for 4 h at 37°C in a humidified incubator with 5% CO_2.
5. Top up the wells with 80 μl of complete cell culture medium. Incubate the cells for 20 h in the humidified CO_2 incubator.

3.3 Qualitative Evaluation Using β-Galactosidase Staining Kit

1. Prepare fresh 1× fixing solution immediately before use.
2. Remove the medium from the wells and wash the cells with 100 μl of PBS (see Note 3).
3. Fix the cells with 100 μl of 1× fixing solution for 10 min at room temperature (see Note 4).
4. While waiting for fixation, prepare complete staining solution.
5. Wash the cells twice with 100 μl of PBS.
6. Stain the cells with 150 μl of complete staining solution for 2 h in humidified CO_2 incubator.
7. Check the cells and take images under inverted microscope (see Note 5).
8. For long-term storage of the plate, replace complete staining solution with 70% glycerol in PBS, seal the plate with parafilm and store at 4°C.

3.4 Quantification of β-Galactosidase Using Galacto-Light Plus™ Beta-Galactosidase Reporter Gene Assay System

1. Prepare lysis solution and reaction buffer at room temperature immediately before use.
2. Remove the medium from the wells and wash the cells with 100 μl of PBS (see Note 3).
3. Lyse the cells at room temperature with 30–50 μl of lysis solution (see Note 6).
4. Mix 5–20 μl of cell lysate with 70 μl of reaction buffer in a luminometer tube (see Note 7). Vortex briefly and incubate for 30 min.
5. Add 100 μl of accelerator-II and vortex briefly. Place tube in luminometer and measure chemiluminiscent signal for 0.1–1 s/tube (see Note 8).
6. To determine the actual amount of β-galactosidase, generate a standard curve by serially diluting 5 ng/μl of β-galactosidase in lysis solution, followed by measurement of chemiluminiscent signal (follow steps 4 and 5).
7. For normalization against total protein, serially dilute 2.5 mg/ml of BSA in lysis solution. Pipette the BSA standards and cell lysate samples into a clean 96-well plate. Add reagents A' and B from D_c protein assay kit to the samples. Incubate for 15 min with gentle agitation; measure absorbance at wavelength of 650–750 nm. Generate a standard curve using BSA standards and calculate total protein concentration of cell lysate (see Note 9).

4 Notes

1. Hydrophilic peptides have a tendency to bind more water. Therefore, it is important to measure the concentration of peptide solution after reconstitution of lyophilized peptide.

For peptides containing amino acids with aromatic ring (tryptophan and tyrosine), concentration can be determined by estimating the molar extinction coefficient and measuring absorbance at 280 nm with spectrophotometer. For other peptides without such features, amino acid analysis is required to determine the peptide concentration.

2. To prevent or minimize peptide or protein degradation, always store the stock solution in individual aliquots at −20 or −80°C after use to avoid freeze-thaw cycles.
3. When dealing with multiple wells of cells, use multichannel pipette to avoid morphological changes or cell detachment due to PBS washing.
4. Some mammalian cell lines might be very sensitive and might undergo morphological changes during fixation. To avoid such scenario, the fixation time could be reduced to 5 min.
5. Before imaging, replace complete staining solution with 100 μl of PBS for better contrast.
6. Put on a shaker for 15 min to improve lysis rate. Check under inverted microscope to confirm complete lysis before proceed to next step.
7. The volume of cell lysate required for measurement depends upon the β-galactosidase concentration. Use 15 μl as a starting point for optimization.
8. Due to the light emission kinetics of the reaction, accelerator-II should be added in the sequence of adding reaction buffer in the previous step.
9. Bubbles will affect absorbance measurement. If bubbles form, use a clean, dry tip to pop them.

Acknowledgments

This work was supported by Institute of Bioengineering and Nanotechnology, Agency for Science, Technology and Research (A*STAR), Singapore.

References

1. Campbell PL, McCluskey J, Yeo JP, Toh BH (1995) Electroporation of antibodies into mammalian cells. Methods Mol Biol 48:83–92
2. Marrero MB, Schieffer B, Paxton WG, Schieffer E, Bernstein KE (1995) Electroporation of pp 60c-src antibodies inhibits the angiotensin II activation of phospholipase C-gamma 1 in rat aortic smooth muscle cells. J Biol Chem 270(26):15734–15738
3. Raptis LH, Liu SK, Firth KL, Stiles CD, Alberta JA (1995) Electroporation of peptides into adherent cells in situ. Biotechniques 18(1):104, 106, 108, 110 passim
4. Nolkrantz K, Farre C, Hurtig KJ, Rylander P, Orwar O (2002) Functional screening of intracellular proteins in single cells and in patterned cell arrays using electroporation. Anal Chem 74(16):4300–4305

5. Bar-Sagi D, Feramisco JR (1985) Microinjection of the ras oncogene protein into PC12 cells induces morphological differentiation. Cell 42(3):841–848
6. Rui M, Chen Y, Zhang Y, Ma D (2002) Transfer of anti-TFAR19 monoclonal antibody into HeLa cells by in situ electroporation can inhibit the apoptosis. Life Sci 71(15):1771–1778
7. Abarzua P, LoSardo JE, Gubler ML, Neri A (1995) Microinjection of monoclonal antibody PAb421 into human SW480 colorectal carcinoma cells restores the transcription activation function to mutant p53. Cancer Res 55(16):3490–3494
8. Theiss C, Meller K (2002) Microinjected anti-actin antibodies decrease gap junctional intercellular commmunication in cultured astrocytes. Exp Cell Res 281(2):197–204
9. Narayanan A, Eifert J, Marfatia KA, Macara IG, Corbett AH, Terns RM, Terns MP (2003) Nuclear RanGTP is not required for targeting small nucleolar RNAs to the nucleolus. J Cell Sci 116(Pt 1):177–186
10. Nagahara H, Vocero-Akbani AM, Snyder EL, Ho A, Latham DG, Lissy NA, Becker-Hapak M, Ezhevsky SA, Dowdy SF (1998) Transduction of full-length TAT fusion proteins into mammalian cells: TAT-p27Kip1 induces cell migration. Nat Med 4(12):1449–1452
11. Vocero-Akbani A, Lissy NA, Dowdy SF (2000) Transduction of full-length Tat fusion proteins directly into mammalian cells: analysis of T cell receptor activation-induced cell death. Methods Enzymol 322:508–521
12. Becker-Hapak M, McAllister SS, Dowdy SF (2001) TAT-mediated protein transduction into mammalian cells. Methods 24(3):247–256
13. Zelphati O, Wang Y, Kitada S, Reed JC, Felgner PL, Corbeil J (2001) Intracellular delivery of proteins with a new lipid-mediated delivery system. J Biol Chem 276(37):35103–35110
14. Dalkara D, Chandrashekhar C, Zuber G (2006) Intracellular protein delivery with a dimerizable amphiphile for improved complex stability and prolonged protein release in the cytoplasm of adherent cell lines. J Control Release 116(3):353–359
15. Buchser WJ, Pardinas JR, Shi Y, Bixby JL, Lemmon VP (2006) 96-well electroporation method for transfection of mammalian central neurons. Biotechniques 41(5):619–624
16. Peng PD, Cohen CJ, Yang S, Hsu C, Jones S, Zhao Y, Zheng Z, Rosenberg SA, Morgan RA (2009) Efficient nonviral sleeping beauty transposon-based TCR gene transfer to peripheral blood lymphocytes confers antigen-specific antitumor reactivity. Gene Ther 16(8):1042–1049
17. Bennion BJ, Daggett V (2003) The molecular basis for the chemical denaturation of proteins by urea. Proc Natl Acad Sci USA 100(9):5142–5147
18. Rudolph R, Böhm G, Lilie H, Jaenicke R (1997) Folding proteins. Protein function: a practical approach, 2nd edn. Oxford University Press, New York
19. Szebeni J (2001) Complement activation-related pseudoallergy caused by liposomes, micellar carriers of intravenous drugs, and radiocontrast agents. Crit Rev Ther Drug Carrier Syst 18(6):567–606
20. Szebeni J (2005) Complement activation-related pseudoallergy: a new class of drug-induced acute immune toxicity. Toxicology 216(2–3):106–121
21. Lo SL, Wang S (2010) Intracellular protein delivery systems formed by noncovalent bonding interactions between amphipathic peptide carriers and protein cargos. Macromol Rapid Commun 31(13):1134–1141

Chapter 26

Lipo-oligoarginine-Based Intracellular Delivery

Jae Sam Lee and Ching-Hsuan Tung

Abstract

Efficient cellular delivery, including plasma membrane permeability and intracellular metabolic stability, is a crucial factor determining the success of therapeutic agents. Cell-penetrating peptides (CPPs) have been widely used for the intracellular delivery of various bioactive molecules into cells to modify cellular functions. We have developed an improved CPP-based cellular delivery vector, named lipo-oligoarginine peptide (LOAP), by conjugating an oligoarginine peptide with a fatty acid moiety. The prepared LOAPs were further stabilized by introducing different combinations of D-Arg residues into the peptide backbone and were systematically evaluated for their membrane-penetrating properties and metabolic stabilities in cells.

Key words Cell-penetrating peptide, Oligoarginine, Fatty acid, Cellular uptake, Drug delivery system, Metabolic stability

Abbreviations

MBHA	4-Methylbenzhydrylamine
Fmoc	9-Fluorenylmethyloxycarbonyl
Pbf	2,2,4,6,7-Pentamethyldihydrobenzofuran-5-sulfonyl
Mtt	Methyltrityl
TA	Thioanisole
EDT	Ethandithiol

1 Introduction

Cell-penetrating peptides (CPPs) have been employed to deliver various membrane impermeable bioactive molecules such as proteins, peptides, nanoparticles, and nucleic acids into cells. Among the CPPs studied, the human immunodeficiency virus 1 (HIV-1) Tat protein has gained much attention due to its efficient internalization capacity and relatively low cytotoxicity (1). The functional domain on the Tat protein that controls translocation through the cell membrane is a short 9-mer peptide, RKKRRQRRR (1–3). The positively charged arginine residues in the Tat peptide are crucial

Volkmar Weissig et al. (eds.), *Cellular and Subcellular Nanotechnology: Methods and Protocols*, Methods in Molecular Biology, vol. 991, DOI 10.1007/978-1-62703-336-7_26, © Springer Science+Business Media New York 2013

for membrane penetration; replacing these with other positively charged amino acids such as lysine, histidine, or citrulline abolishes the ability of the protein to penetrate the membrane, suggesting that the cationic guanidine moiety on the arginine side chain plays an important role in this unique membrane translocation property (4). It has been reported that the cellular uptake efficiency of arginine-rich CPPs depends on the number of arginine residues (5 < R < 15) (4, 5). No internalization was observed in the case of peptides with three or less arginine residues (6), and severe cell toxicity was seen in CPPs having 15 or more arginine residues (5). Interestingly, the internalization efficiency of CPPs is increased when L-arginines are replaced by different combinations of a protease resistant D-arginine sequence (5, 7).

A major obstacle in developing potent therapeutic agents is their poor bioavailability, which prevents them from entering the lipid bilayer of the plasma membrane, resulting in a diminished therapeutic efficacy. Hydrophobic enrichment of therapeutic agents has been attempted to achieve improved delivery into cytoplasmic or intracellular compartments (8). Similarly, fatty acid moieties that could advance cellular association with hydrophobic cell membranes have been incorporated into various CPPs, such as proline-rich, arginine-rich, or cationic peptides (9–11). However, studies on a myristoylated random neutral peptide revealed that a myristoyl group alone was not enough to promote cellular uptake of the reference peptide (12). It was also reported that a hydrophobic peptide modified by a single farnesyl group did not associate strongly with membrane (13). These data imply that conjugation of a fatty acid to a peptide in itself is insufficient to promote cellular uptake, and a second signal is required to achieve stable association of CPPs with cell membrane. Neither a fatty acid (e.g., myristic acid) nor a basic peptide (e.g., oligoarginine) alone is adequate for plasma membrane binding; however, both these moieties when combined within the same protein or peptide template might exhibit enhanced association with the cell membrane. We have previously reported that oligoarginine peptides modified with fatty acid moieties exhibit enhanced cellular uptake efficiency (5, 7, 12), confirming that oligoarginine sequences tethered to fatty acid moieties might form effective delivery vectors, because such modifications could enhance membrane association and potentially lead to improved delivery of cargo molecules into the intracellular compartment.

Although the metabolic stability of CPPs is an important factor for their efficient cellular translocation and for sustained release of cargo molecules into target cells, most CPPs are metabolically unstable and are rapidly degraded before reaching their targets (14–18). The degradation process usually begins with the combined action of exoproteases, which cleave the C-terminal and/or N-terminal residues, as well as endoproteases, which cleave internal

Table 1
Sequences of synthesized LOAPs

Name	Sequence[a]
R7	$B(R)_7K(FITC)\text{-}NH_2$
C12R9	Lauroyl-$B(R)_9K(FITC)\text{-}NH_2$
C12dR9	Lauroyl-$B(dR)_9K(FITC)\text{-}NH_2$
C12dR9-1	Lauroyl-$B(dR)_5(R)_4K(FITC)\text{-}NH_2$
C12dR9-2	Lauroyl-$B(R)_4(dR)_5K(FITC)\text{-}NH_2$
C14R11	Myristoyl-$B(R)_{11}K(FITC)\text{-}NH_2$
C14dR11	Myristoyl-$B(dR)_{11}K(FITC)\text{-}NH_2$

[a] *B* β-alanine, *dR* D-arginine, and *FITC* fluorescein isothiocyanate

amide bonds at specific sites. To minimize proteolytic degradation, the structures of these peptides are often modified so that they are no longer labile to proteases. We recently introduced a series of stable lipo-oligoarginine peptide (LOAP) backbone, in which L-Arg residues were replaced with D-Arg residues (Table 1). Although there was no apparent difference in cellular uptake (15, 19), a C14dR11 peptide (where all Arg residues are in D configuration) expressed superior metabolic stability and enhanced intracellular retention without adverse effects on cell viability compared with its unmodified parent C14R11 peptide (5). This modified LOAP might potentially be an excellent delivery vector to transport various payloads.

Conjugation of fluorescence markers to CPPs has emerged as a useful method for analyzing cellular uptake, intracellular distribution, and metabolic stability of CPPs by confocal laser scanning microscopy (5, 7, 12). Furthermore, fluorescence probes enable direct comparison of cellular uptake efficiencies by flow cytometry (5, 7, 20, 21). Here we describe methods for the systematic evaluation of various LOAPs generated by the introduction of different fatty acid moieties at the N-terminus of oligoarginine peptides, focusing on synthesis and purification, cellular uptake, intracellular distribution, cell toxicity, and metabolic stability in Jurkat cells.

2 Materials

Prepare all solutions using ultrapurified water and analytical grade chemicals and reagents.

2.1 Reagents for Lipo-oligoarginine Peptide Synthesis

1. Rink amide MBHA resin (0.72 mmol/g), and amino acids including Fmoc-Arg(Pbf)-OH, Fmoc-D-Arg(Pbf)-OH, Fmoc-Lys (Mtt)-OH, and Fmoc-β-Ala-OH, 2-(1H-benzotriazole-1-yl)-1,

1,3,3-tetramethyluronium hexafluorophosphate (HBTU), *N*-hydroxybenzotriazole (HOBt), piperidine, dimethylformamide (DMF), and *N*-methyl-2-pyrrolidone (NMP). Peptide synthesis is accomplished using the ABI 433A peptide synthesizer (Life Technologies, Grand Island, NY, USA).

2. Fatty acyl chloride (lauroyl chloride, myristoyl chloride, palmitoyl chloride), *N*,*N*-di-isopropylethylamine (DIEA), methanol, and anhydrous dichloromethane (DCM), used in the conjugation of fatty acid with the oligoarginine peptide. Ninhydrin test is performed using the Kaiser test kit (Sigma-Aldrich, Milwaukee, WI, USA). A vacuum manifold reservoir and 15 ml reservoir filter (Fisher Scientific, Waltham, MA, USA) are also needed.
3. For FITC labeling of LOAPs, trifluoroacetic acid (TFA), Wheaton glass 20 ml scintillation vials, and fluorescein isothiocyanate (FITC).
4. For cleavage and purification of LOAPs, cleavage cocktail chemicals can be obtained (Fisher Scientific, Waltham, MA, USA). The Vydac C18 column and reverse-phase high-performance liquid chromatography (HPLC) are available (Varian, Palo Alto, CA, USA).
5. The MW of LOAPs can be measured using an AB4800 matrix-assisted laser desorption/ionization time-of-flight (MALDI-TOF) Analyzer (Applied Biosystems, Norwalk, CT).

2.2 Cell Culture

1. Maintain human Jurkat cells from the American-Type Culture Collection in RPMI-1640 supplemented with 10% heat-inactivated fetal bovine serum and 1% penicillin/streptomycin. 75 cm^2 culture flasks and 24-well culture plates are also needed.
2. Trypsin-ethylenediaminetetraacetic acid (EDTA) and propidium iodide (PI) are used in flow cytometry analysis. Falcon cup cell strainer 70 μm (Becton Dickinson, Franklin Lakes, NJ, USA) is needed to prevent cell clumping. The FACS LSR II instrument (Becton Dickinson, Franklin Lakes, NJ, USA) is used for analysis with WinMDI software (The Scripps Institute).
3. For cell cytotoxicity analysis, CytoTox 96® Non-radioactive Cytotoxicity Assay (Promega, Madison, WI, USA) and Phenol red-free RPMI 1640 are needed. We performed our analysis on the SpectraMax M2 microplate reader (Molecular Devices, Sunnyvale, CA, USA).
4. For microscopic observation, Olympus FluoView™ 1000 laser scanning confocal microscope (Center Valley, PA) can be used. Glass slide and cover slip are also needed.

3 Methods

LOAPs are synthesized by conventional Fmoc solid-phase chemistry with β-alanine and lysine at either end for conjugation of fatty acid and FITC moieties, respectively. All amino acids and fatty acids are pre-activated with HBTU. Starting with the polystyrene resin, peptides are built "backwards," from the C-terminus to the N-terminus. After synthesis, the peptide is cleaved from the resin, yielding the crude form of the peptide (Scheme 1). Generally, yields of crude peptides are ~90% pure, but HPLC purification can be performed if higher purity is desired.

3.1 Lipo-oligoarginine Peptide Synthesis

1. Load an ABI Model 433A peptide synthesizer equipped with UV monitoring of Fmoc deprotection and conditional cycle feedback, with 0.25 mmol Rink amide MBHA resin (0.7 mmol/g) and HBTU/HOBt as the activating and coupling reagents in NMP and DIEA.
2. The amino acids used in the synthesis of the LOAPs are Fmoc-Arg(Pbf)-OH, Fmoc-D-Arg(Pbf)-OH, Fmoc-Lys(Mtt)-OH, and Fmoc-β-Ala-OH. Position the nonnatural amino acid β-Ala at the N-terminus of the oligoarginine peptide to reduce proteolytic degradation.

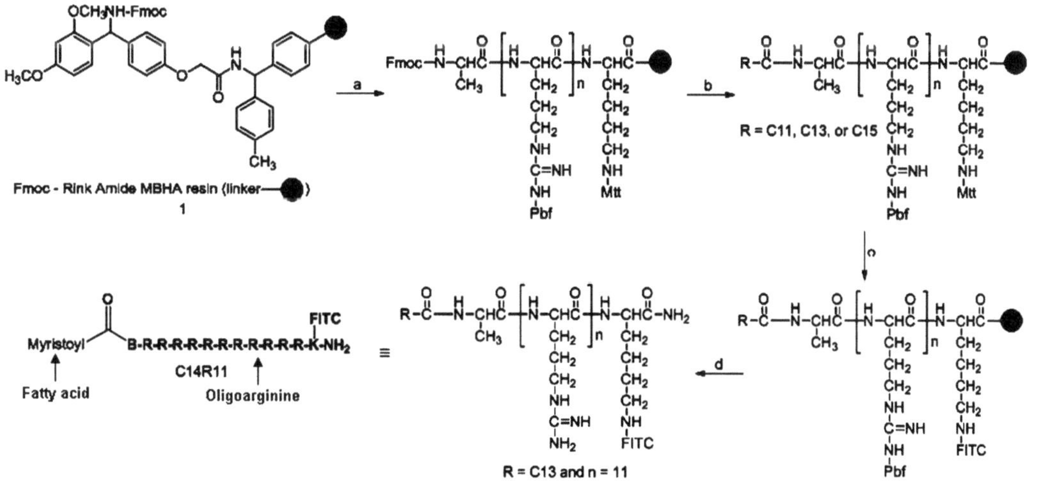

Fmoc - Rink Amide MBHA resin (linker—●)
1

R = C11, C13, or C15

Myristoyl B-R-R-R-R-R-R-R-R-R-R-R-K-NH2 FITC
C14R11
Fatty acid
Oligoarginine

R = C13 and n = 11

Scheme 1 Synthesis of LOAPs. Reagents and conditions. (*a*) Amino acids used are Fmoc-β-Ala-OH, Fmoc-Arg(Pbf)-OH, and Fmoc-Lys(Mtt-OH), with HBTU/HOBt and 20% piperidine/DMF, 0.5 h. (*b*) Conjugation of the fatty acyl chloride in 5 ml of DIEA/NMP solution (2 M) at room temperature for 2 h. (*c*) Removal of Mtt in 5 ml TFA in DCM (1:99, v/v) for 2 min, 9–12 times, followed by labeling with FITC in 5 ml of DIEA/DMF (1:4) at room temperature overnight. (*d*) Cleavage of LOAPs using TFA:TA:EDT:anisole (90%:5%:3%:2%) at room temperature for 3–4 h

3. Deprotect the Fmoc moiety by 20% piperidine/DMF and activate the resin with NMP/DMF (0.4 M) for each cycle (see Note 1).

3.2 Fatty Acid Conjugated Oligoarginine Peptides

1. Suspend the peptide resin (0.08 mmol) in 5 ml of freshly prepared DIEA/NMP solution (2 M).
2. Add fatty acyl chloride (0.32 mmol) dropwise into the resin slurry and stir the mixture gently at room temperature for 2 h.
3. Check the completion of the reaction via the Kaiser test kit. If the coupling of the acyl group to the amino group is not complete, the test will yield a blue color.
4. Collect the raw product by filtering through a vacuum manifold reservoir and wash with DCM (5×), NMP (5×), and methanol (5×).

3.3 FITC Labeling with Lipo-oligoarginine Peptide

1. Selectively remove the 4-methyltrityl (Mtt) protecting group on the lysine side chain using 5 ml of TFA in DCM (1:99, v/v) for 2 min with stirring. Repeat this step 9–12 times, followed by washing with DCM (5×), NMP (5×), and methanol (5×) (see Note 2).
2. React the free lysine-containing peptide with fourfold excess of FITC (0.4 mmol, 155.8 mg) in 5 ml of DIEA/DMF (1:4), and stir the slurry overnight at room temperature in the dark.
3. Wash the bright yellow LOAP resin with NMP and methanol.

3.4 Cleavage and Purification of LOAPs

1. Cleave the fully dried LOAPs from the resin using the cleavage cocktail (90% TFA, 5% TA, 3% EDT, and 2% anisole) for 3–4 h. Longer reaction times will be needed to completely remove the Pbf protecting groups from the longer arginine sequences. Precipitate the yellow LOAPs with diethyl ether or methyl-*t*-butyl ether (see Note 3).
2. Purify the peptide solution by reversed-phase HPLC equipped with a Vydac C18 column using 0.1% TFA and acetonitrile (ACN) as elution buffers. Detect peptide peaks by UV absorption at 220 and 265 nm. Lyophilize the eluted peptide to obtain a yellow cotton-like solid (see Note 4).
3. Characterize the molecular mass of each synthetic LOAP by MALDI-TOF $(M+H)^+$ (see Note 5).

3.5 Cellular Internalization of LOAPs in Jurkat Cells

1. For cellular internalization and cytotoxicity assays, prepare Jurkat cells (3.5×10^5 cells/ml) in a 24-well plate in complete culture medium and incubate overnight at 37°C in a humidified 5% CO_2 incubator.
2. Determine the concentration of each LOAP solution by measuring FITC absorption using a SpectraMax M2 Microplate (Molecular Devices, Sunnyvale, CA) at 490 nm ($\varepsilon = 67{,}000$) in phosphate buffered saline (PBS) buffer (pH = 7.2).

3. Wash the cells with serum-free media and PBS twice, and add freshly prepared LOAP stock solutions (100 μM) to the culture medium in each well, to a final LOAP concentration of 10 μM.
4. Incubate for 10 min at 37°C. Wash with PBS three times, and incubate cells with 200 μl trypsin-EDTA for 3 min at 37°C to remove cell surface-bound peptides (see Note 6).
5. Following trypsin treatment, add 200 μl of culture medium to neutralize the protease and centrifuge at 250×*g* for 3 min. Wash with PBS and incubate with 200 μl ice-cold propidium iodide (1 μg/ml) in PBS+2% FBS to assess cell viability.
6. Cells staining with propidium iodide will be excluded from the flow cytometry analysis. Count and characterize a total of $1–2\times10^4$ events using a FACS LSR II instrument and WinMDI software (Fig. 1).
7. Determine the toxicity of LOAPs to Jurkat cells using the CytoTox 96® Non-radioactive Cytotoxicity Assay (Fig. 2a).
8. Load the LOAP-treated cells (5 μl, 3.5×10^3 cells) onto a regular microscopic glass slide and cover with a cover glass slip.

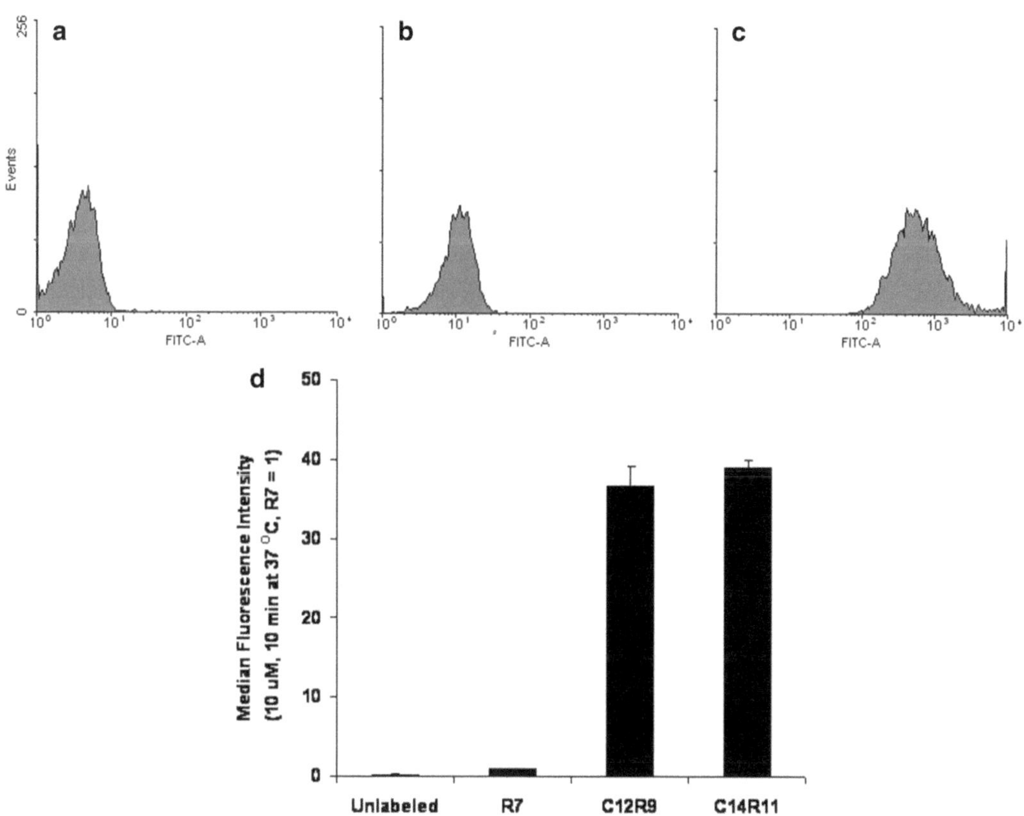

Fig. 1 FACS analysis of LOAP cellular uptake in Jurkat cells. Cells were incubated with 10 μM of LOAPs for 10 min at 37°C. The cells were washed, trypsin treated, PI stained, and analyzed by flow cytometry with $1–2\times10^4$ events. Each histogram indicates (**a**) control, (**b**) R7, and (**c**) C14R11, respectively. (**d**) Cellular uptake of LOAPs. The median fluorescence intensity of R7 in Jurkat cells was set as 1

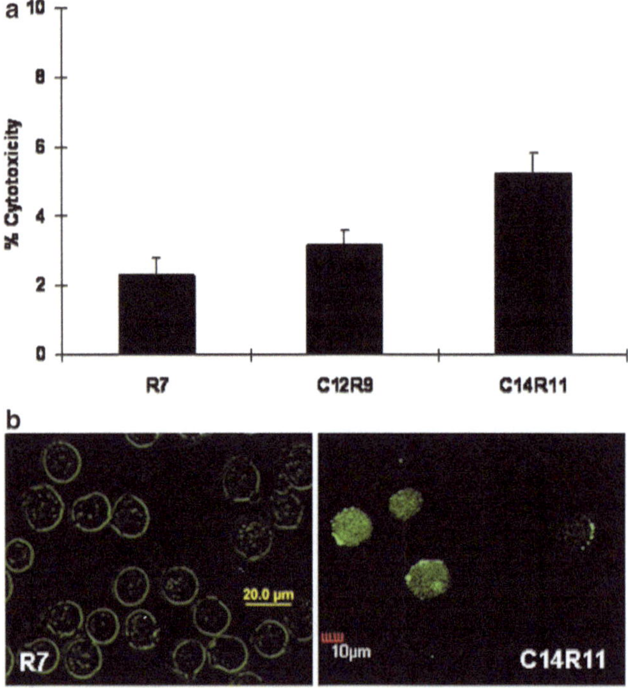

Fig. 2 Cell toxicity and intracellular distribution of LOAPs in Jurkat cells. (**a**) Effects of LOAPs on cell toxicity. Cells were incubated with 10 μM of LOAPs for 60 min at 37°C. Data were normalized to the maximum amount of lactate dehydrogenase (LDH) released from the cultures treated with lysis solution alone and corrected for baseline LDH release from cultures exposed to buffer only. Experiments were performed in triplicate, and data are represented as mean ± SD. (**b**) Visualization of intracellular distribution. Cells were treated with 10 μM of LOAPs for 10 min at 37°C. Cell images were obtained by confocal microscopy

9. Capture fluorescence images using an Olympus FluoView TM 1000 laser scanning confocal microscope equipped with 60× oil immersion objective, with excitation by 488 nm laser line (to excite the FITC tag on the LOAPs; Fig. 2b)

3.6 Metabolic Stability of LOAPs

1. Monitor the metabolic degradation of LOAPs by analytical reversed-phase HPLC (920-LC, Varian) equipped with a fluorescence detector, with excitation at 495 nm and emission at 521 nm (Fig. 3a).
2. At each indicated time point, collect the whole culture media containing the LOAPs for HPLC quantification and mass analysis.
3. Prepare elution solvent A consisting of H_2O (0.1% TFA) and solvent B consisting of ACN (0.1% TFA). Set the linear gradient as 0–60% solvent B over 30 min with 1 ml/min flow rate for a 10 μl sample volume.

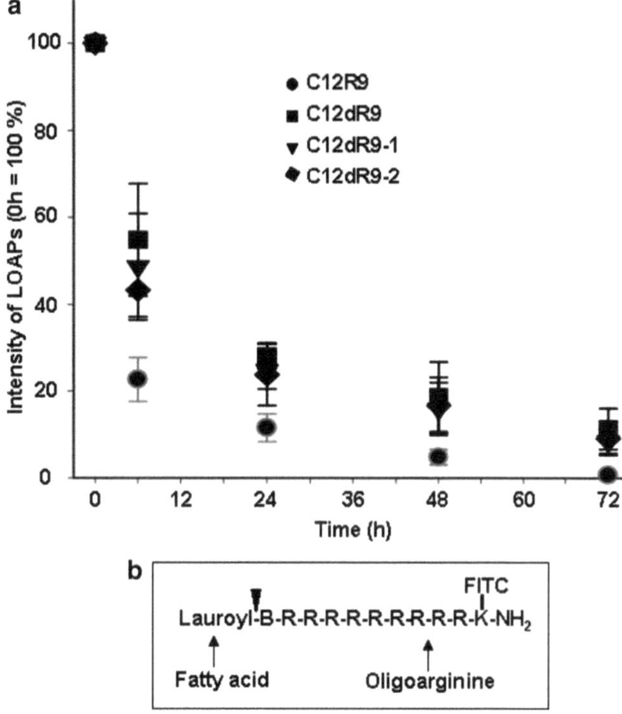

Fig. 3 Release kinetics and cleavage site of LOAPs. (**a**) Jurkat cells were incubated with 10 μM of LOAPs at 37°C for 10 min. Cells were then washed completely; further incubated with a fresh peptide-free culture medium for 6, 24, 48, or 72 h; and harvested for flow cytometry analysis. The fluorescence intensity of LOAPs at 0 h was set as 100%. (**b**) Schematic illustration of the cleavage site in LOAPs. It was found that most internalized LOAPs were hydrolyzed between the lauroyl chain and beta-alanine residue, releasing the intact peptide into the culture medium

4. Collect aliquots of FITC-conjugated sequences (C-terminus) and identify the molecular weight by MALDI-TOF mass (Fig. 3b).

4 Notes

1. Swell Rink amide MBHA resin in 5 ml of DMF for 5 min, and remove the Fmoc group of Rink amide MBHA resin using 20% piperidine/DMF. Dissolve HBTU in a solution of HOBt in DMF and add the first amino acid into this solution along with NMP and DIEA. Transfer the mixture solution directly to the reaction vessel. HBTU activation is highly efficient, especially when difficult couplings are predicted, such as with Arg and the branched side chains of Ile, Val, and Leu. Generally, a coupling time of 2–4 h is sufficient. Remove the Fmoc group of each amino acid using 20% piperidine/DMF. Verify Fmoc

group removal using the ninhydrin test. Modify reaction conditions to compensate by extending the deprotection and/or coupling periods when aggregation occurs. After removing the Fmoc group, wash the resin with NMP to remove the piperidine. Before coupling, the carboxyl group of the amino acid must be activated by adding HBTU, which can be dissolved in a solution of HOBt in DMF.

2. The coupling of the fluorophore (e.g., FITC) to the CPP is usually carried out by covalent linkage to the N-terminal α-amino group or to a lysine ε-amino group in the peptide. The side-chain Mtt group can be selectively removed with 1% TFA in DCM (22). In order to remove the Mtt group completely from the LOAP, the peptide resin should be washed at least 9–12 times using 1% TFA in DCM with gentle shaking for 2 min each time. In washing steps 1–3, the color of the solution will remain unchanged. During washing steps 3–6, the color of the solution will change from pale to dark yellow. In washing steps 6–9, almost all the color will disappear.

3. Wash the bright yellow resin from the previous step with NMP (5 × 5 ml) and methanol (5 × 5 ml), and dry for 60 min in a vacuum. Treat the peptide resin with 10 ml of the deprotection cocktail, and stir gently for 3–4 h. Prepare and assemble columns for the extraction of peptide from the resin. Fill a 50-ml conical tube with 30 ml ether, and place the tube in ice to keep the ether cold. Place the column via a clamp and stand in hood. *Slowly add the cocktail to column.* Once the cocktail has stopped "dripping" from the column, add slight pressure to the top of the column using the air line to remove any remaining peptide from the column. Remove tube from under column, cap, shake, and release pressure. Incubate the ether-peptide mixture for 10 min, then centrifuge the tube at 2,500 rpm (780 × *g*) for 5 min at room temperature. Decant supernatant into a waste carboy. Allow the ether to evaporate under low vacuum drying. Add 15 ml of 10% acetic acid to the centrifuge tube. Freeze and lyophilize. Characterize the molecular weight of the peptide using MALDI mass spectrometry, and store at -20°C.

4. To obtain highly pure peptide, approximately 200 mg/ml crude peptide (the amount depends on the complexity of the HPLC trace of the crude product) can be purified by preparative HPLC (Vydac 218TP1022 column) using a gradient of 60–100% B over 60 min (A = 0.1% TFA/water and B = 0.1% TFA/ACN) at a flow rate of 8 ml/min. Collect all column fractions and pool those containing the ideal product. The amount of pure material recovered depends on the sequence but is usually greater than 90% of that loaded.

5. Dissolve LOAPs in a mixture of 70% ACN and 30% H_2O (0.1% TFA) and desalt using a ZipTipC18 tip (Millipore, Billerica, MA). Spot a 1 μl sample directly onto a MALDI target plate. Apply 1 μl matrix solution (α-cyano-4-hydroxycinnamic acid, 5 mg/ml in 50:50 ACN/H_2O) on the sample spot and allow to dry. Blow the dried MALDI spot with compressed air before inserting into the mass spectrometer (AB4800 MALDI-TOF) Proteomics Analyzer. *Operation conditions*: The instrument must be operated in positive ion reflector mode, with the mass range as required. Acquire 2,000 laser shots and calculate the average for each sample. Perform an external calibration using a calibration mixture including four different masses (904.468, 1296.685, 1570.677, and 2465.199). Applied Biosystems software package 4000 Series Explorer (v. 3.6 RC1) can be used to analyze the acquired data.
6. A very important challenge in quantifying CPP uptake is to distinguish between membrane-associated and intracellular CPPs. For this purpose, trypsin/EDTA is an effective method to detach membrane-associated peptides from the cell surface.

Acknowledgment

This work was supported in part by NIH CA135312, CA114149, and CA126752.

References

1. Vives E, Brodin P, Lebleu B (1997) A truncated HIV-1 Tat protein basic domain rapidly translocates through the plasma membrane and accumulates in the cell nucleus. J Biol Chem 272:16010–16017
2. Frankel AD, Pabo CO (1988) Cellular uptake of the tat protein from human immunodeficiency virus. Cell 55:1189–1193
3. Green M, Loewenstein PM (1988) Autonomous functional domains of chemically synthesized human immunodeficiency virus tat trans-activator protein. Cell 55:1179–1188
4. Mitchell DJ, Kim DT, Steinman L, Fathman CG, Rothbard JB (2000) Polyarginine enters cells more efficiently than other polycationic homopolymers. J Pept Res 56:318–325
5. Lee JS, Tung CH (2010) Lipo-oligoarginines as effective delivery vectors to promote cellular uptake. Mol Biosyst 6:2049–2055
6. Futaki S, Suzuki T, Ohashi W, Yagami T, Tanaka S, Ueda K, Sugiura Y (2001) Arginine-rich peptides. An abundant source of membrane-permeable peptides having potential as carriers for intracellular protein delivery. J Biol Chem 276:5836–5840
7. Lee JS, Tung C-H (2011) Enhanced cellular uptake and metabolic stability of lipo-oligoarginine peptides. Biopolymers. 96:772–779
8. O'Donnell JM, Alpert NM, White LT, Lewandowski ED (2002) Coupling of mitochondrial fatty acid uptake to oxidative flux in the intact heart. Biophys J 82:11–18
9. Fernandez-Carneado J, Kogan MJ, Van Mau N, Pujals S, Lopez-Iglesias C, Heitz F, Giralt E (2005) Fatty acyl moieties: improving Pro-rich peptide uptake inside HeLa cells. J Pept Res 65:580–590
10. Futaki S, Ohashi W, Suzuki T, Niwa M, Tanaka S, Ueda K, Harashima H, Sugiura Y (2001) Stearylated arginine-rich peptides: a new class of transfection systems. Bioconjug Chem 12:1005–1011
11. Koppelhus U, Shiraishi T, Zachar V, Pankratova S, Nielsen PE (2008) Improved cellular activity of antisense peptide nucleic acids by conjugation to a cationic peptide-lipid (CatLip) domain. Bioconjug Chem 19:1526–1534
12. Pham W, Kircher MF, Weissleder R, Tung CH (2004) Enhancing membrane permeability by

fatty acylation of oligoarginine peptides. Chembiochem 5:1148–1151

13. Silvius JR, l'Heureux F (1994) Fluorimetric evaluation of the affinities of isoprenylated peptides for lipid bilayers. Biochemistry 33: 3014–3022
14. Garner P, Sherry B, Moilanen S, Huang Y (2001) In vitro stability of alpha-helical peptide nucleic acids (alphaPNAs). Bioorg Med Chem Lett 11:2315–2317
15. Elmquist A, Langel U (2003) In vitro uptake and stability study of pVEC and its all-D analog. Biol Chem 384:387–393
16. Trehin R, Nielsen HM, Jahnke HG, Krauss U, Beck-Sickinger AG, Merkle HP (2004) Metabolic cleavage of cell-penetrating peptides in contact with epithelial models: human calcitonin (hCT)-derived peptides, Tat(47–57) and penetratin(43–58). Biochem J 382:945–956
17. Rennert R, Wespe C, Beck-Sickinger AG, Neundorf I (2006) Developing novel hCT derived cell-penetrating peptides with improved metabolic stability. Biochim Biophys Acta 1758:347–354
18. Foerg C, Merkle HP (2008) On the biomedical promise of cell penetrating peptides: limits versus prospects. J Pharm Sci 97:144–162
19. Wender PA, Mitchell DJ, Pattabiraman K, Pelkey ET, Steinman L, Rothbard JB (2000) The design, synthesis, and evaluation of molecules that enable or enhance cellular uptake: peptoid molecular transporters. Proc Natl Acad Sci USA 97:13003–13008
20. Vives E, Richard JP, Rispal C, Lebleu B (2003) TAT peptide internalization: seeking the mechanism of entry. Curr Protein Pept Sci 4:125–132
21. Duchardt F, Fotin-Mleczek M, Schwarz H, Fischer R, Brock R (2007) A comprehensive model for the cellular uptake of cationic cell-penetrating peptides. Traffic 8:848–866
22. Bourel L, Carion O, Gras-Masse H, Melnyk O (2000) The deprotection of Lys(Mtt) revisited. J Pept Sci 6:264–270

Chapter 27

Fluorescent Spherical Monodisperse Silica Core–Shell Nanoparticles with a Protein-Binding Biofunctional Shell

Achim Weber, Marion Herz, and Günter E.M. Tovar

Abstract

The production of uniform protein-binding biofunctional fluorescent spherical silica core–shell nanoparticles by a modified Stöber method is described. Fluorescent particle cores with diameters of 100 nm are synthesized in a two-step reaction. Functional shells for subsequent coupling reactions are provided by generating organic shells providing amine and carboxyl groups at the nanoparticles' shell surface. Conjugation of proteins to the nanoparticles is achieved using *N*-(3-dimethylaminopropyl)-*N'*-ethylcarbodiimide hydrochloride (EDC) as coupling agent. The characterization of the nanoparticle systems and their surface functionalization is done by microelectrophoresis, dynamic light scattering (DLS), and a colorimetric detection of the amount of nanoparticle-attached protein via a bicinchoninic acid (BCA) assay. Fluorescently spiked nanoparticle cores with biofunctional shells for molecular recognition reactions may be used as imaging tools or reporter systems.

Key words Silica nanoparticles, Nanoparticles, Core–shell, Surface modification, Monodisperse, Fluorescent dye

1 Introduction

The synthesis of submicron particles with narrow size distribution, uniform shape, and surface modification is of interest in various fields in life science such as molecular biological assays in cell and tissue engineering, molecular biotechnology, biological imaging, or microarrays (1–6). The production of nanoparticulate sol–gel particles is simple, their size can be tailored by a systematic variation of reaction conditions, and the flexible silica chemistry provides versatile possibilities for surface modifications and preparation of a nanoscopically thin organic shell based on the silanol groups that are present on the nanoparticles' surface (7, 8). These silanols can be used to integrate amine groups, carboxyl groups, or other functionalities into the particle shell using organo-functional silanes.

Volkmar Weissig et al. (eds.), *Cellular and Subcellular Nanotechnology: Methods and Protocols*, Methods in Molecular Biology, vol. 991, DOI 10.1007/978-1-62703-336-7_27, © Springer Science+Business Media New York 2013

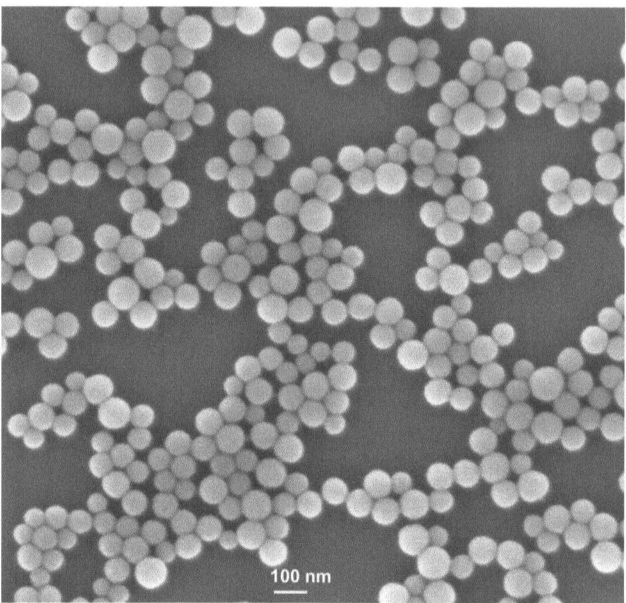

Fig. 1 Typical scanning electron micrograph of silica nanoparticles prepared by the hydrolysis of tetraethoxysilane in the presence of ethanol containing ammonia (Subheading 3.1)

By incorporating a fluorescent dye into the core of these chemically stable and biocompatible particles, you gain a system of high brightness and photostability which can be used for identifying or characterizing biological samples such as cells (4). One of the major advantages of nanoparticles is their small size which leads to a dramatically increased specific surface. A large number of dye molecules can be incorporated into one single silica particle such that much higher optical signals are generated than with isolated dye molecules.

Here, the synthesis of spherical silica nanoparticles with a hydrodynamic diameter of 100 nm is demonstrated via a modified Stober method (see Fig. 1), which consists of the hydrolysis of tetraethoxysilane in a mixture of ethanol and aqueous ammonia (7). After the formation of the nanoparticles, additional layers of reactive functional groups, amine and carboxy, are attached (see reaction schemes in Fig. 2). Fluorescent-dyed particles are synthesized in a two-step reaction. In the first step, the fluorescent dye Alexa Fluor® 488 carboxylic acid succinimidyl ester is chemically bound to (3-aminopropyl)triethoxysilane, an amine containing silane agent (see fluorescence image in Fig. 3). In the second reaction step, the silanized dye and tetraethoxysilane are hydrolyzing and co-condensing to form dye-doped silica nanoparticles (9, 10). A protein is covalently bound to carboxyl-functionalized silica nanoparticles by using carbodiimides that form amide linkages between the carboxyl groups at the particle surface and the amino groups of proteins (11).

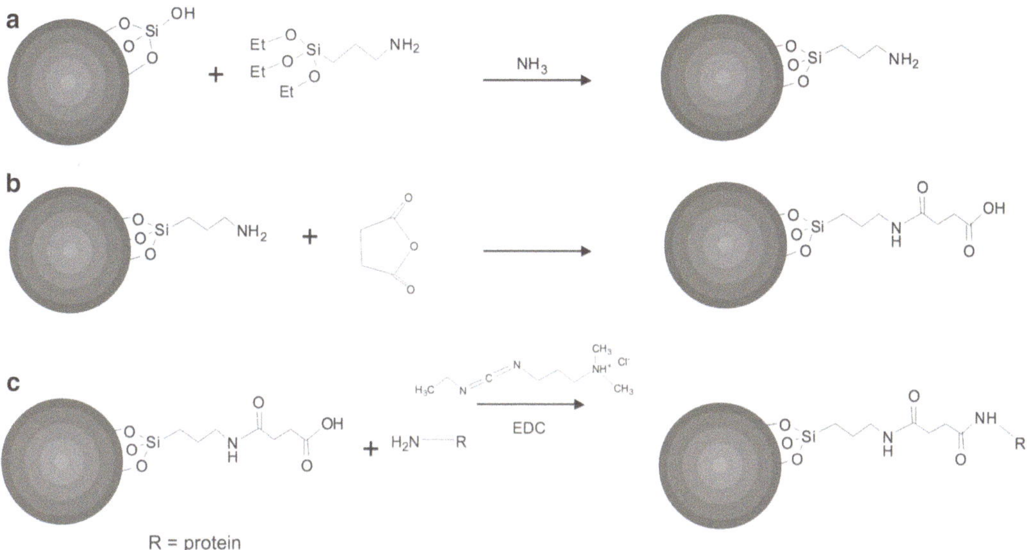

Fig. 2 Reaction scheme of the surface modifications. (**a**) Functionalization of silica particles with (3-aminopropyl)triethoxysilane, (**b**) reaction of silica surface covered by amino groups with succinic anhydride, and (**c**) formation of an amide bond between the carboxylic group of the silica particles and the amine group of the antibody in the presence of N-(3-dimethylaminopropyl)-N′-ethylcarbodiimide hydrochloride (EDC)

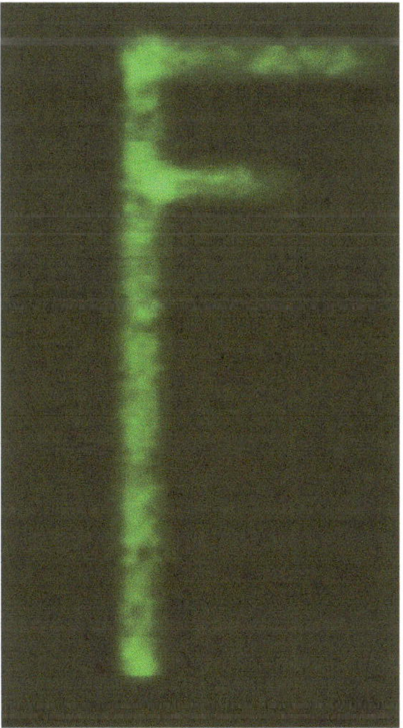

Fig. 3 Fluorescence microscopy image of Alexa Fluor® 488-stained silica nanoparticles attached to a microscope slide, 10× magnification

2 Materials

All solutions are prepared by the use of Milli-Q grade water (<18.2 mΩ/cm) and stored at room temperature (unless described otherwise).

2.1 Synthesis of Unmodified Silica Nanoparticles

1. Magnetic stirrer.
2. Stirring bar.
3. 500 mL closable bottle.
4. Graduated volumetric cylinder, 1 L.
5. Volumetric pipette, 20 mL.
6. Graduated pipette, 20 mL.
7. Piston stroke pipette, 1–10 mL with appertaining tips.
8. Suction bulb.
9. Centrifuge for high sample volume.
10. Centrifugal tubes, 500 mL.
11. Plastic beaker, 5 L.
12. Beaker, 600 mL.
13. Glass funnel.
14. Dialysis tube, MWCO 12–14 kDa, diameter 29 mm, flat width 45 mm, 6.4 mL/cm (volume/length).
15. Dialysis clips.
16. Syringe.
17. Syringe filter, pore size 0.45 μm with a membrane consisting of mixed cellulose esters.
18. Conductivity electrode.
19. Ethanol (HPLC grade, >99.9%).
20. Ammonium hydroxide solution (puriss p.a. plus, ≥25% in water).
21. Milli-Q grade water.
22. Tetraethoxysilane (TEOS, 99%).

2.2 Surface Modification with Amino-Functional Groups

1. Magnetic stirrer.
2. Stirring bar.
3. 100 mL closable bottle.
4. Volumetric pipette, 100 mL.
5. Graduated pipette, 10 mL.
6. Suction bulb.
7. Piston stroke pipette, 100–1,000 μL with appertaining tips.
8. Centrifuge for high sample volume.

9. Centrifugal devices, 30 mL teflon.
10. Centrifugal devices, 30 mL PP.
11. Centrifugal tubes, 1.5 mL.
12. Syringe.
13. Syringe filter, pore size 0.45 μm with a membrane consisting of mixed cellulose esters.
14. Ammonium hydroxide solution (puriss p.a. plus, ≥25% in water).
15. (3-Aminopropyl)triethoxysilane (APTES, 99%).
16. Milli-Q grade water.
17. 100 mM acetate buffer solution, pH 4.7 (ACB47): add approximately 800 mL Milli-Q grade water to a glass beaker. Pipette 11.44 mL of acetic acid to the beaker and then add 3.999 g of sodium hydroxide. Mix and adjust the pH with 1 M caustic soda. Transfer the solution to a 1 L graduated cylinder and make up to 1 L with Milli-Q grade water. Store at room temperature.

2.3 Surface Modification with Carboxyl Functional Groups

1. Magnetic stirrer.
2. Stirring bar.
3. 50 mL closable bottle.
4. Graduated pipette, 10 mL.
5. Suction bulb.
6. Centrifuge for high sample volume.
7. Centrifugal devices, 30 mL PP.
8. Centrifugal devices, 30 mL teflon.
9. Balance.
10. Spatula.
11. Plastic beaker, 5 L.
12. Beaker, 600 mL.
13. Glass funnel.
14. Dialysis tube, MWCO 12–14 kDa, diameter 29 mm, flat width 45 mm, 6.4 mL/cm (volume/length).
15. Dialysis clips.
16. Syringe.
17. Syringe filter, pore size 0.45 μm with a membrane consisting of mixed cellulose esters.
18. Conductivity electrode.
19. 100 mM phosphate buffer solution, pH 7.0 (PB70): add approximately 800 mL Milli-Q grade water to a glass beaker (see Note 1). Weigh 8.8995 g disodium hydrogen phosphate

dihydrate and 6.8995 g sodium dihydrogen phosphate monohydrate and transfer to the beaker. Mix and adjust the pH with 1 M caustic soda. Transfer the solution to a 1 L graduated cylinder and make up to 1 L with Milli-Q grade water.

20. Tetrahydrofuran (>99.9%).
21. Succinic anhydride (>99%).
22. Milli-Q grade water.

2.4 Fluorescent Dye Labeling of Unmodified Silica Nanoparticles

1. Centrifuge for small sample volume.
2. Centrifugal tubes 1.5 mL.
3. Piston stroke pipette 1–10 μL with appertaining tips.
4. Piston stroke pipette 10–100 μL with appertaining tips.
5. Piston stroke pipette 100–1,000 μL with appertaining tips.
6. Shaker for centrifugal tubes.
7. Alexa Fluor® 488 carboxylic acid succinimidyl ester.
8. Ethanol (HPLC grade, >99.9%).
9. (3-Aminopropyl)triethoxysilane (APTES, 99%).
10. Milli-Q grade water.
11. Ammonium hydroxide solution (puriss p.a. plus, ≥25% in water).
12. Tetraethoxysilane (TEOS, 99%).

2.5 Surface Modification of Fluorescent Dye-Labeled Nanoparticles with Amino-Functional Groups

1. Centrifuge for small sample volume.
2. Centrifugal tubes 1.5 mL.
3. Piston stroke pipette 1–10 μL with appertaining tips.
4. Piston stroke pipette 10–100 μL with appertaining tips.
5. Piston stroke pipette 100–1,000 μL with appertaining tips.
6. Shaker for centrifugal tubes.
7. Ammonium hydroxide solution (puriss p.a. plus, ≥25% in water).
8. Tetraethoxysilane (TEOS, 99%).
9. (3-Aminopropyl)triethoxysilane (APTES, 99%).
10. 100 mM acetate buffer, pH 4.7 (ACB47)—*see* preparation noted in Subheading 2.2.
11. Ethanol (HPLC grade, >99.9%).
12. Milli-Q grade water.

2.6 Surface Modification of Fluorescent Dye-Labeled Nanoparticles with Carboxyl Functional Groups

1. Centrifuge for small sample volume.
2. Centrifugal tubes 1.5 mL.
3. Piston stroke pipette 100–1,000 μL with appertaining tips.
4. Balance.
5. Spatula.

6. Shaker for centrifugal tubes.
7. 100 mM phosphate buffer, pH 7.0 (PB70)—see preparation noted in Subheading 2.3.
8. Tetrahydrofuran (>99.9%).
9. Succinic anhydride (>99%).
10. Milli-Q grade water.

2.7 Functionalization of Carboxy-Modified Nanoparticles with Protein

1. Centrifuge for small sample volume.
2. Centrifugal tubes 1.5 mL.
3. Shaker for centrifugal tubes.
4. Piston stroke pipette 100–1,000 μL with appertaining tips.
5. Balance.
6. Spatula.
7. *N*-(3-Dimethylaminopropyl)-*N'*-ethylcarbodiimide hydrochloride (EDC): 8 mM in Milli-Q grade water.
8. Trehalose solution: 5% trehalose in Milli-Q grade water.
9. 100 mM 2-(*N*-morpholino)ethanesulfonic acid sodium salt (MES), pH 4.5 (MES45): Add approximately 800 mL Milli-Q grade water to a glass beaker (see Note 1). Weigh 21.72 g MES and transfer to the beaker. Mix and adjust the pH with HCl. Transfer the solution to a 1 L graduated cylinder and make up to 1 L with Milli-Q grade water.
10. 100 mM phosphate buffer saline, pH 7.4 (PBS74): add approximately 800 mL Milli-Q grade water to a glass beaker (see Note 1). Weigh 8 g sodium chloride, 0.2 g potassium chloride, 0.2 g potassium dihydrogen orthophosphate, and 1.44 g disodium hydrogen phosphate dihydrate and transfer to the beaker. Mix and adjust the pH with 1 M caustic soda. Transfer the solution to a 1 L graduated cylinder and make up to 1 L with Milli-Q grade water.
11. Triton X-100 (PBS/Tr): 0.1% Triton X in PBS74.
12. Protein, e.g., gelatin, other proteins or antibodies are also usable.

3 Methods

All reactions are done at room temperature (unless described differently).

3.1 Synthesis of Unmodified Silica Nanoparticle Cores with a Hydrodynamic Diameter of 100 nm

1. Measure out 400 mL of ethanol in a 1 L volumetric cylinder and pour it into a 500 mL closable bottle.
2. Put a stirring bar into the bottle and place it onto a magnetic stirrer. Stir at 500 rpm.

3. Add 20 mL of Milli-Q grade water, 10.3 mL of ammonium hydroxide solution, and 5.35 mL (corresponding to 5 g) of TEOS while stirring. Then close the bottle and let the reaction run for at least 12 h (see Notes 2 and 3).
4. Transfer the resulting particle suspension into two 500 mL centrifugal devices and tare. Centrifuge at 10,000 rpm for 60 min (Beckman Coulter Avanti J, Rotor JA-10, 17,700 × *g*). Discard the supernatant. Resuspend the particle pellets thoroughly in 100 mL of Milli-Q grade water with the help of an ultrasonic bath (see Note 4).
5. Preincubate a 30 cm piece of dialysis tube in Milli-Q grade water for approximately 5 min.
6. Fill a 5 L plastic beaker with Milli-Q grade water, put a stirring bar into it, and place the beaker onto a magnetic stirrer.
7. Seal the dialysis tube at one end using two clips and fill in the particle suspension with the help of a glass funnel. Rinse the centrifugal devices and the funnel with Milli-Q grade water and close the dialysis tube at the top with again two clips. Place it into the beaker filled with Milli-Q grade water (see Note 5).
8. Gently stir at about 200 rpm such that the water inside the beaker will be moved while the dialysis tube floats motionless.
9. Exchange water daily. Dialysis may be stopped when the conductivity of the water is below 1.3 μS/cm (see Note 6). This usually takes 5–7 days.
10. Finally pass the particle suspension through syringe filters with a pore size of 0.45 μm.
11. A SEM image of these particles is seen in Fig. 1.

3.2 Surface Modification with Amino-Functional Groups

1. Suspend 100 mg of silica particles in 100 mL Milli-Q grade water and pour it into a closable bottle.
2. Put a stirring bar inside the bottle and place the bottle onto a magnetic stirrer. Stir at about 500 rpm.
3. Add 8 mL of ammonium hydroxide solution and 200 mg of APTES (20% w/w of the used particle mass) while you keep on stirring (see Notes 2 and 3).
4. Close the bottle and make sure that the reaction time is kept at exactly 1 h (see Note 7).
5. Split the particle suspension into four centrifugation devices made of teflon (see Notes 8 and 9). Centrifuge and resuspend the particles in ACB47. Transfer suspension to centrifugal devices made of PP and repeat washing several times (see Note 10). Finally pass the particle suspension through syringe filters with a pore size of 0.45 μm.

3.3 Surface Modification with Carboxylic Groups

1. Centrifuge 500 mg of amino-modified silica nanoparticles at 20,000 rpm for 20 min in 30 mL centrifugal devices made of PP (Beckman Coulter Avanti J, Rotor JA-25.50, 48,384 × *g*).
2. Throw away the supernatant and resuspend the particle pellets in 25 mL of PB70 with the support of an ultrasonic bath. The particle suspension should also be transferred into centrifugal devices made of teflon because of the ongoing working step.
3. Centrifuge at 10,000 rpm for 30 min (Beckman Coulter Avanti J, Rotor JA-25.50, 12,096 × *g*). Aspire the supernatant again and resuspend the particle pellet in 25 mL of tetrahydrofuran whereby the suspension is poured into a 50 mL closable bottle (see Note 11).
4. Add 875 mg of succinic anhydride to the reaction suspension and homogenize it by using the ultrasonic bath for 1 h while you sonify for 5 min and let it rest for 10 min. Repeat that for four times and make sure that the bath is ice cooled at all time as tetrahydrofuran has a high vapor pressure and the reaction suspension is getting warm very fast.
5. Shake the particle suspension for at least 16 h.
6. Pour the modified particle suspension into centrifugal devices made of teflon and rinse carefully to make sure that all particles are collected.
7. Wash the particles with centrifugation and resuspension into 50 mL of Milli-Q grade water as mentioned above.
8. Then fill the thoroughly resuspended particles inside a dialysis tube as described in Subheading 3.1 and exchange again the water until the conductivity is below 1.3 μS/cm. That will usually take 3–5 days.

3.4 Fluorescent Labeling of Unmodified Silica Nanoparticles

1. Add 79 μL of ethanol to 1 mg of Alexa Fluor® 488 fluorescent dye in a 1.5 mL centrifugal tube and homogenize the mixture.
2. Add 3.47 mg of APTES to 1 mL of ethanol in a separate 1.5 mL centrifugal tube and mix it carefully.
3. For silanization of the dye, pipette 79 μL of the APTES solution to the Alexa dye solution and shake the mixture constantly for 2 h (see Note 12).
4. Centrifuge 10 mg silica particles in a 1.5 mL centrifugal device at 15,000 rpm for 20 min (Hermle Z216 MK, Rotor 220.87 V11, 21,380 × *g*).
5. Aspire the supernatant and resuspend the particle pellet in 354 μL of Milli-Q grade water.
6. Add 35 μL of ammonium hydroxide solution, 16.7 μL of the silanized Alexa dye solution, and 3 μL of TEOS to the carefully resuspended particles (see Notes 2 and 3).

7. Shake the suspension for 24 h at room temperature.
8. Centrifuge the particles at 15,000 rpm for 20 min.
9. Throw away the supernatant and wash the dye-doped particles three times with ethanol and one time with Milli-Q grade water as described above.

3.5 Surface Modification of Fluorescent Dye-Labeled Nanoparticles with Amino-Functional Groups

1. Pipette the corresponding volume of 10 mg dye-doped nanoparticle suspension into a 1.5 centrifugal tube and add 100 μL of ammonium hydroxide solution.
2. Furthermore add 3.7 μL TEOS and 11 μL APTES and shake the reaction mixture for 24 h at room temperature.
3. Wash the amino-modified particles with centrifugation and resuspension two times with 1 mL ethanol, one time with 1 mL ACB47, and two times with 1 mL Milli-Q grade water as mentioned in Subheading 3.4.
4. Characterize the particles with zeta potential measurements explained in Subheading 3.8.

3.6 Surface Modification of Fluorescent Dye-Labeled Nanoparticles with Carboxy-Functional Groups

1. Centrifuge 10 mg of dye-doped and amino-modified silica nanoparticles at 15,000 rpm for 20 min in a 1.5 mL centrifugal device.
2. Aspire the supernatant and resuspend the particle pellet in 1 mL of PB70 with the support of an ultrasonic bath.
3. Again centrifuge the particle suspension at 15,000 rpm for 20 min. Throw away the supernatant and resuspend the particle pellet in 1 mL of tetrahydrofuran (see Note 11).
4. Add 17.5 mg of succinic anhydride to the reaction suspension and homogenize it by using the ultrasonic bath for 1 h while you sonify for 5 min and let it rest for 10 min. Repeat that reaction cycle for four times; make sure that the bath is ice cooled at all time as tetrahydrofuran has a high vapor pressure and the reaction suspension is getting warm very fast.
5. Shake the particle suspension for at least 16 h.
6. Wash the particles three times with centrifugation and resuspension into 1 mL of Milli-Q grade water as described in Subheading 3.4.

3.7 Functionalization of Carboxy-Modified Silica Nanoparticles with Proteins

1. Add 45 μg of protein and 100 μL of EDC solution to 1 mg of carboxy-modified silica nanoparticles (see Note 13).
2. Fill up to 1 mL with MES45.
3. Shake the suspension for 16 h at 4°C.
4. Centrifuge the suspension at 15,000 rpm for 10 min and aspirate the supernatant very carefully (Hermle Z216 MK, Rotor 220.87 V11, 21,380 × *g*).

5. Resuspend the particle pellet in 1 mL of PBS/Tr and centrifuge the suspension again at 15,000 rpm for 10 min (see Note 14).
6. Repeat the washing procedure with PBS74 and at least resuspend the pellet in 1 mL of 5% trehalose solution.
7. Storage of the produced particles is possible at 4°C for several weeks.

3.8 Particle Characterization

1. The particle size distribution is measured by dynamic light scattering (DLS) using a Zetasizer 3000HSA (Malvern Instruments, UK) as well as the measurement of the zeta potential via microelectrophoresis. Only distilled/deionized water (Milli-Q-grade water purification system, TKA, Germany) is used for all analytical measurements. The values shown in Table 1 are an average of five measurements. The hydrodynamic particle diameter is determined at a fixed scattering angle of 90°. Measurements are performed with 500 μg particles dispersed in 3 mL Milli-Q grade water for size distribution and buffer solution for zeta potential distribution.
2. Electron microscopic investigation is done with a field emission scanning electron microscope Leo 1530 VP Gemini (Carl Zeiss AG, Germany). The picture shown in Fig. 1 is recorded with a 1,000× magnification. For gaining a particle monolayer, the suspension is diluted to 100 μg/mL with Milli-Q grade water.
3. The fluorescence image is obtained using an Olympus BX60 fluorescence microscope fitted with a 2.5× lens, an HQ-filter set for Cy2 sel, Alexa 488, and a soft imaging system color view camera. Illumination of the nanoparticles is done at 495 nm with a mercury lamp; emission is collected at 519 nm. The concentration of the sample is 10 mg/mL in Milli-Q grade water.

Table 1
Hydrodynamic diameter determined by dynamic light scattering and zeta potential measured via microelectrophoresis both of initial silica nanoparticles, after surface modification with (3-aminopropyl)triethoxysilane (APTES) and after ongoing functionalization of amino-modified nanoparticles with succinic anhydride

	$\langle d_{DLS} \rangle$ (nm)	Zeta potential (mV)	
		pH = 4.7	pH = 7.0
Silica	108	−14	−21
Amino-silica	94	+48	−16
Carboxyl-silica	123	−24	−34

4. The detection of protein concentration is done spectrophotometrically by comparison with known protein standards via bicinchoninic acid (BCA) assay. A set of protein standards has to be prepared with concentrations between 2 and 0 mg/mL (blank) preferably using the same diluent as the sample. 20 μL of each standards and sample are pipetted into individual wells of a 96-microwell plate to which 200 μL of "BCA" reagent is added. Incubation time is 30 min at 37°C. After cooling down to room temperature, absorbance is measured at 562 nm (see Note 15).

4 Notes

1. Putting some water inside of the beaker helps to dissolve the salts very quickly.
2. Holding ammonium hydroxide solution close to the reaction bottle while you are pipetting helps avoiding spillage and makes sure that the correct amount will be introduced into the reaction mixture. Because of the high vapor pressure of the ammonium hydroxide solution, it will drop out of a graduated pipette.
3. Work under a hood while pipetting the ammonium hydroxide solution. It is harmful when inhaling it and will burn your anatomical airway.
4. Join the two pellets to one by using only the amount of 100 mL Milli-Q grade water. By doing that, you are able to keep the working volume small. So put only about 50 mL of water into the first centrifugal device and resuspend the pellet by using ultrasonic sound. Make sure that there are no clusters left in the suspension. Always check that by overturning the device and observing the suspension by bright light. If there are no clusters left, add the suspension to the second pellet and start resuspending that as well.
5. Use the remaining 50 mL of Milli-Q grade water to rinse the centrifugal devices. It is better to use smaller volumes for rinsing and repeat the procedure 2–3 times.
6. It is not necessary to change the water of the dialysis more than two times a day. The system needs approximately 8–10 h to equilibrate.
7. One hour reaction time is perfect for saturating the particles' surface with amine groups. If the reaction time is exceeded, the particles will start to agglomerate because of the excess amount of silane inside the reaction mixture.
8. Centrifugation devices made of teflon are necessary because of the high basic pH inside the suspension, which would destroy other plastic material like polypropylene.

9. By working with less volume in many centrifugal devices, the purification effect is better. You will use more buffering solution for the same amount of particles. Again after your last centrifugation, join the four pellets with little amount of ACB47 to gain one particle suspension.
10. Working with teflon centrifugal devices makes it harder to resuspend particle pellets with ultrasonic sound. The material is soft and the energy can hardly penetrate the devices. Switching to harder centrifugal devices allows centrifugation at higher speed velocity and makes resuspension much easier.
11. The particles are totally discharged inside of this media, so it is normal that the suspension is not stable and the particles look as they seem to agglomerate. Make sure that no particle suspension will touch the inner surface of the bottle when pouring the suspended particles inside. The solvent will dry immediately, and there will be no way to bring the dried particles back into the reaction suspension after that. So they will not be modified.
12. Storage of the prepared dye is possible at −18°C for several months.
13. EDC should be dissolved in Milli-Q grade water not until it is needed for the coupling trial. Otherwise hydrolysis of EDC will take place, and the activation of the carboxy groups may be prevented.
14. For destabilizing the particle pellet, knock the centrifugal device at a table several times. That will reduce the time needed for resuspension of the pellet in an ultrasonic bath and preserve the coupled antibody at the particles' surface.
15. For additional information about the BCA assay, please visit www.thermo.com/pierce.

References

1. Elwing H, Clark SR, Billsten P (1994) A fluorescence technique for investigating protein adsorption phenomena at a colloidal silica surface. Colloid Surf B Biointerfaces 2:457–461
2. Künzelmann U, Böttcher H (1997) Biosensor properties of glucose oxidase immobilized within SiO_2 gels. Sensor Actuat B Chem 38–39:222–228
3. Böttcher H, Slowik P, Suess W (1998) Sol–gel carrier systems for controlled drug delivery. J Sol–gel Sci Technol 13:227–281
4. Scheurich P et al (2005) Tumor necrosis factor (TNF)-functionalized nanostructured particles for the stimulation of membrane TNF-specific cell responses. Bioconjugate Chem 16: 1459–1467
5. Borchers K et al (2005) Microstructured layers of spherical biofunctional core-shell nanoparticles provide enlarged reactive surfaces for protein microarrays. Anal Bioanal Chem 383:738–746
6. Vaccarin A, Pifferi PG (1978) New solvents for paper and silica Gel thin-layer chromatography of anthocyanins. Chromatographia 11:193–196
7. Stöber W, Fink A (1968) Controlled growth of monodispersed silica spheres in the micron size range. J Colloid Interface Sci 26:62–69
8. Schiestel T, Brunner H, Tovar GEM (2004) Controlled surface functionalization of silica nanospheres by covalent conjugation reactions and preparation of high density streptavidin nanoparticles. J Nanosci Nanotechnol 4: 504–5011

9. Wang L, Zhao W, Tan W (2008) Bioconjugated silica nanoparticles: development and applications. Nano Res 1:99–115
10. Herz M et al. (2009) In vitro study of mouse fibroblast tumor cells with TNF coated and Alexa 488 marked silica nanoparticles with an endoscopic device for real time cancer visualization. Mater Res Soc Symp Proc 1190
11. Hermanson GT (2008) Bioconjugate techniques, 2nd edn. Elsevier, Amsterdam

Chapter 28

Direct Quantification of PTD Transduction Using Real-Time Monitoring

Mi-Sook Lee and Song Her

Abstract

Protein transduction domains (PTD or cell-permeable proteins) have attracted much attention as drug carriers because of their ability to penetrate cellular membranes. Although numerous PTD have been identified and their properties elucidated, their mechanism of action has not been fully understood due to the absence of a reliable quantification method. This chapter provides a direct method for quantifying cellular transduction of PTD in vitro and in vivo using bioluminescence imaging (BLI). This methodology exploits noninvasive techniques to create an environment suitable for the real-time imaging of PTD transduction and is therefore a promising tool for studying the mechanism of PTD transduction and the in vivo application of new therapeutic candidates.

Key words Drug carrier, Protein transduction domains, Cell-permeable proteins, Cellular transduction, Bioluminescence imaging, Real-time imaging

1 Introduction

Protein transduction domains (PTD) have attracted increasing attention in intracellular therapeutic protein delivery (1), and quantifying cellular transduction of PTD is important for comprehending both their mode of transduction and the efficacy of new drugs (2). Current approaches for quantifying PTD transduction, including flow cytometry and confocal fluorescence microscopy, are based on the fluorescent labeling of peptides, but flow cytometry can lead to false-positive results originating from cell surface-bound peptides (3). Also, fluorescence imaging by confocal fluorescence microscopy, which can monitor the subcellular localization of peptides and discriminate between internalized and extracellular fluorescent peptides, is limited by statistical problems (4). Alternatively, direct peptide detection by MALDI-TOF using isotope-labeled peptide was recently reported (5, 6). Although this method allows for the direct quantification of cell-permeable

Volkmar Weissig et al. (eds.), *Cellular and Subcellular Nanotechnology: Methods and Protocols*, Methods in Molecular Biology, vol. 991, DOI 10.1007/978-1-62703-336-7_28, © Springer Science+Business Media New York 2013

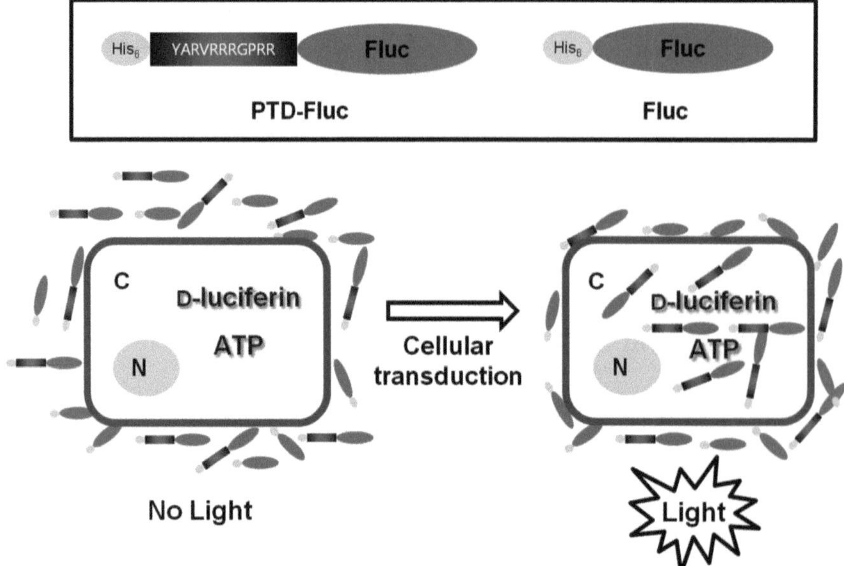

Fig. 1 Schematic illustration of bioluminescence in living cells following PTD-Fluc transduction. *Inset*: Structures of PTD-Fluc containing 11 amino acids (YARVRRRGPRR) and Fluc as a control. After PTD-Fluc transduction, luminescence is emitted by an ATP-dependent luciferin–luciferase reaction (*C* cytosol, *N* nucleus)

proteins and the discrimination of extracellular membrane-bound and intracellular peptides, it can only be used with cell lysates and is not suitable for in vivo measurements.

Bioluminescence imaging (BLI) based on luciferase activity is a rapid and sensitive method for in vitro and in vivo studies of ongoing biological phenomena. The most commonly used luciferase for BLI is firefly luciferase (Fluc), which requires ATP-Mg^{2+} and oxygen in the presence of the substrate, D-luciferin, to produce bioluminescence (7). Using firefly luciferase-tagged PTD (PTD-Fluc), we quantified internalized PTD by the real-time monitoring of ATP-dependent luciferase activity in vitro and in vivo (Fig. 1). This chapter provides a basic protocol for the real-time quantification of PTD transduction using BLI.

2 Materials

2.1 Luciferase Assay of Purified Protein

1. Purified PTD-Fluc (8) (see Note 1).
2. Purified Fluc (8) (see Note 1).
3. Luciferin substrate solution: Combine 1 mM D-luciferin (Xenogen, Alameda, CA, USA) (see Subheading 2.2), 3 mM ATP (ATP disodium salt), and 15 mM $MgSO_4$ in 30 mM HEPES (pH 7.8) using fresh, deionized ATP-free water. Store the solution at −20°C in polypropylene or glass tubes.
4. 96-well black microplate with a clear bottom.

2.2 PTD-Fluc Transduction In Vitro

1. HeLa cells or other biologically relevant cell lines of interest.
2. Dulbecco's modified Eagle's medium (DMEM) complete media: DMEM supplemented with 10% (v/v) fetal bovine serum and 1× (v/v) pen/strep (100× antibiotic solution: 10,000 U of penicillin and 10,000 U of streptomycin).
3. 1× Phosphate-buffered saline without Mg^{2+} and Ca^{2+}.
4. D-Luciferin firefly (potassium salt, MW = 318.42, #XR-1001; Xenogen) stock solution: Reconstitute a 1.0 g of D-luciferin in 33.3 ml of sterile 1× PBS without Mg^{2+} and Ca^{2+} to make a 30 mg/ml (0.1 M) stock solution. Filter sterilize through a 0.2-μm syringe filter. Store in aliquots at −20°C and protect from light (see Note 2).
5. Syringe filter, 0.2 μm.

2.3 PTD-Fluc Transduction In Vivo

1. Male 4-week-old ICR mice weighing approximately 30 g each (SPF grade).
2. 0.5-cc, 27-gauge insulin syringes.

2.4 BLI Equipment (Xenogen)

1. Isoflurane anesthesia chamber.
2. IVIS-200.
3. Caliper/Xenogen IVIS® Living Image software, version 3.0.

3 Methods

3.1 Cell-Free Luciferase Assay

Prior to the in vitro or in vivo analysis of PTD transduction, the following protocol using PTD-Fluc and Fluc should be carried out to accurately determine the activity and concentration of each protein.

1. Bring the luciferin substrate solution to room temperature before starting.
2. Transfer 10 μl each of PTD-Fluc and Fluc in Ni-NTA elute buffer to a 96-well black microplate. Commercially available purified Fluc can be used as a standard for calibration (see Note 3).
3. Add 90 μl of D-luciferin substrate solution and use the substrate solution without purified protein as a blank.
4. Immediately acquire an image using the IVIS-200 imaging system (*see* Subheading 3.2, step 7).
5. Generate a luciferase standard curve for light emission (photons/second, p/s) vs. concentration (ng/ml).
6. Based on the standard curve, determine the concentration of each purified protein (see Note 4).

3.2 Transduction of PTD-Fluc In Vitro

The following quantification protocol based on the ATP-dependent luciferase reaction allows the direct and real-time measurement of transduction in live cells without interfering with surface-bound PTDs (Fig. 1).

1. Seed HeLa cells in a 96-well black microplate in 100 μl of DMEM complete media at 10,000 cells per well.
2. Culture the cells in a humidified atmosphere at 37°C under 5% CO_2 for 20–24 h.
3. Prepare 100 μl of fresh PTD-Fluc dissolved in DMEM complete media to a final concentration of 1–500 nM. Also, prepare fresh Fluc solution in DMEM complete media (see Note 5).
4. Remove the media from the cells and wash them once with pre-warmed DMEM complete media, taking care not to detach the cells.
5. Add 100 μl of PTD-Fluc or control Fluc simultaneously using a multichannel pipette to the appropriate wells (see Notes 6 and 7).
6. At suitable time points, wash the cells twice with 1× PBS without Mg^{2+} and Ca^{2+} to remove surface-bound proteins (see Note 8).
7. To image Fluc, add 100 μl of D-luciferin stock solution to each well at a final concentration of 150 μg/ml using a multichannel pipette.
8. Image initially at 1–5 s, 10 bin, f/4 using Living Image software, version 3.0. The imaging times and binning can then be adjusted accordingly (see Note 9).

3.3 Transduction of PTD-Fluc In Vivo

The results of real-time in vivo monitoring of PTD transduction are illustrated in Fig. 3.

1. Prepare 100 μl of fresh PTD-Fluc dissolved in PBS (50 μM stock solution) to final dose range of 0.1–10 μM/kg body weight. Also prepare fresh Fluc in PBS at the same concentration as PTD-Fluc (see Note 5).
2. Inject 100 μl of PTD-Fluc or Fluc intraperitoneally into unanesthetized male ICR mice using a 27-gauge needle (see Notes 10 and 11).
3. Immediately after injection of each protein, administer 30 mg of D-luciferin/kg body weight intraperitoneally using a 27-gauge needle (see Note 12).
4. Ten minutes after the injection of D-luciferin, anesthetize the mice in an isoflurane induction chamber for 1–2 min (mean time until loss of the righting reflex) and then transfer to an IVIS-200 imaging chamber (see Note 13).
5. Take a brief test image at 1 min, 10 bin, f/8. Use this image to estimate the exposure time and binning needed for subsequent

images (usually a 1–5-min exposure). Quantify the bioluminescence (p/s) within a region of interest encompassing the specific area of bioluminescence.

6. Repeat steps 3–5 every 2 h (see Note 14).

4 Notes

1. Hexahistidine (His_6)-tagged PTD-Fluc and Fluc expressed in *Escherichia coli* BL21 (DE3) cells harboring pRSET-PTD-Fluc or pRSET-Fluc should be purified using an Ni-NTA Fast Start Kit (#30600, Qiagen) (8, 9). Although affinity tags using His_6 are powerful and convenient tools for the purification of recombinant proteins, you can also use genetically engineered fusion partners such as glutathione *S*-transferase (GST), FLAG, or maltose-binding protein (MBP) (10); however, you should check whether the tag has a deleterious effect on the biological properties of the recombinant protein (e.g., effects on solubility and biological activity).
2. D-Luciferin (potassium salt) can be dissolved up to a concentration of 50 mg/ml (62.8 mM); it precipitates out at 60 mg/ml.
3. To establish a standard luciferase calibration curve, make a series of luciferase solutions (10 μl each) ranging in concentration from 1 pg/ml to 1 ng/ml in elution buffer. Alternatively, each sample can be serially diluted by a factor of 10.
4. Keep the purified proteins (if possible, maintain at >10 μM since a high concentration will help prevent a loss of protein stability) in an Ni-NTA elute buffer supplemented with 10% glycerol (#30600, Qiagen) at −80°C in small aliquots for up to a few days. Do not thaw and refreeze; whenever possible, use either fresh or one-time-thawed proteins.
5. Make sure that the activity and concentration of the PTD-Fluc protein are accurate and essentially no different from those of the control (Fluc). You should always perform a cell-free luciferase assay (see Subheading 3.1) and check the activity and concentration of each sample prior to running PTD transduction experiments.
6. Treating the PTD proteins at the same time is important. PTD-Fluc may appear to be internalized very rapidly (within minutes), although the response to treatment is dependent on cell type. Figure 2 illustrates the real-time monitoring of PTD transduction in living cells.
7. Check the viability of the cells under a microscope before proceeding to the next step. The criteria for determining the maximal PTD-Fluc concentration depend on cellular viability.
8. If you wish to test for PTD transduction in a matter of minutes, this step can be skipped (i.e., go on to BLI) because

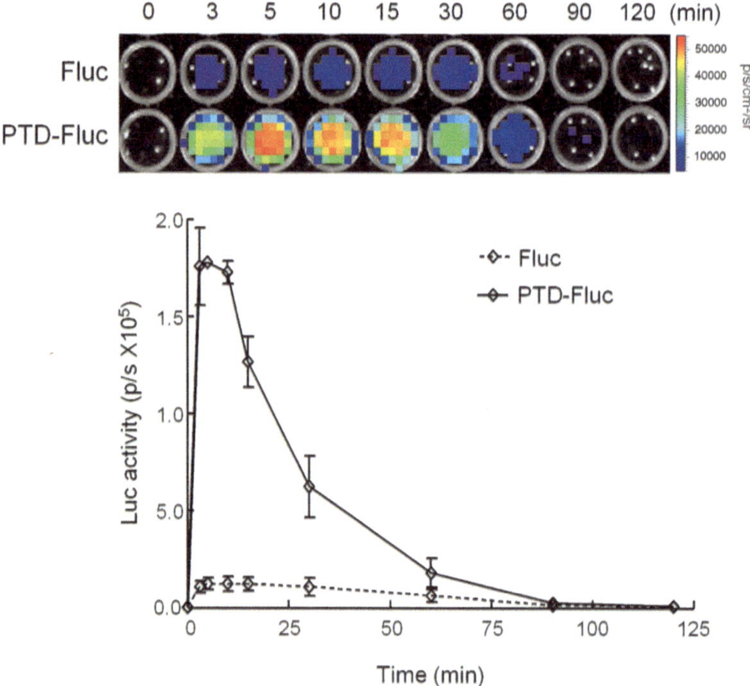

Fig. 2 Real-time imaging and quantification of PTD-Fluc transduction. Time-course representative image (*upper panel*) and quantification of luciferase activity in the presence of 40 nM Fluc or PTD-Fluc (*lower panel*). The data represent the mean ± SEM for each assay, performed in triplicate with three independent experiments. *Colored bars* indicate the bioluminescence signal intensity (p/s/cm^2/sr) (Reproduced with permission from ref. 8)

ATP-free extracellular or surface-bound PTDs may not affect the luciferase reaction. This may be confirmed by comparing the results of PTD-Fluc and control (Fluc) transduction (Fig. 2).

9. The level of bioluminescence is directly proportional to the exposure time, depending on the level of PTD-Fluc transduction. If you are unsure of what exposure time to use, start with low-sensitivity settings and increase as necessary (typical range, 0.5–1 min).

10. In general, the maximum intraperitoneal injection volume should not exceed 10 ml/kg body weight for adult mice.

11. Intraperitoneal administration is the injection of a substance into the peritoneal cavity. Note that intraperitoneal delivery is difficult to perform correctly, as you can easily misplace the dose into the intestine, gut, urinary bladder, muscle, or other organs. To avoid puncturing the abdominal viscera, hold the animal with its head tilted downward and insert the needle quickly into the lower left of the midline umbilicus.

12. The first image capture after PTD-Fluc injection should be delayed by approximately 15 min because it takes time to reach peak luminescence output from intact cells after D-luciferin injection (Fig. 3).

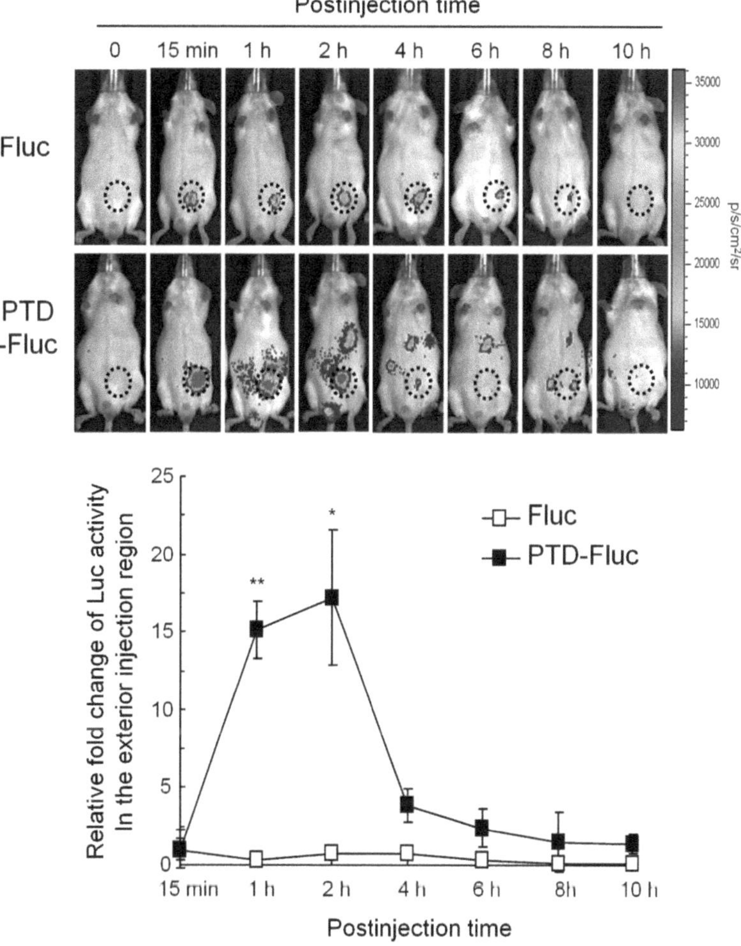

Fig. 3 In vivo real-time imaging and quantification of PTD-Fluc transduction. ICR mice were injected intraperitoneally once with 10 μM/kg PTD-Fluc or Fluc. Next, the mice were given an injection of D-luciferin at 30 mg/kg after 15 min, 1, 2, 4, 6, 8, and 10 h. Anesthesia and BLI were performed as in Subheading 3.3, steps 4 and 5. Representative images from the same mouse in the Fluc-treated group ($n=4$) and PTD-Fluc-treated group ($n=6$) are displayed as pseudocolor images of peak bioluminescence, with variations in color representing the light intensity at a given location (*upper panel*). *Red* represents the most intense light emission, while *blue* corresponds to the weakest signal. The *colored bar* indicates the bioluminescence signal intensity (p/s/cm^2/sr). The mice were imaged with an integration time of 3 min at a binning of 10. The *dotted circle* indicates the region of injection. Quantification of in vivo tracking by measuring the exterior luciferase activity (*lower panel*). Exterior activity was calculated by subtracting the activity in the *dotted circle* from the total activity. *Error bars* represent the SEM. *Significantly different from the Fluc-injected control group ($p<0.05$); **$p<0.01$ (Reproduced with permission from ref. 8)

13. The mice should be anesthetized with 3.0% isoflurane delivered in 100% oxygen in a gas anesthesia induction chamber prior to imaging. Once the mice have been anesthetized, they should be moved onto the imaging stage; anesthesia should be maintained with 1.0–3.0% isoflurane in 100% oxygen inside the IVIS-200.
14. One of the advantages of BLI is that a mouse can be imaged repetitively (e.g., before and after drug administration). This strategy allows each mouse to serve as its own control, which reduces experimental variation.

Acknowledgment

This study was supported by a KBSI grant (T31901) to S.H.

References

1. Johnson RM, Harrison SD, Maclean D (2011) Therapeutic applications of cell-penetrating peptides. Methods Mol Biol 683:535–551
2. Jarver P, Mager I, Langel U (2010) In vivo biodistribution and efficacy of peptide mediated delivery. Trends Pharmacol Sci 31:528–535
3. Richard JP, Melikov K, Vives E et al (2003) Cell-penetrating peptides. A reevaluation of the mechanism of cellular uptake. J Biol Chem 278:585–590
4. Fischer R, Waizenegger T, Kohler K et al (2002) A quantitative validation of fluorophore-labelled cell-permeable peptide conjugates: fluorophore and cargo dependence of import. Biochim Biophys Acta 1564:365–374
5. Burlina F, Sagan S, Bolbach G et al (2005) Quantification of the cellular uptake of cell-penetrating peptides by MALDI-TOF mass spectrometry. Angew Chem Int Ed Engl 44:4244–4247
6. Balayssac S, Burlina F, Convert O et al (2006) Comparison of penetratin and other homeodomain-derived cell-penetrating peptides: interaction in a membrane-mimicking environment and cellular uptake efficiency. Biochemistry 45:1408–1420
7. Greer LF 3rd, Szalay AA (2002) Imaging of light emission from the expression of luciferases in living cells and organisms: a review. Luminescence 17:43–74
8. Lee MS, Kwon EH, Choi HS et al (2010) Quantification of cellular uptake and in vivo tracking of transduction using real-time monitoring. Biochem Biophys Res Commun 394:348–353
9. Choi JM, Ahn MH, Chae WJ et al (2006) Intranasal delivery of the cytoplasmic domain of CTLA-4 using a novel protein transduction domain prevents allergic inflammation. Nat Med 12:574–579
10. Murphy MB, Doyle SA (2005) High-throughput purification of hexahistidine-tagged proteins expressed in *E. coli*. Methods Mol Biol 310:123–130

Chapter 29

Genotoxic Assessment of Carbon Nanotubes

Olivera Nešković, Gordana Joksić, Ana Valenta-Šobot, Jelena Cvetićanin, Djordje Trpkov, Andreja Leskovac, and Sandra Petrović

Abstract

Carbon nanotubes are unique one-dimensional macromolecules with promising application in biology and medicine. Since their toxicity is still under debate, here we describe an investigation of genotoxic properties of purified single-walled carbon nanotubes (SWCNT), multiwall carbon nanotubes (MWCNT), and amide-functionalized purified SWCNT. We used two different cell systems: cultured human lymphocytes where we employed cytokinesis-block micronucleus test and human fibroblasts where we investigate the induction of DNA double-strand breaks (DSBs) employing H2AX phosphorylation assay.

Key words Carbon nanotubes, Genotoxicity, Human cells, Micronuclei, γ-H2AX foci

1 Introduction

Carbon nanotubes (CNT) are unique, one-dimensional macromolecules, whose outstanding properties have sparked an abundance of research since their discovery in 1991 (1). Single-walled carbon nanotubes (SWCNT) are constructed of a single sheet of graphite (diameter 0.4–10 nm), while multiwall carbon nanotubes (MWCNT) consist of multiple concentric graphite cylinders of increasing diameter (10–100 nm) (2). Both SWCNT and MWCNT possess high tensile strengths, are ultralight weight, and have excellent thermal and chemical stability. In combination with their metallic and semiconductive electronic properties, this remarkable array of features has seen a plethora of applications proposed.

One of the major areas of CNT research is the field of biomedical materials and devices. Many applications for CNT have been proposed including biosensors, drug and vaccine delivery vehicles, and novel biomaterials (2). CNT can be used as nanofillers in existing polymeric materials to both dramatically improve mechanical properties and create highly anisotropic nanocomposites

Volkmar Weissig et al. (eds.), *Cellular and Subcellular Nanotechnology: Methods and Protocols*, Methods in Molecular Biology, vol. 991, DOI 10.1007/978-1-62703-336-7_29, © Springer Science+Business Media New York 2013

(3, 4). They can also be used to create electrically conductive polymers and tissue engineering constructs with the capacity to provide controlled electrical stimulation (5–7). However, before such materials can be incorporated into new and existing biomedical devices, the toxicity and biocompatibility of CNT needs to be thoroughly investigated. Within the realm of biotechnology, carbon nanotubes, a major class of carbon-based tubular nanostructures, have been utilized as platforms for ultrasensitive recognition of antibodies (8) as nucleic acids sequencers (9) and as bioseperators, biocatalysts (10), and ion channel blockers (11) for facilitating biochemical reactions and biological processes. Towards nanomedicine, an emerging field of utilizing nanomaterials for novel and alternative diagnostics and therapeutics has been developed. CNT have been utilized as scaffolds for neuronal and ligamentous tissue growth for regenerative interventions of the central nervous system and orthopedic sites (12), substrates for detecting antibodies associated with human autoimmune diseases with high specificity (13), and carriers of contrast agent aquated Gd^{3+}-ion clusters for enhanced magnetic resonance imaging (14). When coated with nucleic acids (DNA or RNA) or proteins, CNT have been shown as effective substrates for gene sequencing and as gene and drug delivery vectors to challenge conventional viral and particulate delivery systems (15–19). Consequently, efforts to take advantage of the physical and chemical properties of CNT in biological settings must first circumvent the hydrophobicity of these nanomaterials.

Research over the past decade has shown that CNT and fullerenes can be readily modified, either covalently or non-covalently, by incorporating chemical and biological functional groups for much enhanced solubility and bioavailability. The covalent modification of single-walled carbon nanotubes, for example, normally involves esterification or amidation of acid-oxidized nanotubes and sidewall covalent attachment of functional groups (20–24). However, these covalent schemes are often marred by undesirable modifications to the physical and chemical properties of SWCNT (25). Furthermore, such functionalized SWCNT often have dangling bonds at the defective sites and are prone to generating free radicals. In comparison, the non-covalent modifications of SWCNT employ adsorption of proteins, biopolymers and synthetic polymers (DNA, RNA, polyvinyl pyrrolidone, polystyrene sulfonate), and surfactants (sodium dodecyl sulfate or SDS) to form supramolecular assemblies (8, 26–34). For both covalent and non-covalent solubilization schemes, the introduction of surfactants, surface charges, organic solvents, and residues may induce additional genotoxicity. Developing well-characterized solubilization schemes on the base of functionalized nanotubes is thus crucial for facilitating the full range biological and biomedicinal applications of nanotubes and their derivatives (35).

Structural characterization is necessary while investigating the influence of carbon nanotubes on the living systems, since previous investigations showed that the genotoxicity level is different depending of the length, diameter, total surface, irregularities in the lattice organization, purity and presence of the functional groups, and the solubility of carbon nanotubes (36).

Considering numerous possible interactions between CNT and biomolecules, as well as their existing and potential applications in biological sciences and other fields of science and technology, it is of great interest to determine their influence on living systems.

Here we present a study investigating genotoxic properties of carbon nanotubes: purified SWCNT, MWCNT, and amide-functionalized purified SWCNT on cultured human lymphocytes employing cytokinesis-block micronucleus test and enumeration of γ-H2AX foci in human fibroblast cell line.

2 Materials

2.1 Laboratory Equipment

1. Laminar flow cabinet.
2. Ultrasonic bath.
3. Incubator.
4. Water bath.
5. Centrifuge.
6. Vortex mixer.
7. Micropipette.
8. Microscope.
9. Microscope slide staining jar.
10. Microscope with illuminator for fluorescence microscopy.

2.2 Solutions

1. Carbon nanotubes dispersion: Disperse 5 mg of each carbon nanotube sample (SWCNT, MWCNT, and amide-functionalized purified SWCNT) into 1 ml of 96% ethanol. Sterilize in an autoclave. Sonicate prior to use for 30 min.
2. Cytochalasin B stock solution: Dissolve 10 mg cytochalasin B into 33.3 ml DMSO in laminar hood. Filter DMSO with single-use syringe filter with pore sizes 0.20 μm. Aliquot in 2 ml safeseal micro tubes (see Note 1).
3. Hypotonic solution: Mix equal volumes of 0.56% KCl and 0.9% NaCl water solutions. Calculate final volume according to number of specimens. Keep in wash bottle in water bath at 37°C (see Note 2).

4. Fixative solution: In 3× volume of methanol (CH_3OH), add 1× volume of acetic acid (CH_3COOH).
5. SØRENSEN'S phosphate buffer: (Stock solutions: (a) 0.2 M Na_2HPO_4 and (b) 0.2 M NaH_2PO_4) To prepare 100 ml of working buffer (0.1 M, pH 6.8), mix 24.5 ml of (a) with 25.5 ml of (b). Dilute to 100 ml with ddH_2O.
6. 10% Giemsa in SØRENSEN'S buffer: To prepare 100 ml, add 10 ml Giemsa stain to 90 ml of SØRENSEN'S buffer.
7. 1% HCl ethanol solution: To prepare 1 l, add 10 ml 37% HCl to 990 ml 95% ethanol.
8. 90% ethanol solution: To prepare 100 ml, add 10 ml ddH_2O to 90 ml 100% ethanol.
9. 70% ethanol solution: To prepare 100 ml, add 30 ml ddH_2O to 70 ml 100% ethanol.

2.3 Cell Culture Components

1. Human dermal fibroblasts HDMEC (PromoCell GmbH, Heidelberg, Germany).
2. T25 tissue culture flasks.
3. Dulbecco's modified Eagle's medium (DMEM).
4. Fetal bovine serum.
5. 0.05% tripsin-EDTA.
6. Polyprep glass slides.
7. Disposable Petri dishes (100 mm).
8. Lithium heparin BD Vacutainer® tubes with BD Hemogard® closure (Becton-Dickinson, Franklin Lakes, NJ, USA).
9. PB-MAX karyotyping medium (Invitrogen-Gibco, Paisley, UK).

2.4 Immunofluorescence (γ-H2AX Assay) Components

1. 1× phosphate-buffered saline (PBS): To prepare 1 l, add 8 g NaCl, 0.2 g KCl, 1.44 g Na_2HPO_4, and 0.24 g KH_2PO_4 to 800 ml deionized H_2O. Adjust pH to 7.4. Adjust volume to 1 l with deionized H_2O. Sterilize by autoclaving and store at room temperature.
2. 4% formaldehyde: 10 ml 37% formaldehyde dilute in 90 ml PBS. Prepare fresh (see Note 3).
3. 0.2% Triton-X: Add 200 μl Triton-X (Sigma-Aldrich Co., Steinheim, Germany) to 100 ml deionized PBS. Store at 4°C.
4. 1× TBS-T (Tris-buffered saline Tween-20) buffer: To prepare 1 l, add 8.8 g NaCl, 0.2 g KCl, 3 g Tris-base, and 500 μl Tween 20–800 ml ddH_2O. Adjust pH to 7.4. Adjust volume to 1 l with ddH_2O. Sterilize by autoclaving. Store at 4°C (see Note 4).
5. 0.5% BSA: Dissolve 0.05 g bovine serum albumin (BSA, Sigma-Aldrich) into 10 ml deionized PBS.

6. Primary antibody: Anti-phospho-H2AX (Ser 139) mouse monoclonal antibody (Upstate Cell Signaling Solutions), dilute in 0.5% BSA (v/v = 1:500).
7. Secondary antibody: Antimouse fluorescein isothiocyanate (FITC) antibody, dilute in 0.5% BSA (v/v = 1:400).
8. 4,6-Diamidino-2-phenylindole (DAPI)-containing antifade solution (Vector Laboratories Inc., Burlingame, CA, USA).
9. Coverslips (22 × 50 × 0.13 mm).
10. Clear, nonfluorescent nail varnish.

3 Methods

3.1 Cell Culture-Human Blood Cells

1. Collect fresh blood by venepuncture in 6 ml lithium heparin BD Vacutainer® tubes with BD Hemogard® closure. Store blood at 4°C prior to procedure.
2. Aliquot PB-max karyotyping medium (Invitrogen-Gibco, Paisley, UK), 4.5 ml in each 10 ml sterile tissue/culture test tube in laminar flow hood. Add 0.5 ml of whole blood. Close the cap of the test tube and put into the incubator.
3. Keep cell culture in an incubator at 37°C. One hour after the stimulation of cells, add the agent of interest to cultures: different volumes of carbon nanotubes dispersion (25, 50, 100, and 150 μl per 5 ml of total cell culture volume) and one untreated specimen as a control.

3.2 Micronuclei Preparation: Micronucleus Assay

1. Add 0.1 ml of the cytochalasin B solution after 44 h of culture, and then incubate for next 28 h.
2. Centrifuge cell suspension the next day (after 72 h of cell harvesting) at 500 × *g* for 10 min.
3. Remove the supernatant.
4. Resuspend pellet on vortex mixer and add pre-warmed hypo tonic solution.
5. Keep in water bath for 5 min at 37°C.
6. Centrifuge at 500 × *g* for 10 min.
7. Remove supernatant up to 1 ml and add fixative solution on a vortex mixer to 12 ml.
8. Leave samples at room temperature for 30 min.
9. Repeat steps 6 and 7 until the suspension is clear.
10. After last centrifugation, aspirate up to 0.5 ml. Resuspend pellet with Pasteur pipette and prepare slides (37).

3.3 Preparing Slides for Micronucleus Assay

1. Degrease slides with detergent, wash thoroughly with distilled water, and keep over night in 1% HCl ethanol solution. Prior to use, wash slides with distilled water and bi-distilled water.
2. Onto clean, dry slides, put three drops of the cell suspension.
3. Air-dry the slides.
4. Stain slides (in staining jar) with 10% Giemsa in SØRENSEN's buffer (pH 6.8) for 10 min.

3.4 Slide Scoring for Micronucleus Assay

1. Score at least 1,000 binuclear (BN) cells per sample. Analyze slides with a microscope using magnification 400× or 1,000× when necessary.
2. Score a minimum of 1,000 binucleated cells to evaluate the percentage of cells with one, two, three, four, or more than four micronuclei.
3. Calculate a cytokinesis-block proliferation index (CBPI) as follows: CBPI = MI + 2MII + 3(MIII + MIV)/N, where MI–MIV represents the number of cells with one to four nuclei, respectively, and N is the number of cells scored (38).

3.5 Cell Culture-Human Fibroblasts

1. Incubate normal human dermal fibroblasts HDMEC in tissue culture flask for 48 h under standard tissue culture conditions in DMEM, supplemented with 10% of fetal bovine serum at 37°C and in the atmosphere of 10% CO_2.
2. When reach 80–90% confluence, remove growth medium from the flask by aspiration.
3. Incubate the flask with 1 ml 0.05% Trypsin-EDTA at 37°C for 4–7 min. Examine the flasks microscopically to make sure the cells begin to round. The cells should detach from the flask surface after 7 min (see Note 5).
4. Tighten cap and lightly tap the side of the flask to lift the remaining cells from the flask. Wash the sides of the flask with growth medium to inactivate the trypsin (see Note 6). Gently mix cells and medium. Pipette the cell suspension up and down so as to obtain a suspension of individual cells.

3.6 Immunofluorescence (γ-H2AX Assay)

1. Distribute 1 ml aliquots of the cell suspension to polyprep slides (see Note 7).
2. Transfer polyprep slides into Petri dishes (see Note 8).
3. Add appropriate dose of carbon nanotubes (0.5–30 μl per/ml) into cell suspension seeded on the polyprep slide. Close the dish and incubate at 37°C in a humid atmosphere for the next 24 h (see Note 9).
4. Wash slides in PBS for 5 min.

5. Fix cells in fresh 4% formaldehyde (see Note 3) for 15 min.
6. Permeabilize cells in 0.2% Triton-X at 4°C for 10 min.
7. Block reaction by transferring slides in 0.5% BSA/PBS for 30 min, at room temperature.
8. While blocking, prepare primary antibody.
9. Aspirate blocking solution and apply 100 μl diluted primary antibody to each slide. Incubate for 1 h in a light-tight damp container.
10. Wash slides three times in TBS-T for 3 min each.
11. Incubate slides with 100 μl secondary anti-goat antibody conjugated with Fluorescein isothiocyanate (FITC) for 2 h in a light-tight damp container.
12. Wash slides three times in TBS-T for 3 min each.
13. Fix cells with 70, 90, and 100% ethanol, 5 min each, and air-dry in the dark.
14. Counter stain cells with 15 μl 4′,6′-diamidino-2-phenylindole (DAPI)-containing antifade solution and cover with coverslips. Apply nail varnish to seal the samples.

    For the best results, examine specimens immediately. For long-term storage (several weeks), store slides at 4°C protected from light.
15. At least 200 cells should be analyzed to evaluate the number of γ-H2AX positive foci by using the microscope with illuminator for fluorescence microscopy and the computer software ImageJ.

4 Notes

1. Store at −20°C.
2. Make it prior to use. Use double distilled water.
3. Formaldehyde is toxic; use only in fume hood.
4. Adjustment of pH should be performed with 1 N HCl.
5. If the cells do not become detached after 7 min, incubate an additional 1–2 min.
6. Add growth medium at volume equal to or greater than volume of trypsin added.
7. Sterile pipette must be used.
8. One polyprep slide per Petri dish.
9. All subsequent incubations should be carried out at room temperature unless otherwise noted.

Acknowledgments

This work was financially supported by the Ministry of Education and Science of the Republic of Serbia, under Project No. III 45005 and Project No. 173046.

References

1. Iijima S (1991) Helical microtubules of graphitic carbon. Nature 354:56–58
2. Lin Y, Taylor S, Li H, Fernando SKA, Qu L, Wang W, Gu L, Zhou B, Sun YP (2004) Advances toward bioapplications of carbon nanotubes. J Mater Chem 14:527–541
3. Koerner H, Price G, Pearce NA, Alexander M, Vaia RA (2004) Remotely actuated polymer nanocomposites—stress-recovery of carbon-nanotube-filled thermoplastic elastomers. Nat Mater 3:115–120
4. Sen R, Zhao B, Perea D, Itkis ME, Hu H, Love J, Bekyarova E, Haddon RC (2004) Preparation of single-walled carbon nanotube reinforced polystyrene and polyurethane nanofibers and membranes by electrospinning. Nano Lett 4:459–464
5. Grunlan JC, Mehrabi AR, Bannon MV, Bahr JL (2004) Water-based single-walled-nanotube-filled polymer composite with an exceptionally low percolation threshold. Adv Mater 16:150–153
6. Huang JE, Li XH, Xu JC, Li HL (2003) Well-dispersed single-walled carbon nanotube/polyaniline composite films. Carbon 41:2731–2736
7. Supronowicz PR, Ajayan PM, Ullmann KR, Arulanandam BP, Metzger DW, Bizios R (2002) Novel current-conducting composite substrates for exposing osteoblasts to alternating current stimulation. J Biomed Mater Res A 59:499–506
8. Chen RJ, Bangsaruntip S, Drouvalakis KA, Kam NWS, Shim M, Li Y, Kim W, Utz PJ, Dai H (2003) Noncovalent functionalization of carbon nanotubes for highly specific electronic biosensors. Proc Natl Acad Sci USA 100:4984–4989
9. Wang J, Liu G, Jan MR, Zhu Q (2003) Electrochemical detection of DNA hybridization based on carbon-nanotubes loaded with CdS tags. Electrochem Commun 5:1000–1004
10. Mitchell DT, Lee SB, Trofin L, Li N, Nevanen TK, Soderlund H, Martin CR (2002) Smart nanotubes for bioseparations and biocatalysis. J Am Chem Soc 124:11864–11865
11. Park KH, Chhowalla M, Iqbal Z, Sesti F (2003) Single-walled carbon nanotubes are a new class of ion channel blockers. J Biol Chem 278:50212–50216
12. Hu H, Ni Y, Montana V, Haddon RC, Parpura V (2004) Chemically functionalized carbon nanotubes as substrates for neuronal growth. Nano Lett 4:507–511
13. Wang J, Liu G, Jan MR (2004) Ultrasensitive electrical biosensing of proteins and DNA: carbon-nanotube derived amplification of the recognition and transduction events. J Am Chem Soc 126:3010–3011
14. Sitharaman B et al (2005) Superparamagnetic gadonanotubes are high-performance MRI contrast agents. Chem Commun 31:3915–3917
15. Pantarotto D, Briand JP, Prato M, Bianco A (2004) Translocation of bioactive peptides across cell membranes by carbon nanotubes. Chem Commun 1:16–17
16. Pantarotto D, Singh R, McCarthy D, Erhardt M, Braind JP, Prato M, Kostarelos K, Bianco A (2004) Functionalized carbon nanotubes for plasmid DNA gene delivery. Angew Chem Int Ed Engl 43:5242–5246
17. Kam NWS, Jessop TC, Wender PA, Dai HJ (2004) Nanotube molecular transporters: internalization of carbon nanotube - protein conjugates into mammalian cells. J Am Chem Soc 126:6850–6851
18. Liu Y, Wu DC, Zhang WD, Jiang X, He CB, Chung TS, Goh SH, Leong KW (2005) Polyethylenimine-grafted multiwalled carbon nanotubes for secure noncovalent immobilization and efficient delivery of DNA. Angew Chem Int Ed Engl 44:4782–4785
19. Lu Q, Moore JM, Huang G, Mount AS, Rao AM, Larcom LL, Ke PC (2004) RNA polymer translocation with single-walled carbon nanotubes. Nano Lett 4:2473–2477
20. Sano M, Kamino A, Okamura J, Shinkai S (2001) Self-organization of PEO-graft-single-walled carbon nanotubes in solutions and langmuir - blodgett films. Langmuir 17:5125–5128
21. Banerjee S, Wong SS (2002) Structural characterization, optical properties, and improved solubility of carbon nanotubes functionalized

with Wilkinson's catalyst. J Am Chem Soc 124:8940–8948
22. Pompeo F, Resasco DE (2002) Water solubilization of single-walled carbon nanotubes by functionalization with glucosamine. Nano Lett 2:369–373
23. Bahr JL, Mickelson ET, Bronikowski MJ, Smalley RE, Tour JM (2001) Dissolution of small diameter single-wall carbon nanotubes in organic solvents. Chem Commun 1:193–194
24. Sun Y, Wilson SR, Schuster DI (2001) High dissolution and strong light emission of carbon nanotubes in aromatic amine solvents. J Am Chem Soc 123:5348–5349
25. Pantarotto D, Partidos CD, Hoebeke J, Brown F, Kramer E, Briand JP, Muller S, Prato M, Bianco A (2003) Immunization with peptide-functionalized carbon nanotubes enhances virus-specific neutralizing antibody responses. Chem Biol 10:961–966
26. Shim M, Kam NWS, Chen RJ, Li Y, Dai H (2002) Functionalization of carbon nanotubes for biocompatibility and biomolecular recognition. Nano Lett 2:285–288
27. Matarredona O, Rhoads H, Li Z, Harwell JH, Balzano L, Resasco DE (2003) Dispersion of single-walled carbon nanotubes in aqueous solutions of the anionic surfactant NaDDBS. J Phys Chem B 107:13357–13367
28. Yurekli K, Mitchell CA, Krishnamoorti R (2004) Small-angle neutron scattering from surfactant-assisted aqueous dispersions of carbon nanotubes. J Am Chem Soc 126:9902–9903
29. O'Connell MJ et al (2002) Band gap fluorescence from individual single-walled carbon nanotubes. Science 297:593–596
30. O'Connell MJ, Boul P, Ericson LM, Huffman C, Wang Y, Haroz E, Kuper C, Tour J, Ausman KD, Smalley RE (2001) Reversible water-solubilization of single-walled carbon nanotubes by polymer wrapping. Chem Phys Lett 342:265–271
31. Rao R, Lee J, Lu Q, Keskar G, Freedman KO, Floyd WC, Rao AM, Ke PC (2004) Single-molecule fluorescence microscopy and Raman spectroscopy studies of RNA bound carbon nanotubes. Appl Phys Lett 85:4228–4230
32. Strano MS, Zheng M, Jagota A, Onoa GB, Heller DA, Barone PW, Usrey ML (2004) Understanding the nature of the DNA-assisted separation of single-walled carbon nanotubes using fluorescence and Raman spectroscopy. Nano Lett 4:543–550
33. Wang H, Zhou W, Ho DL, Winey KI, Fischer JE, Glinka CJ, Hobbie EK (2004) Dispersing single-walled carbon nanotubes with surfactants: a small angle neutron scattering study. Nano Lett 4:1789–1793
34. Zheng M, Jagota A, Semke ED, Diner BA, McLean RS, Lustig SR, Richardson RE, Tassi NG (2003) DNA-assisted dispersion and separation of carbon nanotubes. Nat Mater 2:338–342
35. Smart SK, Cassady AI, Lu GQ, Martin DJ (2006) The biocompatibility of carbon nanotubes. Carbon 44:1034–1047
36. Zhu Y, Li WX (2008) Cytotoxicity of carbon nanotubes. Sci China Ser B Chem 51(11):1021–1029
37. Fenech M (2000) The in vitro micronucleus technique. Mutat Res 455:81–95
38. Fenech M (2003) Human micron nucleus project. Mutat Res 534(1–2):65–75

Chapter 30

Separation Science: Principles and Applications for the Analysis of Bionanoparticles by Asymmetrical Flow Field-Flow Fractionation (AF4)

Alexandre Moquin, Françoise M. Winnik, and Dusica Maysinger

Abstract

Field-flow fractionation is an analytical technique that allows the separation of particles over a size range, from a few nanometers to several microns in diameter. The separation takes place under mild conditions and is suited for the analysis of neutral or charged particles. A single measurement yields the size and concentration of each component of a mixture. However, developing a suitable fractionation method can be tedious and time-consuming. In this chapter, we present asymmetrical flow field-flow fractionation (AF4) conditions that have proven their reliability for the analysis of quantum dots and other nanoparticles in the 5–50 nm size range. Common pitfalls are emphasized together with strategies to overcome them.

Key words Asymmetrical flow field-flow fractionation, Quantum dots, Aggregation, Nanoparticles, Dynamic light scattering

1 Introduction

Field-flow fractionation (FFF) offers great versatility in the types of sample to be analyzed, and it provides a full sample characterization in a single measurement typically within 30 min (1–3). Various FFF subtechniques are available depending on the field applied to the sample (4–9). They each have specific advantages and drawbacks. For example, sedimentation FFF (SdFFF), in which the field is gravitational, has a much greater resolving power than the other subtechniques, but it is only applicable for particles larger than 100 nm that have a density significantly different from that of the carrier liquid. In flow field-flow fractionation (FlFFF), the field is provided by a crossflow applied perpendicularly to the elution flow. The acronym AF4 (asymmetrical flow field-flow fractionation) refers to FlFFF for which the channel is of trapezoidal shape instead of an elongated hexagon (10). Most commercial FlFFF systems are using AF4, in view of its versatility, relative simplicity, and its wide range in

Volkmar Weissig et al. (eds.), *Cellular and Subcellular Nanotechnology: Methods and Protocols*, Methods in Molecular Biology, vol. 991, DOI 10.1007/978-1-62703-336-7_30, © Springer Science+Business Media New York 2013

terms of particle size (11–13). In the following section, we present the principle of the fractionation and review the importance of each parameter as well as describe step-by-step the process to obtain a fractionation of quantum dots (QDs) and optimize the fractionation conditions.

Quantum dots are one of the most representative nanoparticles used in the nanosciences. They are nanometer-sized semiconductor crystals, which can be made with core or core-shell architecture, and have been extensively used for imaging applications both in vitro, at the single-cell level (14–16), and in vivo (17–23), mainly in animals bearing tumors. Their main attractiveness relies on the fact that their luminescence emission can be tuned by controlling the size and size distribution of the particles. Because, the emission maxima are directly related to the sizes, it is possible to have a qualitative appreciation of the size distribution of a QD sample by measuring its luminescent emission spectrum. FlFFF can provide quantitative data on the QD's (and other nanoparticles) size distribution, presence of aggregates, adsorbed proteins (opsonized particles), and nanoparticle decomposition (24).

1.1 Principle of Flow Field-Flow Fractionation

In FFF, the separation of the sample takes place inside a narrow ribbonlike channel clamped between two parallel surfaces through which a field can be applied (Fig. 1). A carrier liquid is pumped through the channel from the inlet (sample injection) to the outlet (detector). A parabolic flow profile (Newtonian flow) is established inside the channel, as in a capillary tube. Flow velocities vary from 0 on the walls to a maximum value in the center of the channel. A field is applied perpendicularly to the flow direction, while the carrier liquid containing the sample flows through the channel. This induces the transport of the sample towards one wall, creating a concentration gradient. A diffusion flux in the opposite direction is induced according to Fick's law leading to a steady state where each component of the sample reaches a position at a unique distance from the wall. Due to the parabolic flow profile, the components are transported in the direction of the longitudinal channel axis at varying velocities, depending on their distance from the channel walls. Since smaller objects diffuse at a faster rate than larger ones, the elution from the channel outlet proceeds from the smaller species to the larger ones. In AF4, three flows are used (Chapter 18 in (4, 10)): (1) a tip flow is introduced from the top of the channel, and it is through this flow that the sample is injected and carried down the channel; (2) a focus flow is introduced from the midsection of the channel or from the output of the channel, depending on the instrument. This flow creates a focusing zone as it meets the tip flow at the top of the channel. The distance between the input and the focusing zone is controlled by adjusting the rate of each flow; (3) a crossflow is applied perpendicularly to the direction of the channel, either passively or with a pump that sucks the

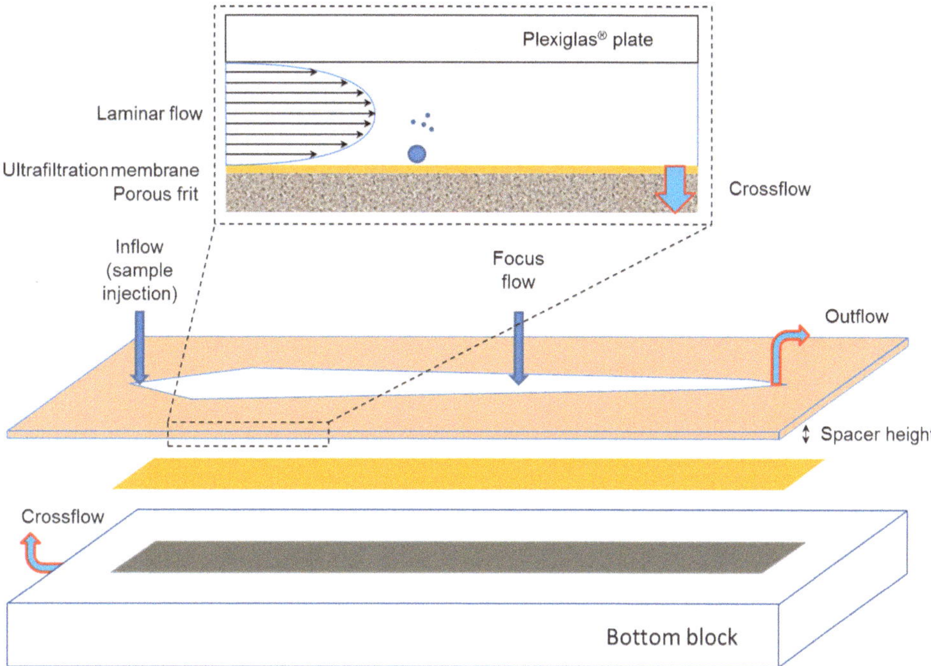

Fig. 1 Schematic representation of the channel used in asymmetrical flow field-flow fractionation. The *inset* illustrates the laminar flow in the channel

carrier liquid out through an ultrafiltration membrane placed on the bottom wall of the channel.

1.2 Features of the Channel

The channel is held in place by two blocks. At the channel bottom, there is a stainless steel block fitted with a permeable frit serving as a support for the membrane. A spacer sheet, 190–800 μm thick, is placed on top of the membrane. A trapezoidal area is cut out within this sheet to form the channel. The thickness of the spacer sheet determines the thickness of the channel, although the channel is slightly thinner than the spacer itself due to the compression imposed on the sheet by the two blocks. The actual channel volume can be determined experimentally by measuring the elution time of a sample which has a well-known diffusion coefficient at a given temperature (Chapter 18 in (4)). The second block, made of Plexiglas, forms an impermeable wall on top of the channel. The advantage of Plexiglas is that it offers a smooth surface and is transparent, allowing the user to measure the distance of the focusing line from the top of the channel and to check for air bubbles, leaks, disturbances in the flow, or sample adsorption on the membrane.

1.3 Features of the Membrane

Several membranes are available commercially (Table 1). The selection of the most suitable membrane for a specific sample is

Table 1
List of membrane materials available for analysis of different sample types in aqueous media

Membrane name	Use	MW cutoff (kDa)	Supplier	Catalog number
Regenerated cellulose (RC)	Analysis of proteins/ peptides/antibodies/ viruses/nanoparticles/ carbon nanomaterials	1/10/30	Postnova	Z-AF4-MEM-612
RC amphiphilic	Analysis of amphiphilic or cationic polymers	10	Postnova	Z-MEM-AQU-631
Polyethersulfone (PES)	Negatively charged samples/biopolymers/ polyelectrolytes	0.3/1/5/10/20	Postnova	Z-AF4-MEM-611
Cellulose triacetate (CTA)	Proteins/peptides/ antibodies/viruses	5/10/20	Postnova	Z-AF4-MEM-613
PVDF hydrophilic	Large hydrophilic particles/nanoparticles/carbon nanomaterials	30/200	Postnova	Z-AF4-MEM-614

based on the following considerations: (1) the membrane material must be compatible with the solvent (no swelling or dissolution); (2) the sample analyzed should not adsorb on the membrane; (3) the membrane surface should be flat, smooth, and devoid of ridges; (4) thick membranes should be avoided as they may protrude into the channel upon compression of the spacer, thereby causing channel clogging; the optimal membrane thickness is 250 μm; and (5) the molecular weight cutoff (MWCO) of the membrane is commonly set at 10 kDa. The use of membranes with smaller MWCO values results in an increase in the system pressure. If membranes of low MWCO are needed, the crossflow value should be decreased, keeping the ratio of the crossflow to detector flow (V_c/V_{out}) constant. Membranes with higher MWCO values may lead to significant loss of small species in a sample. The use of an inappropriate membrane may lead to loss of signal upon sample injection, an indication that the sample is adsorbed onto the membrane. This problem can be solved by conditioning the membrane: several injections of the same sample are performed until the membrane surface is saturated, such that the subsequent runs will lead to full sample recovery. It is also possible to choose a carrier liquid with a different ionic strength and pH or with an added surfactant. For charged samples, a membrane with the same charge should be used, to prevent electrostatically induced adsorption of the sample to the membrane.

2 Materials

2.1 Apparatus

AF4 experiments were carried out on an AF2000-MT (Postnova Analytics GmbH, Landsberg, Germany), equipped with a channel oven for temperature control in the mid-temperature range (MT: 5–80°C), which was left at ambient temperature for this experiment. Three pumps were used: two isocratic pumps (Postnova, PN1130) to provide the tip flow and focus flow, while a Kloehn syringe pump (Postnova, PN1610) provided the crossflow. The degasser (Postnova, PN7505) was placed between the carrier liquid reservoir and the isocratic pumps. Two 0.1 μm VVPP filters (Durapore, VVLP04700, Millipore) were placed in-line after each pump. They were changed regularly, as soon as the pressure increased above 20 bar for a flow rate of 1 mL/min of aqueous carrier liquid (see Note 1).

The AF4 was equipped online with a UV absorbance detector from Shimadzu (SPD-20A, sold by Postnova PN3211, operating from 190 to 700 nm), measuring absorbance at the wavelength of 300 nm, which is strongly absorbed by quantum dots. A spectrofluorometric detector, also from Shimadzu (RF-10A_{XL}, sold by Postnova PN3410), was placed after the UV absorbance detector.

For the negatively charged QDs, a polyethersulfone membrane (Z-AF4-MEM-611-10KD) was installed in the channel.

2.2 Reagents and Chemicals

1. Deionized water was obtained from a Millipore Milli-Q water system (18.2 MΩ cm, 25°C) and was further filtered using 0.1 μm VVPP filters for aqueous solutions. The 10 mM NaCl solution was prepared from the same Milli-Q water and was filtered using the 0.1 μm VVPP filter. The carrier liquid was degassed by sonication for 15 min.
2. Bovine serum albumin was purchased from Sigma-Aldrich (≥98%, A7906) as lyophilized powder.
3. Chemicals used to prepare the QDs were purchased from Sigma-Aldrich and PCI Synthesis and were of technical grade. The end product was purified by precipitation in methanol and resuspension in chloroform, before surface ligand exchange using 3-mercaptopropionic acid (MPA). After suspension in water, the QDs were separated from the excess MPA by precipitation in the presence of 1:1 (v/v) water to ethanol and resuspension in deionized water.

2.3 Sample Preparation (Rewrite: Shorten and Adapt to Materials + Add a Section in Met.)

If possible, the sample should be prepared in the same liquid as the carrier solution. It should be freshly prepared prior to analysis. The volume to be injected is determined by the volume of the injection loop. It can be tuned by modifying the length or the inner diameter of the loop tubing. In general, volumes between 3 and 100 μL are selected. The sample load per injection should be small (1–100 μg/injection) (25). This will decrease particle-particle and/or particle-wall interactions that can occur with sample overload, resulting in reduced resolution or sample loss. The optimal sample load has to be adjusted to the sensitivity of the detectors in order to obtain acceptable signal to noise ratios (see Note 2 about Sample overloading). Surfactants may be used to increase the solubility and improve the dispersion of the sample (4). Certain samples can also be evenly dispersed using vigorous mechanical agitation or sonication; however, care must be taken to avoid damaging or causing changes to the surface of the samples.

We have chosen as an example, a QD sample with MPA as a surface ligand. The CdSe/CdZnS QDs were prepared by following the protocol presented by Pons et al. (26). QDs have been analyzed previously by FFF in environmental studies (27, 28), as proof of concept (29, 30), or in Zattoni et al. (31) for polymeric coated QDs.

Two modes exist in FlFFF: the normal mode, for small particles with diameter between 2 nm and 1 μm, and the steric mode for particles larger than the micrometer. Depending on the size of the particles contained in the sample to be analyzed, the method will have to be adapted. A higher sample load (50–1,000 μg) is usually required in the steric mode.

2.4 Eluent

Virtually any liquid compatible with the sample and the channel/tubing materials can be used in AF4. Stainless steel channels are also available for fractionation analysis requiring organic solvents. All liquids should be degassed and filtered with 0.1–0.2 μm filters to remove any particulate material before entering the pumps. The use of an online degasser is recommended to remove residual microbubbles, which can cause baseline noise in light scattering detection.

Aqueous buffers should be prepared from deionized water to which salts are added. Bactericides, such as sodium azide (0.01–0.02%, 0.05% in case of long-term disuse), may be added to prevent bacterial growth (see Note 3). The ionic strength of the buffer has to be selected carefully as it affects sample retention time, stability against aggregation, and adsorption on the membrane. If the ionic strength is too low, repulsion forces between charged particles will cause the particles to equilibrate further from the membrane, and consequently, the sample will elute too quickly. Hupfeld et al. have studied how the ionic strength and osmotic pressure of the carrier liquid affects the retention time of liposomes (32). The pH of the solution affects the retention time of samples carrying pH-sensitive groups (carboxylic acids, amines, etc.). For protein analysis, the buffer pH has to be different from the isoelectric point of the protein in order to avoid adsorption of the protein on the membrane. Ethanol is not recommended to use in the long term as it may cause cracks in the Plexiglas plate (see Note 4).

Viscosity should also be considered when choosing a proper solvent, as the relationship between the hydrodynamic radius of the particles and the retention time is dependent on the viscosity. It is recommended to use solvents of low viscosity (see Note 5).

2.5 Miscellaneous Instrumental Features

Filters and Tubings: Filters should be placed in-line after each pump to prevent contamination of the sample by particulate matter from the pumps. The filters should be changed frequently (when the pump pressure reaches 20 bar for 1 mL/min, see Note 5). Selection of the proper tubing is important in the FlFFF system, since the inner diameter of the tubing regulates the pressure in different regions of the system. Tubing with an inner diameter of 0.01 in is usually satisfactory. The length of the tubing should be kept as small as possible, mainly in the critical regions such as the tubing linking the injector to the channel, the channel to the detectors, and in between detectors. The dead volume contained in this tubing is responsible for band broadening because in the absence of field, the sample diffuses freely. One should keep in mind that small inner diameters can introduce shear stress on the sample. For analysis of shear sensitive samples, it is possible to use larger diameter tubing, at the expense of resolution. PEEK (polyetheretherketone) tubing is used for all of the applications of FlFFF in aqueous media. For organic solvents, such as tetrahydrofuran, PEEK cannot be used; stainless steel is recommended in this case.

Detectors: For complete characterization of a sample, it is useful to have more than one detection method. The concentration of the eluting sample is determined with a UV/Vis absorption detector or a refractive index detector (see Note 6). A multiangle light scattering (MALS) detector yields the molecular weight and the root-mean-square (rms) radius of each sample component as it elutes from the channel. A photon correlation spectroscopy (PCS) or dynamic light scattering (DLS) detector measures the diffusion coefficient of the eluting components. The diffusion coefficients can also be calculated from the species retention times using the FFF theory. Therefore DLS detection provides an independent method to confirm the validity of the theory. Other detectors, such as a fluorescence detector, can be linked to the instrument for enhanced characterization of samples subjected to analysis. Finally, an added feature of FlFFF is the ability to collect fractions as they elute out of the channel. The collected fractions can be analyzed by complementary techniques, such as scanning electron microscopy, transmission electron microscopy, inductively coupled plasma, and mass spectrometry, or used for in vitro biological experiments.

3 Methods

Have ready before sample analysis:

3.1 Channel Preparation

The channel should be prepared in advance with the appropriate membrane and spacer for the sample to be analyzed. It is recommended to change the membrane after 30 runs.

1. Rinse all the elements and soak the membrane and the frit for a few minutes in Milli-Q water, before assembling the channel.
2. Place the frit in the bottom block and lay the membrane above it with the smooth side facing up, making sure it stays aligned with the frit.
3. Install the spacer above the membrane.
4. Close the channel with the block of Plexiglas.
5. Use a torque wrench to bolt the channel together. The torque wrench is used to provide a precise and uniform pressure throughout the channel. Tighten the bolts from the center of the channel moving outwards in a spiral fashion. The amount of torque to apply to the bolts depends on the channel type and should be specified in the manufacturer's guide (it varies from 2 to 9 N.m).

After the channel is assembled and placed in a vertical position:

1. Connect the focus flow tubing, and pump the filtered carrier liquid into the channel at a rate of 1–2 mL/min to fill the channel and to flush air bubbles towards the remaining channel openings.
2. Connect the tubings located at the bottom of the channel. This will force the carrier liquid to move up the channel, evacuating the large air bubbles with it.
3. Once the carrier liquid reaches the top of the channel, connect the tip flow input tubing to the channel, and pump liquid through it, thereby filling the remainder of the channel until carrier liquid comes out from the crossflow output. Once the carrier liquid reaches the top of the channel, the crossflow tubing can be connected to the channel. Crossflow should be switched on at 1–1.5 mL/min with a tip flow of 2 mL/min. This will remove the remaining air bubbles in the channel through the membrane and the frit. Carrier liquid should be pumped through the channel at 1 mL/min for 1–2 h to equilibrate the channel and remove any air bubbles trapped in the channel. If the frit has been allowed to dry completely, carrier liquid should be pumped continuously through the channel for 2–4 days approximately, with a flow rate of 1–2 mL/min.

Table 2
Set of flow rates used for the analysis of mercaptopropionic-coated CdSe/CdZnS quantum dots

V_c/V_{out}	2	3	4	5
Detector flow V_{out} (mL/min)	0.5	0.5	0.5	0.5
Crossflow V_c (mL/min)	1.0	1.5	2.0	2.5
Focus flow V_{foc} (mL/min)	1.3	1.8	2.3	2.8
Tip flow during focusing (mL/min)	0.2	0.2	0.2	0.2

The crossflow was linearly decreased for all methods from the initial value to 0 mL/min over a period of 20 min, before being let constant at 0 mL/min for 10 min. The channel used had a length of 27.5 cm, a width of 2.0 cm, and a thickness of 350 micrometers.

Setting Up the Instrument for Sample Fractionation, Sample Preparation, and Sample Injection

Setting up the instrument (before the sample preparation and separation):

1. Turn on the detectors before the sample preparation (you need about 1 h to establish the baseline conditions).
2. Set the tip flow rate (same as detector flow rate to be used: i.e., 0.5 mL/min) using AF2000 software (Postnova).
3. Open a "New Run" in the same software and create the method using the flow rates presented in Table 2.
4. Preset the focusing flows according to the selected method.
5. Open the light scattering software.
6. Create a new template and put all parameters for the sample and separation conditions (as requested by the light scattering software and seen on the monitor).
7. Keep the instrument on "stand by" while you prepare your sample.

3.2 Sample Preparation and Injection

1. Have QDs (prepared commercially or in-house, at least 35 μL per injection and concentration above 1 mg/mL) in the medium as the carrier liquid.
2. If required, sonicate or vortex the sample to obtain a homogeneous suspension.
3. Use Hamilton microsyringe to withdraw the sample (at least 10 μL more than the volume of the sample loop) from the tube. Remove any air bubble from the syringe.
4. The inject lever on the instrument should be in the "Inject" position. Insert the syringe through the septum and release 2–3 μL of the sample (to eliminate any risk of injecting air bubbles in the sample loop).

5. Change the lever to the "Load" position and inject your sample (add 10 μL more than the sample loop volume to fully load it).
6. Change the lever position back to the "Inject" position. This will start the sample fractionation.

3.3 Sample Analysis

1. The signal from the detectors (UV or refractive index, fluorescence, light scattering) will appear on the monitor.
2. During the elution, the user can already start analyzing the elution profile and determine how to improve the elution.
   (a) The first thing to look at is the following: Is there a reasonable signal? If no, then the reason might be because of total adsorption of the sample on the membrane; if that is the case, see Note 7. Another possibility is that the sample load is too small. Third possibility is the detectors are not sensitive enough.
   (b) If there is not enough separation between the main peak eluting while the crossflow is kept constant and the residual peak eluting after the crossflow has been decreased, the user can either increase the time the crossflow is maintained constant or can decrease the V_c/V_{out}.
3. Once the run is finished (with or without the rinsing step), it is good to either rerun the same method using a blank injection, to rinse the channel and verify if there was any sample adsorption during the previous run which could have detached during the subsequent run, thereby contaminatng it. The other option is to open the purge valve and flush the channel with a fast axial flow rate (2 mL/min tip flow) for 10 min. One of the advantages of FFF techniques is the absence of stationary phase; therefore, there is no risk of interaction between the sample and the stationary phase, and flow rates can be changed rapidly without risking erosion of the stationary phase material.
4. For the next run, the user can start by opening the same template as the previous run and modifying the ratio of crossflow/outflow, either decreasing it if the sample was retained too much in the channel or increasing it if the sample eluted too rapidly. In the first step, outflow rate should be kept the same. Later on, to improve the resolution, the outflow rate can be increased; it is better to start with low outflow rates (<0.5 mL/min). Once an appropriate ratio of crossflow/outflow rate has been determined, for which there is no sign of peak splitting and the retention time is short enough, then it is interesting to increase the crossflow as much as possible in between runs to determine the best one, as this will increase the retention level (t_r/t_0) and lead to a better resolution between the components (see Note 8).

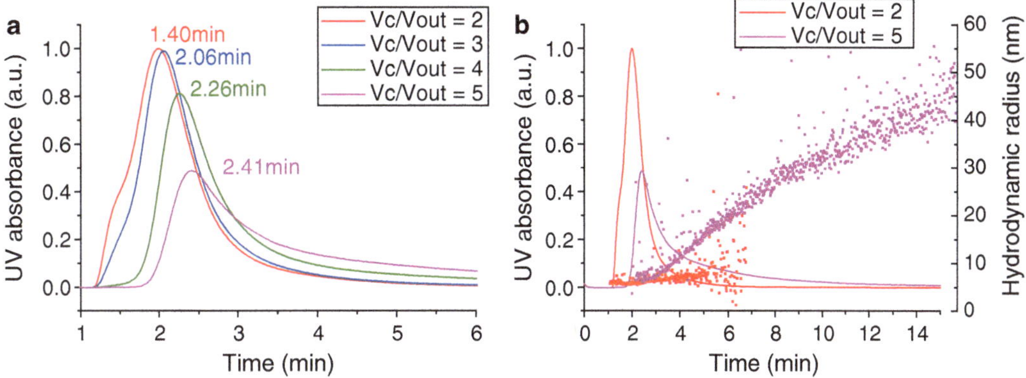

Fig. 2 (**a**) AF4 fractogram of the UV/Vis signal at $\lambda = 300$ nm plotted vs. time using increasing crossflow/outflow ratios. (**b**) Plot of ratios 2 and 5 showing the UV absorbance on the left y-axis and the hydrodynamic radius (Rh) on the right y-axis plotted as a function of elution time

Another way to visually determine at which crossflow/outflow rate the components will elute out, one can position the time of elution on the method template and see during the crossflow decrease, at which crossflow value the samples start to elute out (see Notes 9–12).

5. As a result of the QD separation by AF4, one can obtain the fractogram illustrated in Fig. 2:

The fractogram of QD-MPA was obtained using deionized water as a carrier liquid and a 10 kDa cut-off polyethersulfone (PES) membrane as an accumulation wall. The ratio of crossflow to channel flow was increased from 2 to 5 to observe the effect on retention time. The UV signals were all normalized to the signal from the $V_c/V_{out} = 2$ condition. By integrating the UV signal as a function of elution time for each peak, we determined that the areas were all equal, meaning that there was no loss of sample material in between the different flow rate conditions presented in Table 2.

In the first step, 5 mg/mL solution was injected, with an injection loop of 21.5 μL, which represents a mass of 107.5 μg. The detector flow rate was set at 0.5 mL/min to avoid excessive flow rates in the channel. Since the QDs are relatively small particles (8–16 nm in diameter), we used a higher crossflow rate than the detector flow rate. We tested four V_c/V_{out} ratios: 2, 3, 4, and 5, V_c being the crossflow rate and V_{out} being the flow rate at the output of the channel, also called flow rate at the detectors. The fractograms presented in Fig. 2 show the relative retention time for the various crossflow to detector flow ratios. Retention time was increased progressively with the increase of crossflow/channel flow ratio. Bellow values of $V_c/V_{out} = 3$, the peak of the QDs coelutes with the

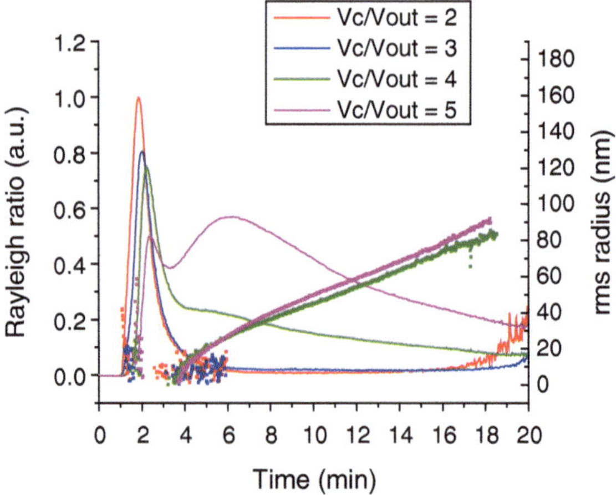

Fig. 3 AF4 fractogram showing the Rayleigh ratio signal as a function of elution time for increasing ratios of V_c/V_{out}. The rms radius (geometrical radius) is reported on the right y-axis

void peak. By increasing to 4 and 5, we obtain good separation between the main peak and the void. However, a shoulder to the right appears.

By looking at the QELS and MALS signals (Figs. 2 and 3, respectively), we observe that this shoulder corresponds to larger aggregates which have formed because of the strong flow rates in the channel. The Rh plot (Fig. 2) shows aggregates between 10 and 50 nm in hydrodynamic radius. This corresponds to a few QDs agglomerating together under the influence of the strong crossflow rates. In Fig. 3, showing the Rayleigh ratio as a function of elution time, a secondary peak appears for the V_c/V_{out} ratios 4 and 5, even though the UV/Vis signal intensity is low, because of the stronger scattering intensity of larger particles. The agglomerated QD particles formed are large enough to give precise information on the rms radius of the eluting particles, as seen in Fig. 3. Non-agglomerated QDs were too small compared to the wavelength of the laser, to give useful information on the rms size. They were smaller than the limit of detection of the MALS detector. Aggregates, of up to 200 nm in geometric diameter, are formed under the influence of the crossflow rate.

The effect of increasing the rate of the crossflow for a given detector flow on the nanoparticle recovery was also assessed, by integrating the elution peak of the UV/Vis signal as a function of elution time to obtain the area under the curve (as seen in Fig. 2). Results show that the recovery, as determined by the area under the curve, was the same over several injections using the same sample concentration and while increasing the crossflow/outflow ratio. Therefore, the increased flow rates in the channel induced

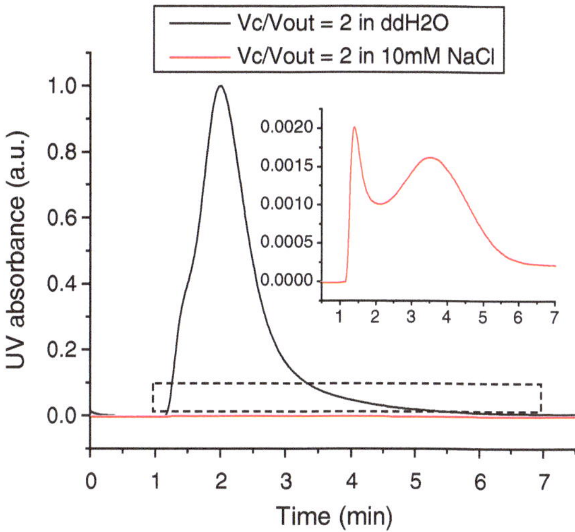

Fig. 4 Effect of 10 mM NaCl on the fractionation of QD-MPA on a polyethersulfone membrane. The UV/Vis absorbance at 300 nm is plotted as a function of elution time

aggregation of the QD-MPA; however it did not lead to increased sample adsorption to the membrane material.

As observed earlier, the V_c/V_{out} of 4 and 5 lead to a shoulder. The Rayleigh ratio shows a more important signal compared to the UV/Vis detector, as the signal read corresponds to larger particle, which scatter more light even at lower concentration. The rms radius is sensitive to particles which are approximately 1/10th of the wavelength of the laser used. Here, individual QD-MPA are not seen, as they are too small (10–16 nm); however, the crossflow-induced aggregation leads to a strong and useful signal.

Keeping the same membrane material (PES), we changed the carrier liquid, to assess the effect of ionic strength on the stability of these small ligand-coated QDs. The carrier liquid was changed to a solution of sodium chloride (10 mM, in deionized water), which was properly filtered on a 0.1 μm VVPP (Durapore) filter and degassed. The same fractionation method was used ($V_c/V_{out}=2$, detector flow = 0.5 mL/min) for both carrier liquid conditions.

Figure 4 shows that there is almost complete adsorption of the QD sample to the membrane material. The sizes using the sodium chloride carrier liquid cannot be reported accurately because of the too low concentration of the sample exiting the channel, but from the peak shape, it doesn't appear that the salts caused aggregation of the eluting QDs.

Through this case study, we have briefly observed the effect of crossflow/detector flow ratio on the retention of small spherical nanoparticles; we have seen that by increasing this ratio, we were able to shift the elution of the particles to longer retention times.

However, we have also observed the limits of increasing the crossflow/detector flow, as this affected the stability of the particles being analyzed. This effect was accentuated by the nature of the sample to be analyzed, which relies uniquely on electrostatic repulsion between particles for its stability in suspension. The higher crossflows were able to force the particles in vicinity which accelerated their precipitation. However, it didn't seem to affect negatively the particle-wall interactions, as no loss of material was observed when increasing the crossflow rates. The strength of the hyphenated technique is that whatever happens in the channel is picked up by the changes in elution time but also the light scattering detectors, which provide crucial information for the thorough characterization of the state of the sample. In this case study, we were able to quantify and characterize the sizes of the aggregates formed in the channel, through the combined use of the UV/Vis, MALS, and the QELS detectors.

The importance of the carrier liquid composition on the analysis was also underlined. In this study, the addition of a salt to the carrier liquid induced particle-wall interactions. By screening the surface charges of the QD-MPA as well as the PES ultrafiltration membrane (pI~2.4) (33), the salts in the carrier liquid caused the QDs to come in proximity to the membrane. This is why the user should always do verifications of possible interactions when studying charged samples, as the carrier liquid composition, the membrane material, and the surface properties of the sample play a role in the good process of the fractionation. The interactions may not always cause complete adsorption of the sample to the membrane material; they may slow the elution resulting in overestimation of the size if using the retention time theory alone. This is an example why the hyphenation of AF4, by mounting light scattering detectors, offers such a powerful solution to the thorough analysis of samples of broad size distribution, as the in-line detectors do not rely on the retention time to measure accurately the sizes. The fractionation is useful too to separate the species by size and present them as monodispersed samples to the light scattering detectors, therefore solving the problem of sizes being biased by the presence of a few large particles.

4 Notes

1. If available, read the pressure at the tip flow pump and the system pressure when the tip flow rate is set at 1 mL/min. For a given carrier liquid, the pressure should not vary too much. If it has increased significantly, this is sign that there is blockage in the channel or, more frequently, that the filter placed after the pump needs to be replaced.
2. Sample overloading usually occurs for sample loads above 100 µg. It is caused by the relatively small volume in which

the separation occurs (usually in a 5 μm high layer above the channel wall). Each particle requires a certain volume to reach equilibrium; if the sample is too concentrated, then the particles cannot reach equilibrium because of steric interference with each other, leading to cases of precipitation and aggregation if the particles are attracted to one another. For charged particles in which there is significant particle-particle repulsion, the particles need even more volume to equilibrate; therefore, the sample loads have to be smaller (about 1 μg). Signs of overloading are peak fronting and digitation. To test for sample overloading, several runs with varying sample loads are necessary; if the retention time does not vary with the sample load, then there is no overloading.

3. The amount added must be taken into consideration in the calculation of the ionic strength. Solutions containing sodium azide should be disposed of appropriately.
4. Ethanol should be avoided as a carrier liquid if it is used for prolonged periods of time as it can cause the Plexiglas to crack.
5. If buffers are used, the usual procedures should be used to flush the pump pistons to remove the crystalline materials, from the salts, that may damage the pump seals.
6. Refractive index detectors are sensitive to pressure and cannot withstand pressures in excess of 100 psi; therefore, they should be placed last in-line of the detectors. Large inner-diameter tubing should be used as output to reduce as much as possible back pressure.
7. To determine if there is sample adsorbing to the membrane, inject the same sample using the same method and measure the concentration eluting out; if there is increase or decrease of signal intensity and area under the peak, then it is possible that the sample is interacting with the membrane. Possible solutions: Membrane pretreatment; changing the membrane; changing the salt concentration of the carrier solution; injecting more sample to have enough at the detectors: use preconcentration of the sample by injecting several times the same sample during the focusing period, or use a slot pump to remove the excess of solvent contained above the sample layers.
8. Increasing the crossflow rate/outflow rate ratio will also lead to a proportional increase of system pressure.
9. It is important not to decrease the crossflow rate too quickly, as it may cause sample to be released by the channel prematurely, because of disturbance caused in the channel laminar flow by rapid crossflow decrease. This can also cause the system pressure to increase rapidly, causing significant pressure drop once the crossflow reaches zero, which often causes a small depression in the signal read at the detectors, as well as the

unknown complex effects occurring in the channel when the system's pressure changes abruptly.

10. It is usually not recommended to keep the sample retained in the channel for too long with a too high crossflow rate/outflow rate, if it is not required by the sample size, as sample can interact with itself (particle-particle) causing aggregation and appearance of larger particles or can interact with the membrane, causing sample loss due to reversible/irreversible adsorption and therefore poor recovery and signal intensity.
11. If bubbles appear in the syringes of the crossflow pump, this may be caused by two problems. Either, upon using high crossflow rates, there is cavitation occurring in the syringes which may be caused by an air leak. The solution to this is to change the whole syringe by ordering a new part or to find the leaky gasket and change it. The other possibility is that air bubbles were introduced in the channel and they have been pumped out of the channel by the crossflow pump. This may be because of a too weak axial flow rate and a too strong crossflow rate. If this is the case, the system pressure will indicate null pressure in the system, which means that solvent is backing up from the output back into the channel and out by the crossflow pump. To resolve this problem, it is necessary either to disassemble the channel and rewet the membrane and frit which may be dry or to pump solvent at a high flow rate while opening the purge valve
12. If an acute peak is observed during the elution of the sample, this may have been caused by the passage of an air bubble through the detectors; this usually means that the bubble will be seen on the signals of several detectors in the order of their placement in-line of the instrument. If this peak reappears when repeating the same run, then it's possible that it is caused by a drop in the pressure from the channel to the detectors and can be solved by using a small length of thinner tubing between the channel and the detector, therefore compensating for the drop of pressure.

References

1. Giddings JC (1966) A new separation concept based on a coupling of concentration and flow nonuniformities. Separation Sci 1(1): 123–125
2. Giddings JC (1989) Field-flow fractionation of macromolecules. J Chromatogr 470(2):327–35
3. Giddings JC et al (1980) Analysis of biological macromolecules and particles by field-flow fractionation. Methods Biochem Anal 26:79–136
4. Schimpf ME, Caldwell K, Giddings JC (2000) Field flow fractionation handbook, vol xviii. Wiley-Interscience, New York, p 592
5. Thompson GH, Myers MN, Giddings JC (1969) Thermal field-flow fractionation of polystyrene samples. Anal Chem 41(10):1219–1222
6. Giddings JC, Yang FJF, Myers MN (1974) Sedimentation field-flow fractionation. Anal Chem 46(13):1917–1924
7. Caldwell KD, Gao YS (1993) Electrical field-flow fractionation in particle separation. 1. Monodisperse standards. Anal Chem 65(13): 1764–1772
8. Giddings JC, Yang FJ, Myers MN (1976) Flow-field-flow fractionation: a versatile new separation method. Science 193(4259):1244–1245
9. Vickrey TM, Garcia-ramirez JA (1980) Magnetic field-flow fractionation: theoretical basis. Sep Sci Technol 15(6):1297–1304
10. Wahlund KG, Giddings JC (1987) Properties of an asymmetrical flow field-flow fractionation

channel having one permeable wall. Anal Chem 59(9):1332–1339
11. Wahlund KG, Litzen A (1989) Application of an asymmetrical flow field-flow fractionation channel to the separation and characterization of proteins, plasmids, plasmid fragments, polysaccharides and unicellular algae. J Chromatogr 461:73–87
12. Fraunhofer W, Winter G (2004) The use of asymmetrical flow field-flow fractionation in pharmaceutics and biopharmaceutics. Eur J Pharm Biopharm 58(2):369–383
13. Weers JG, Arlauskas RA (2004) Particle size analysis of perfluorocarbon emulsions in a complex whole blood matrix by sedimentation field-flow fractionation. Colloids Surf B Biointerfaces 33(3–4):265–269
14. Pinaud F et al (2010) Probing cellular events, one quantum dot at a time. Nat Methods 7(4): 275–285
15. Maysinger D, Lovric J (2007) Quantum dots and other fluorescent nanoparticles: quo vadis in the cell? Adv Exp Med Biol 620:156–167
16. Pierobon P, Cappello G (2011) Quantum dots to tail single bio-molecules inside living cells. Adv Drug Deliv Rev 64(2):167–178
17. Huang HC et al (2011) Inorganic nanoparticles for cancer imaging and therapy. J Control Release 155(3):344–357
18. Mahmoudi M, Serpooshan V, Laurent SS (2011) Engineered nanoparticles for biomolecular imaging. Nanoscale 3(8):3007–3026
19. Algar WR et al (2011) The controlled display of biomolecules on nanoparticles: a challenge suited to bioorthogonal chemistry. Bioconjug Chem 22(5):825–858
20. Rosenthal SJ et al (2011) Biocompatible quantum dots for biological applications. Chem Biol 18(1):10–24
21. Choi HS, Frangioni JV (2010) Nanoparticles for biomedical imaging: fundamentals of clinical translation. Mol Imaging 9(6):291–310
22. Hutter E, Maysinger D (2011) Gold nanoparticles and quantum dots for bioimaging. Microsc Res Tech 74(7):592–604
23. Gaponik N et al (2010) Progress in the light emission of colloidal semiconductor nanocrystals. Small 6(13):1364–1378
24. Smith MH et al (2010) Monitoring the erosion of hydrolytically-degradable nanogels via multiangle light scattering coupled to asymmetrical flow field-flow fractionation. Anal Chem 82(2):523–530
25. Giddings JC (1993) Field-flow fractionation: analysis of macromolecular, colloidal, and particulate materials. Science 260(5113): 1456–1465
26. Pons T et al (2009) Synthesis of near-infrared-emitting, water-soluble CdTeSe/CdZnS core/shell quantum dots. Chem Mater 21(8):1418–1424
27. Bouby M, Geckeis H, Geyer FW (2008) Application of asymmetric flow field-flow fractionation (AsFlFFF) coupled to inductively coupled plasma mass spectrometry (ICPMS) to the quantitative characterization of natural colloids and synthetic nanoparticles. Anal Bioanal Chem 392(7–8):1447–1457
28. Hassellov M et al (2008) Nanoparticle analysis and characterization methodologies in environmental risk assessment of engineered nanoparticles. Ecotoxicology 17(5):344–361
29. Rameshwar T et al (2006) Determination of the size of water-soluble nanoparticles and quantum dots by field-flow fractionation. J Nanosci Nanotechnol 6(8):2461–2467
30. Al Hajaj et al. (2011) ACS Nano, 5(6): 4909–4918
31. Zattoni A et al (2009) Asymmetrical flow field-flow fractionation with multi-angle light scattering detection for the analysis of structured nanoparticles. J Chromatogr A 1216(52):9106–9112
32. Hupfeld S et al (2010) Liposome fractionation and size analysis by asymmetrical flow field-flow fractionation/multi-angle light scattering: influence of ionic strength and osmotic pressure of the carrier liquid. Chem Phys Lipids 163(2):141–147
33. Ricq L et al (1997) Electrokinetic characterization of polyethersulfone UF membranes. Desalination 109(3):253–261

Chapter 31

Polymersomes-Mediated Delivery of Fluorescent Probes for Targeted and Long-Term Imaging in Live Cell Microscopy

Irene Canton and Giuseppe Battaglia

Abstract

Fluorescent microscopy becomes an essential tool for live imaging analysis of complex biological pathways and events as it enables noninvasive real-time/real-space imaging. The design of fluorescent probes to provide dynamic information and long-term tracking of samples without altering physiological and structural integrity is critical in live imaging. In recent years, nanotechnology has produced a new generation of imaging probes with promising applications in live imaging. In particular, we describe the use of pH-sensitive amphiphilic block copolymer $PMPC_{25}$-$PDPA_{70}$. This polymer forms biomimetic nanometer-sized vesicles (known as polymersomes) that are readily uptaken by a wide variety of cell types. The pH sensitivity confers much needed endolysomal escape capability without inducing cellular toxicity or stress. Two different characteristic compartments in the polymersomes (hydrophilic core and hydrophobic membrane) allow for encapsulation of different labeling cargoes such as lipidic cell membrane probes, quantum dots, fluorescent dyes, and fluorescent biomolecules such as nucleic acid and protein probes.

Key words Live imaging, Polymersomes, Fluorescent probes, Endocytic uptake, Intracellular delivery

1 Introduction

The use of fluorescence microscopy has become essential in biology and biomedical sciences for the study of functional and structural aspects at the cellular, tissue, or whole animal level. Due to recent technical advances (1, 2) and compared to many other imaging techniques, fluorescence microscopy today achieves high spatial resolution; it is very sensitive and specific with safe and relatively easy detection procedures. One of the biggest advantages is perhaps that it allows compatibility with living specimens, thus providing real-time dynamic studies of biological events. However, live imaging is a fairly recent technology prompted by the introduction of video signal techniques in the 1980s. Fluorescence imaging has traditionally been performed on fixed samples, and at present, improvements are very much needed, especially in the design of fluorescent probes and chemo-sensors suitable to perform live imaging in cells.

Volkmar Weissig et al. (eds.), *Cellular and Subcellular Nanotechnology: Methods and Protocols*, Methods in Molecular Biology, vol. 991, DOI 10.1007/978-1-62703-336-7_31,

The ability to bring together multi-functionalities within a single particle gives an advantage to nanotechnology in the creation of such revolutionary imaging devices. These functionalities mainly include coatings for biocompatibility (3) and/or prolonged half-life (4), targeting sequences (5) combined with drug encapsulation for theragnostic applications (6), powerful bioimaging fluorescent dyes (7), activating fluorescent probes in response to bio-stimuli (8), and also combination of fluorescent and magnetic labels (9).

To gain the most relevant information in live cellular imaging, the probe has often to target the inside of cells. Small hydrophobic molecules can permeate cell membranes with relative ease, but hydrophilic molecules and especially large macromolecules such as proteins and nucleic acids require a vector to assist their transport across the cell membrane. Nanotechnologists have exploited endocytosis as a successful and reliable gate to enter cells in live bioimaging (10, 11). Understanding the uptake process of the fluorescent nanoparticle from the particular endocytic mechanism to intracellular dynamics of fluorescent nanoparticles is imperative to its biological application. Furthermore, to enable rational design of nanovectors, interactions of the probe with the cell as well as with the biological environment that surrounds the cells need to be properly understood (i.e., effect of cell seeding density, effect of cell trypsinization on receptor-mediated uptake, or presence of serum in the media). Indeed, there are a number of underlying mechanisms that govern the success/failure of the fluorescent probe. Efficient uptake characterization protocols must demonstrate three important parameters in nanoformulations: how fast the signal is detected, how much of the formulation is required to obtain a good enough signal, and how long can be this signal tracked. The ultimate challenge is to unify all these parameters into an efficient intracellular delivery protocol that additionally causes no unwanted effects on the cells (i.e., toxicity and cellular stress).

We have developed an efficient nanovector for the delivery of both hydrophilic and hydrophobic fluorescent imaging compounds (Massignani PLoS ONE) as well as bioactives within cells (12). This is based on the self-assembly of pH-sensitive poly(2-methacry loxyethylphosphorylcholine)-block-(2-(diisopropylamino)ethylmethacrylate) (PMPC-PDPA) (13). The PMPC block is highly biocompatible, while the PDPA block imparts pH sensitivity to the copolymer. At physiological pH, the copolymer forms colloidally stable nanometer-sized vesicles, whereas below the pK_a of the PDPA block, rapid dissociation of the vesicles occurs, as the tertiary amine groups on the PDPA chains become protonated. We have demonstrated that PMPC-PDPA polymersomes are efficiently uptaken by cells and the cargo is able to escape the endocytic pathway. The delivery of the fluorophores and bioactives is achieved without affecting the metabolic activity of the cell or triggering proinflammatory pathways (14).

2 Materials

Follow these general aseptic guidelines when using PMPC-PDPA vesicles for intracellular DNA delivery into mammalian cells. Always use sterile solutions and, when possible, endotoxin/RNAse-free reagents/materials. This is especially important when encapsulating RNA (RNAse-free) or when using highly sensitive stimulation experiments (cell cultures and/or animal models).

Use low-passage cells, and ensure that cells are healthy and greater than 90%viable. Transfect cells with PMPC-PDPA polymersomes (both fluorophore and/or bioactive-containing polymersomes) at 70–80%confluence. To increase accuracy and reduce assay variability, we recommend performing triplicate wells for each sample condition.

2.1 Polymer Film Production

1. $PMPC_{25}$-$PDPA_{70}$ is poly(2-(diisopropylamino) ethyl methacrylate)-b-poly(2-methacryloyloxy-ethyl phosphorylcholine) (PMPC25-b-PDPA70) diblock copolymer (13).
2. Chloroform to methanol solution (2:1).
3. Glass vials (28 ml clear universal vial capped, Reagecon Diagnostics Limited, Shannon, Co. Clare, Ireland).

2.2 Polymersome Formation

HCl 1 M (sterile).
NaOH 1 M (sterile).
PBS pH 2 (sterile).
Sepharose 4B GPC column (sterile).

3 Methods

Procedures are to be carried out at room temperature unless otherwise stated. Figure 1 illustrates few examples of probes that can be delivered by PMPC-PDPA polymersomes:

3.1 Polymer Film Making and Encapsulation of Water-Insoluble Agents via Stirring Method

This step is necessary to facilitate subsequent solubilization of the polymer in aqueous solution. The formation of a thin film is required to successfully solubilize the polymer. Maintaining polymer to solvent mixture ratios is crucial to obtain a thin film as well as the diameter of the glassware used to evaporate the polymer to solvent mixture. It is recommended to scale up accordingly.
For a 20 mg polymer film:

1. In a clean glass vial weight 20 mg of $PMPC_{25}$-$PDPA_{70}$.
2. In a fume cupboard, mix 6 ml of chloroform and 3 ml of methanol (ratio 2:1) and add it to the polymer glass vial (see Note 1) gently stirring until the polymer is totally dissolved. The solution must be clear at this point; if it appears cloudy,

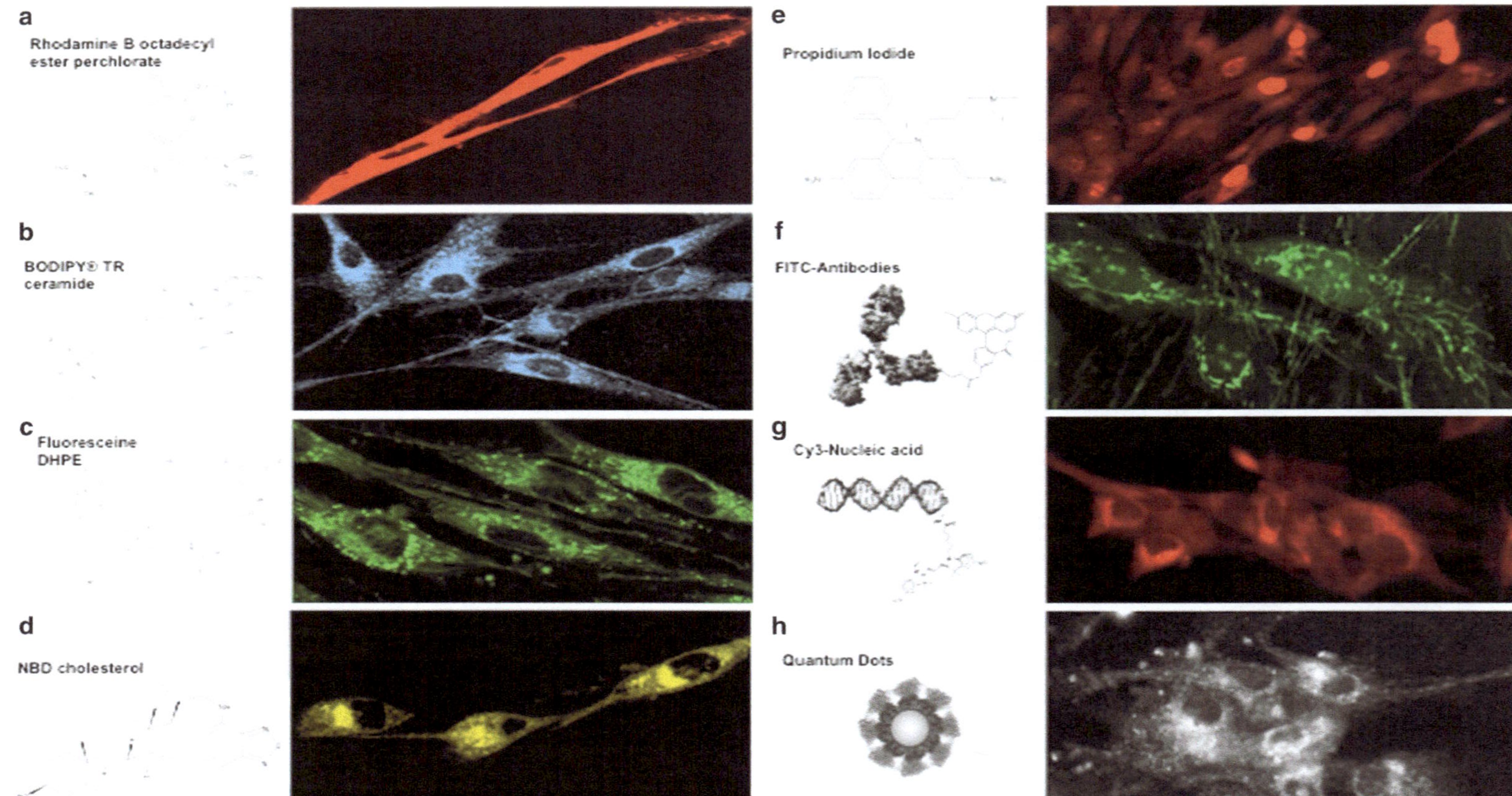

Fig. 1 Confocal micrographs of live primary human dermal fibroblast after 24 h incubation with PMPC-PDPA polymersomes loaded with (**a**) Rhodamine B octadecyl ester perchlorate, (**b**) BODIPY TR ceramide, (**c**) fluorescein 1,2-dihexadecylphosphatidylethanolamine (DHPE), (**d**) labeled NBD cholesterol, (**e**) DNA-staining membrane-impermeable propidium iodide, (**f**) FITC-labeled antibody antitubulin, (**g**) labeled nucleic acids, or (**h**) *quantum dots*

discard the solution due to the presence of impurities. Concentrations of polymer in methanol to chloroform solvent mixture above 4 mg/ml will produce thicker films upon evaporation and will be difficult to solubilize.

3. Water-insoluble dyes or agents to be encapsulated are then added to the polymer to solvent solution. In general, a 5%molar ratio (dye to polymer) is necessary.
4. Evaporate the solution overnight under vacuum to obtain a uniform film (see Note 2).
5. To form the polymersomes and encapsulate hydrophobic dyes, add 0.1 ml pH 7.4 PBS/mg of polymer in the film and stir with a magnetic stirrer for 10 days. This ensures complete and homogeneous solubilization.
6. Once a day for ten consecutive days, sonicate the sample for 30 min to allow "onion-like" structures (multilamellar structures) to break.

3.2 Hydrophilic Dyes and Bioactive Agents Encapsulation: Polymersome Formation via pH Switch

1. Dissolve the film completely in sterile PBS pH 2, making a 10 mg/ml polymer solution (see Note 3). Ensure a pH of 2 in the polymer solution to optimize the solubilization by adding dropwise 1 M HCL.
2. Once the polymer film is completely dissolved, filter sterilize the solution using a 200 nm diameter pore filter (see Note 4).
3. Increase the pH of the solution slowly to pH 6 (see Note 5) by adding dropwise sterile NaOH 1 M with the help of a 10 μl Gilson pipette while stirring vigorously with a vortex to avoid supramolecular aggregations and precipitation of the polymer.
4. When encapsulating water-soluble dyes or agents, add the solution of the dye or agent at this point (see Note 6). We recommend the following concentrations of hydrophilic agents:
   (a) Hydrophilic dyes (i.e., cascade blue, propidium iodide): 5–10%molar ratio (dye to polymer).
   (b) DNA: 5 μg/mg polymer in the solution.
   (c) siRNA/ODNs (with/without fluorescent labels): 4 μg/mg polymer in the solution.
   (d) Protein (including antibodies): 5 μg/mg of polymer in the solution.
5. Increase further the pH to 7 to allow the formation of vesicles and the encapsulation of dyes/agents by adding dropwise sterile NaOH 1 M with the help of a 10 μl Gilson pipette while stirring vigorously with a vortex.
6. Sonicate the sample for 30 min to allow "onion-like" structures to break (see Note 7).

3.3 Separation of Encapsulated Fraction by Size Exclusion Chromatography

1. Harvest the vesicles from the solution by molecular-size fractioning in aseptic conditions with a clean and sterile GPC column filled 2/3 of the length with Sepharose 4B (see Note 8).
2. Once the PBS has completely entered the column, add your sample onto the sepharose carefully (see Note 9) (also, save a small volume of sample aside, for calculating the encapsulation efficiency).
3. Straight after the sample has gone inside the column, top the bed volume up with sterile PBS and start collecting the fractions (see Note 10).
4. An increased turbidity of the collected fractions will reveal presence of polymersomes. Pull the turbid samples together. Annotate the final volume of polymersomes.

3.4 Calculation of Polymersome Size Distribution and Size Control Procedures

Calculate the size and distribution of the polymersomes using dynamic light scattering (see Note 11). It is always recommendable to use TEM as a complementary technique to confirm polymersome morphology.

3.5 Calculation of the Bioactive/Agent Encapsulation Efficiency

1. Measure the dye/agent concentration by fluorescence or to identify the ex/em spectra at pH 6.
2. Create a standard curve of the agent dissolved in polymersome solution (pH 6) (see Note 12).
3. Measure the absorbance or fluorescence of the samples collected (before column (BC) and after column (AC)) and calculate the concentration from the standard curve data.
4. Calculate the % encapsulation efficiency:

$$(\text{Concentration AC})/(\text{Concentration BC}) \times 100$$

3.6 Cellular Uptake of Polymersomes

Polymersomes should be added directly in normal cell medium. Cell types that grow in suspension (i.e., lymphocytes) as well as adherent monolayers can actively uptake PMPC-PDPA polymersomes, provided that cells are viable and able to do endocytosis.

1. Seed cells on required well size tissue culture plates at an appropriate density to achieve 80–90%confluent monolayers after 1–2 days (see Note 13).
2. Add then the polymersomes containing the desired cargo onto the cells typically in a 1:10 dilution (v/v) directly into the cell monolayers (see Note 14).
3. Incubate the cells at 37°C in a humidified CO_2 incubator for 24 h (see Note 15).
4. Afterwards, image the cells in imaging medium (when using imaging dyes) or assay the cells for transfection efficiency (plasmidic, siRNA, and protein cargoes).

4 Notes

1. Plasticware made of polystyrene will dissolve in a solution with a high concentration of chloroform. This will add impurities to the film that will sometimes form a cloudy precipitate. Use chloroform-resistant plastic or preferably glassware.
2. Ensure that all solvent has evaporated prior to solubilization of the film in aqueous solution. Failure to remove solvent will produce precipitates and reduce significantly encapsulation efficiency of hydrophobic compounds.
3. The concentration of 10 mg/ml is required to maintain the colloidal stability of the vesicles; above this concentration, the vesicles aggregate.
4. From this point onwards, the solution is sterile and aseptic techniques should be followed. The procedures should be carried out in a class II sterile cabinet. Use of sterile or aseptic accurate pH meter is also necessary from this point.
5. pH 6 is an adequate (not too low) pH for ensuring stability when encapsulating hydrophilic and bioactive cargoes that are normally maintained at physiological pH. Molecules that are negatively charged (i.e., nucleic acids) at pH 6 can interact with the positive tertiary amine groups, and this can result in increased encapsulation efficiency. However, strong interactions polymer/encapsulates can also promote precipitation of the sample. It is recommended to work at lower polymer/encapsulate concentration if precipitation events occur.
6. We recommend adding the hydrophilic cargoes to the polymer solution at pH 6 in a maximum volume of 10%(v/v). This is to avoid dilution of the polymer and maximize encapsulation efficiency.
7. The sonication time may be tailored to the particular bioactive to encapsulate. Some cargoes are much more sensitive than others to sonication, and standardization assays testing stability of the cargo over sonication time are strongly recommended. In any case, we recommend also maintaining the sonication bath at a constant room temperature (or below room temperature if the samples are sensitive) to avoid overheating and degradation of bioactive samples/dyes.
8. GPC columns can sometimes resist high-pressure sterilization via autoclave (consult with manufacturer). Otherwise, cleaning the columns profusely with sterilizing agents (laboratory use surfactants) followed by overnight sterilization in 70%EtOH is recommended. Several washes with sterile deionized water should be performed in a laminar flow class II cabinet. Sepharose 4B is not autoclavable; it should be sterilized by

suspension (or washing) in 70%EtOH overnight at 4°C. The next day, wash the sepharose 4B in sterile PBS (at least 10 volumes) to ensure that no ethanol remains. Adapt the length of the column to the volume of sample intended to put through. Follow manufacturer's instructions for best results.

9. During the size exclusion chromatography procedure, ensure that the surface on the sepharose in the column does not dry out; this will result in clotting of the sample onto the column.
10. Do not add PBS to push the sample through until your sample has completely entered the column. This will result in unwanted dilution of your sample and will decrease the efficiency of recovery.
11. Size distribution can be controlled using an extruder with a defined set of pore size membrane. For optimal cellular uptake, polymersomes should not be bigger in size than 200 nm. Hydrophilic agents cannot be encapsulated inside hydrophobic cores of polymeric micelles; in these cases the lower limit should be thoroughly controlled, rejecting column fractions with size below 60 nm.
12. Work within the correct maxima wavelengths of ex/em (equally for UV/Vis absorption) for your dye/bioactive agent to calculate the encapsulation efficiency. You can use additional dyes to enhance detection or modify your agents (i.e., DAPI to label DNA, PicoGreen® to label double stranded RNA, bicinchoninic acid to detect protein via UV/Vis absorption). To allow for effective detection, the standard curve should start at the same theoretical concentration of the cargo in the sample before passing through the column (original sample) and should contain serial dilutions down to 1:100 of that original concentration in the sample.
13. We strongly recommend allowing cells to grow for 2 days in the tissue culture plates before adding the polymersomes to the culture, especially after the use of enzymatic products (i.e., trypsin) to detach the monolayers. An easy way to increase cellular uptake is starving the cells overnight (no serum for 16 h) prior to adding the polymersomes in serum containing medium. This, in general, increases cellular uptake of polymersomes.
14. The 1:10 dilution is an arbitrary value based on the nontoxic effect of these nanoparticles on most cell types. We advise to perform toxicity curves for each cell type using MTT tests or similar. As the nanoparticles are in PBS solution, concentrations above 10%in the medium may produce stress due to the PBS effect. If required to go above 10 %, it is necessary to add a toxicity control with the same volume of PBS and no polymersomes.

15. Removal of the polymersome medium is not required; however growth medium may be changed after 24 h of cell contact with polymersomes. When using plasmidic cargoes, it is also recommended assaying the target gene 24–48 h after transfection (48–72 h after incubation).

References

1. Rust MJ, Bates M, Zhuang X (2006) Sub-diffraction-limit imaging by stochastic optical reconstruction microscopy (STORM). Nat Methods 3:793–795
2. Schermelleh L, Heintzmann R, Leonhardt H (2010) A guide to super-resolution fluorescence microscopy. J Cell Biol 190:165–175
3. Massignani M, Canton I, Sun T, Hearnden V, Macneil S, Blanazs A, Armes SP, Lewis A, Battaglia G (2010) Enhanced fluorescence imaging of live cells by effective cytosolic delivery of probes. PLoS One 5:e10459
4. Yang Z, Zheng S, Harrison WJ, Harder J, Wen X, Gelovani JG, Qiao A, Li C (2007) Long-circulating near-infrared fluorescence core-cross-linked polymeric micelles: synthesis, characterization, and dual nuclear/optical imaging. Biomacromolecules 8:3422–3428
5. Kim J, Cao L, Shvartsman D, Silva EA, Mooney DJ (2011) Targeted delivery of nanoparticles to ischemic muscle for imaging and therapeutic angiogenesis. Nano Lett 11:694–700
6. Nasongkla N, Bey E, Ren J, Ai H, Khemtong C, Guthi JS, Chin SF, Sherry AD, Boothman DA, Gao J (2006) Multifunctional polymeric micelles as cancer-targeted, MRI-ultrasensitive drug delivery systems. Nano Lett 6:2427–2430
7. Little LE, Dane KY, Daugherty PS, Healy KE, Schaffer DV (2011) Exploiting bacterial peptide display technology to engineer biomaterials for neural stem cell culture. Biomaterials 32:1484–1494
8. Park C, Im MS, Lee S, Lim J, Kim C (2008) Tunable fluorescent dendron-cyclodextrin nanotubes for hybridization with metal nanoparticles and their biosensory function. Angew Chem Int Ed Engl 47:9922–9926
9. Erogbogbo F, Yong KT, Hu R, Law WC, Ding H, Chang CW, Prasad PN, Swihart MT (2010) Biocompatible magnetofluorescent probes: luminescent silicon quantum dots coupled with superparamagnetic iron(III) oxide. ACS Nano 4:5131–5138
10. Slowing I, Trewyn BG, Lin VS (2006) Effect of surface functionalization of MCM-41-type mesoporous silica nanoparticles on the endocytosis by human cancer cells. J Am Chem Soc 128:14792–14793
11. Liong M, Lu J, Kovochich M, Xia T, Ruehm SG, Nel AE, Tamanoi F, Zink JI (2008) Multifunctional inorganic nanoparticles for imaging, targeting, and drug delivery. ACS Nano 2:889–896
12. Lomas H, Massignani M, Abdullah KA, Canton I, Presti C, MacNeil S, Du J, Blanazs A, Madsen J, Armes SP, Lewis AL, Battaglia G (2008) Non-cytotoxic polymer vesicles for rapid and efficient intracellular delivery. Faraday Discuss 139:143–159, discussion 213–128, 419–120
13. Du J, Tang Y, Lewis AL, Armes SP (2005) pH-sensitive vesicles based on a biocompatible zwitterionic diblock copolymer. J Am Chem Soc 127:17982–17983
14. Massignani M, LoPresti C, Blanazs A, Madsen J, Armes SP, Lewis AL, Battaglia G (2009) Controlling cellular uptake by surface chemistry, size, and surface topology at the nanoscale. Small 5:2424–2432, Weinheim an der Bergstrasse, Germany

Chapter 32

Protocol for the Preparation of Stimuli-Responsive Gold Nanoparticles Capped with Elastin-Based Pentapeptides

Vincent Lemieux, P. Hans H.M. Adams, and Jan C.M. van Hest

Abstract

Stimuli-responsive materials are playing an increasingly important role in a wide range of applications such as drug delivery, diagnostics, sensors, and tissue engineering. Among them, gold nanoparticles responding to changes in their surrounding environment are of particular interest due to their size-related optical properties. Here, we present a novel strategy for the preparation of gold nanoparticles exhibiting a stimuli-responsive behavior. We rely on the use of a ligand consisting of only a single repeat of the elastin-based pentapeptide VPGVG. In this contribution, we describe a protocol for the solid-phase peptide synthesis of thiol-terminated VPGVG ligand, and for the preparation of gold nanoparticles covered with the pentapeptide through a ligand-exchange reaction.

Key words Elastin, VPGVG, Gold nanoparticles, LCST, Solid-phase peptide synthesis, Ligand exchange

1 Introduction

Recently, nanomaterials sensitive to changes in their environment have been at the core of many research programs in the materials science community. Among this class of materials, gold nanoparticles (Au NPs) with surface coatings responsive to external stimuli such as light exposure and variations in pH or temperature have received much attention. This is not surprising since they are promising candidates in the development of many new applications in the field of sensory sciences and for the controlled release of active agents (1–6).

To date, the primed strategy for the preparation of thermosensitive particles is the grafting of linear thermally responsive polymers onto the surface of Au NPs (1–4). By far the most commonly used polymer is poly(*N*-isopropylacrylamide) (pNIPAM). When heated at around 32°C, aqueous solutions of pNIPAM undergo a phase transition leading to the precipitation of the polymeric chains (7). However, even on gold surfaces, this lower critical solution temperature (LCST) can only be varied within a 10°C range, limiting the number of potential applications (5). In order to broaden the scope of thermally

Volkmar Weissig et al. (eds.), *Cellular and Subcellular Nanotechnology: Methods and Protocols*, Methods in Molecular Biology, vol. 991, DOI 10.1007/978-1-62703-336-7_32, © Springer Science+Business Media New York 2013

responsive nanoparticles, novel strategies are required to obtain nanoparticles with flexibly tunable transition temperatures. This could be achieved by interfacing nanoparticles with stimuli-responsive polypeptides/proteins such as elastin-like polypeptides (ELPs).

ELPs are synthetic polypeptides derived from the structural protein elastin, whose most prominent amino acid sequence is VPGVG (where V=valine, P=proline, and G=glycine) (8). When heated, synthetic polymers made of the VPGVG sequence, poly(VPGVG), undergo a transition from a hydrophilic random coil conformation to a hydrophobic β-spiral; the ELPs then aggregate and precipitate (9). Linear poly(VPGVG) is not the only ELP that exhibits this LCST behavior. The phenomenon has also been observed with polymers having VPGVG side chains (10–13) and even with single repeats (14) of the pentapeptide. Moreover, the transition temperature of these elastin-based structures can be tuned within a wide temperature range by varying various parameters such as the molecular weight, concentration, and pH (15, 16). Replacing the fourth residue with any other amino acid (VPGXG), except proline, also has an important influence on the LCST (15). The idea of coating gold nanostructures with ELPs has been exploited in the past (17, 18). Thermally and optically responsive gold nanoassemblies have been prepared using cysteine-containing ELPs. However, the preparation of such polypeptides involves genetically encoded recombinant methods and is nontrivial to most chemists.

We recently described a novel strategy to prepare thermally responsive materials based on the use of gold nanoparticles capped with a layer of only a single repeat unit of thiol-functionalized VPGVG pentapeptide (19) (Fig. 1). This simple 5-mer peptide offered the advantages of a facile, efficient, multi-gram compatible synthesis, and ease of derivatization. In this contribution, we present the extended protocol for the preparation of these thermoresponsive gold nanoparticles, including the solid phase peptide synthesis of the thiol-terminated VPGVG ligand and the preparation of DMAP-capped nanoparticles used for the ligand-exchange reaction. Figure 2 shows an overview of the protocol.

2 Materials

All solvents and reagents were purchased from Aldrich and used as received unless indicated otherwise.

2.1 Components for the Preparation of the Elastin-Based Ligand Using Solid-Phase Peptide Synthesis

1. Resin: *p*-alkoxybenzyl alcohol “Wang” resin with a loading of 1.14 mmol/g.
2. Solvent: Dimethylformamide (DMF).
3. Protected amino acids: 9-Fluorenylmethoxy carbamate protected valine (Fmoc Val-OH, >99%), glycine (Fmoc Gly-OH, >99%), and proline (Fmoc Pro-OH, >99%).

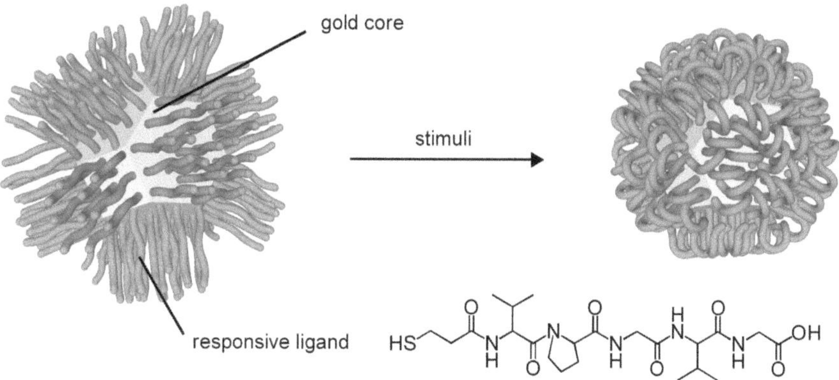

Fig. 1 Schematic of the stimuli-responsive gold nanoparticles and molecular structure of the thiol-terminated VPGVG ligand

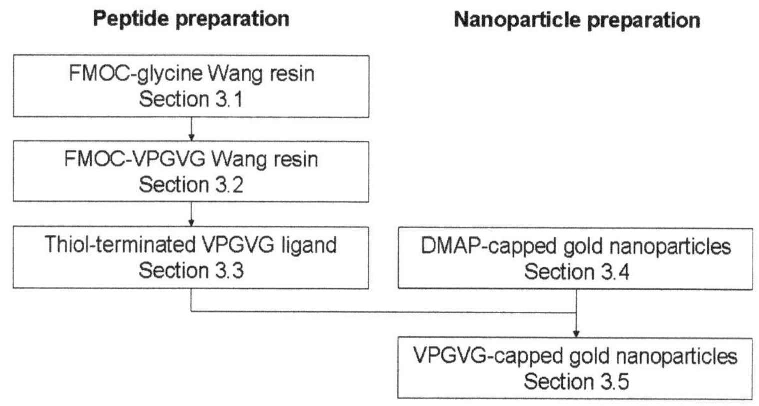

Fig. 2 Overview of the steps for the preparation of VPGVG-capped gold nanoparticles

4. Terminal thiol unit: 3-Mercaptopropionic acid (≥99%).
5. Coupling reagents: 1-Hydroxybenzotriazole hydrate (HOBt, >98%) and *N,N*-di-isopropylcarbodiimide (DIPCDI, >98%).
6. Deprotection solution: Dimethylformamide (DMF) (J.T. Baker) solution containing 20% v/v piperidine.
7. Cleavage reagents and scavenger: Trifluoroacetic acid (TFA, 99%) (Acros); ethane dithiol (EDT, >99%).

2.2 Components for the Kaiser (Ninhydrin) Tests

1. Solution A: Ninhydrin, ethanol.
2. Solution B: Phenol, ethanol.
3. Solution C: Potassium cyanide (KCN), pyridine (seccosolv).

2.3 Components for the Preparation of the Gold Nanoparticles

1. Gold source: Hydrogen tetrachloroaurate trihydrate ($HAuCl4 \cdot 3H_2O$, 99.9+%).
2. Reducing agent: Sodium borohydride (AF granules, 10–40 mesh, 98%).
3. Ligands: Tetraoctylammonium bromide (TOAB, ≥98%); 4-(dimethylamino)pyridine (DMAP, 99%).
4. Solvents: Toluene; deionized water with a typical resistivity of 18.2 MΩ/cm was obtained using a Labconco Water Pro PS purification system.
5. Dialysis membrane: Spectra/Por molecular porous membrane tubing with a MWCO of 12,000–14,000 g/mol.

3 Methods

3.1 Synthesis of Fmoc-Glycine-Functionalized Resin

1. Prepare a suspension of Wang resin (30 g) in DMF (300 mL) and cool in an ice bath.
2. Add Fmoc Gly-OH (13.5 g, 45 mmol), HOBT (9.20 g, 60 mmol), and DIPCDI (4.30 g, 34.2 mmol).
3. Shake the mixture for 6 h.
4. Filter the functionalized resin and wash repeatedly with dichloromethane (DCM) (3×50 mL), DMF (3×50 mL), and isopropyl alcohol (3×50 mL).
5. Resuspend the resin in DCM (300 mL) and cool in an ice bath.
6. Add benzoyl chloride (10.2 mL) and pyridine (8.4 mL) in order to cap the unfunctionalized groups.
7. Shake the mixture for 30 min
8. Filter, and wash repeatedly with DCM (3×50 mL), DMF (3×50 mL), and isopropyl alcohol (3×50 mL).
9. Dry the resin in air and then under vacuum.
10. Determine the loading of the resin (see Note 1).

3.2 Synthesis of Fmoc-VPGVG-Functionalized Resin

The Fmoc-VPGVG resin was synthesized by standard solid-phase methods using an Fmoc-glycine-functionalized "Wang" resin (20, 21).

1. Mix the Fmoc-Gly functionalized resin (5.0 g, loading 0.64 mmol/g) with DMF (45 mL) and allow the resin to swell for 20 min.
2. Filter the resin.
3. Add a DMF solution containing 20% v/v piperidine (45 mL).
4. Shake the mixture for 20 min to remove the Fmoc group.

5. Perform a Kaiser test (22) to verify the presence of free primary amines (positive result) (see Note 2). Repeat steps 2–5 if the Kaiser test is negative.
6. Filter, and wash repeatedly with DMF (3×45 mL).
7. Coupling of the next amino acid: Add to the resin a solution of Fmoc Val-OH (3.26 g, 9.6 mmol), a 1 M HOBt solution in DMF (11.5 mL, 11.5 mmol), and a 1 M DIPCDI solution in DMF (10.6 mL, 10.6 mmol). Dilute to about 45 mL with DMF.
8. Shake the mixture for 45 min.
9. Filter, and wash repeatedly with DMF (3×45 mL).
10. Perform a Kaiser test to verify the completeness of the reaction (negative result) (see Note 2). Repeat steps 7–10 if the Kaiser test is positive.
11. Repeat steps 2–10 with the following three amino acids in this order: Fmoc Gly-OH (2.85 g, 9.6 mmol), Fmoc Pro-OH (3.24 g, 9.6 mmol), and Fmoc Val-OH (3.26 g, 9.6 mmol).
12. Wash repeatedly with DCM (3×40 mL), DMF (3×40 mL), and isopropyl alcohol (3×40 mL).
13. Dry the resin in air and then under vacuum.
14. Determine the loading of the resin (see Note 1).

3.3 Synthesis of Thiol-Functionalized VPGVG Peptide

1. Mix the Fmoc-VPGVG functionalized Wang resin (2.0 g, loading 0.52 mmol/g) with DMF (45 mL) and allow it to swell for 20 min.
2. Filter the resin.
3. Add a DMF solution containing 20% v/v piperidine (45 mL).
4. Shake the mixture for 20 min to remove the Fmoc group.
5. Perform a Kaiser test to verify the presence of free primary amines (positive result) (see Note 2). Repeat steps 2–5 if the Kaiser test is negative.
6. Filter, and wash repeatedly with DMF (3×45 mL).
7. Coupling of the thiol-containing acid: add to the resin a solution of 3-mercaptopropionic acid (0.26 mL, 3.0 mmol), a 1 M HOBt solution in DMF (3.6 mL, 3.6 mmol), and a 1 M DIPCDI solution in DMF (3.3 mL, 3.3 mmol). Dilute to about 45 mL with DMF.
8. Shake the mixture for 60 min.
9. Filter, and wash repeatedly with DMF (3×45 mL).
10. Perform a Kaiser test to verify the completeness of the reaction (negative result) (see Note 2). Repeat steps 7–10 if the Kaiser test is positive.
11. Wash repeatedly with DCM (3×40 mL), DMF (3×40 mL), and isopropyl alcohol (3×40 mL).

12. Allow the resin to dry in air.
13. Cleave the thiol-VPGVG peptide from the resin: Add to the resin 10 mL of a solution containing 95% of TFA, 2.5% of water, and 2.5% of EDT to reduce any disulfide complexes involving the VPGVG peptide.
14. Stir the mixture for 60 min.
15. Precipitate the peptide by adding a third of the TFA solution (3.333 mL) dropwise to diethyl ether (~40 mL).
16. Centrifuge and decant.
17. Resuspend the solid in diethyl ether, and add dropwise a further 3.33 mL of TFA solution.
18. Repeat steps 16 and 17 with the last remaining portion of TFA solution (3.33 mL).
19. Resuspend the solid in diethyl ether (~40 mL).
20. Centrifuge and decant.
21. Repeat steps 19 and 20 twice.
22. Dry the white solid in air.
23. Purify the peptide by column chromatography using silica gel and $CHCl_3$/MeOH/water (65:25:4) as the eluent. ($R_f = 0.24$).
24. From 2.0 g of Fmoc-VPGVG functionalized Wang resin, 461 mg of peptide was obtained (see Note 3).

3.4 Synthesis of DMAP-Stabilized Gold Nanoparticles

The DMAP-stabilized gold nanoparticles were prepared according to the method proposed by Curasso (23) as described by Lennox (24, 25).

1. Mix an aqueous solution of hydrogen tetrachloroaurate trihydrate (30 mM, 30 mL, 0.9 mmol) with a solution of tetraoctylammonium bromide (TOAB) (25 mM, 80 mL, 2.0 mmol) in toluene.
2. Stir vigorously the biphasic mixture until all the tetrachloroaurate has transferred into the toluene layer, giving a deep orange organic layer and colorless aqueous layer.
3. Add to the stirring mixture a freshly prepared aqueous solution of sodium borohydride (0.4 M, 25 mL, 10 mmol) over a period of 2 s. The organic phase should immediately become deep red as the TOAB-capped gold nanoparticles are formed.
4. Stir the mixture for 90 min.
5. Separate the organic layer from the aqueous layer and wash it three times with deionized water (3 × 100 mL).
6. Dry the solution over anhydrous sodium sulfate, and filter it.
7. To this deep red colored solution, add an aqueous solution of DMAP (0.1 M, 80 mL, 8 mmol). The color of the aqueous

phase progressively becomes deep ruby as the phase transfer of the particles takes place and the ligand exchange from TOAB to DMAP occurs.

8. Stir the mixture for 60 min.
9. Isolate the aqueous layer containing the nanoparticles.
10. Adjust the concentration of the gold nanoparticles with an aqueous solution of DMAP (0.1 M, 97 mL) to obtain a solution with a gold content of 1 mg/mL (see Note 4).
11. Store the DMAP-stabilized gold nanoparticles solution at 4°C (see Notes 5 and 6).

3.5 Synthesis of VPGVG-Capped Gold Nanoparticles

1. Add the thiol-functionalized VPGVG peptide (15.5 mg, 0.03 mmol) to a stirring aqueous solution of the DMAP-stabilized gold nanoparticles (1 mg/mL gold content, 10.8 mL, 0.054 mmol of Au).
2. Stir the solution overnight at ambient temperature.
3. Place the solution in a dialysis bag and dialyze the solution against deionized water for 24 h, changing the water periodically.
4. Freeze-dry the solution to obtain the VPGVG-capped gold nanoparticles as a deep purple solid (yield: 14.8 mg) (see Notes 7–9).

4 Notes

1. The loading of the resin can be evaluated by mass difference before and after the peptide coupling.
2. The presence or absence of terminal amino groups on the resin can be confirmed by a Kaiser (ninhydrin) test. Prepare three stock solutions: (1) 500 mg of ninhydrin in 10 mL ethanol, (2) 80 g phenol in 20 mL ethanol, and (3) 2 mL 0.001 M solution of KCN diluted to 100 mL with pyridine. Add 10–20 mg of resin to a 12×75 mm test tube. Rinse the resin with DMF, centrifuge, and decant. Repeat twice more. Add 2–3 drops of each stock solution to the resin in the test tube. Place the test tube in a boiling water bath for 5 min. If the resin beads remain white/yellow (negative test), the reaction is complete. If the resin beads become dark blue (positive test), some amino groups are deprotected.
3. Characterization of the thiol-terminated VPGVG ligand: ^{1}H NMR (DMSO-d_6, 400 MHz): δ 12.50 (bs, 1H), 8.30 (t, J=5.8 Hz, 1H), 8.18 (t, J=6.0 Hz, 1H), 8.08 (d, J=8.5 Hz, 1H), 7.61 (d, J=9.0 Hz, 1H), 4.30 (m, 2H), 4.18 (m, 1H), 3.80 3.40 (m, 6H), 2.63 (t, J=7.5 Hz, 2H), 2.44 (q, J=7.5 Hz, 2H), 2.22 (t, J=7.5 Hz, 1H), 2.10–1.90 (m, 4H), 1.90–1.75 (m, 2H), 0.90 (d, J=6.7 Hz, 3H), 0.87 (d, J=6.7 Hz, 6H),

0.82 (d, J=6.7 Hz, 3H). ^{13}C NMR (DMSO-d$_6$, 100 MHz): δ 171.82, 171.16, 170.59, 170.35, 169.87, 168.38, 59.87, 59.22, 55.82, 47.50, 43.24, 42.16, 38.85, 30.86, 30.11, 29.12, 24.52, 20.14, 19.12, 18.90, 18.73, 18.53. IR (solid): ν 3,284 (NH); 3,071, 2,965, 2,924, 2,872 (CH); 1,622 (C=O amide I); 1,530 (amide II); 1,443, 1,410, 1,313, 1,233, 1,208, 1,037 (unassigned). LC-MS for $C_{22}H_{37}N_5O_7S$ in order of decreasing intensity: 538.5 ($M+Na^+$), 516.4 ($M+H^+$), 554.4 ($M+K^+$), 560.5 ($M-H^++2\ Na^+$), 279.2 ($M+2\ Na^+$).

4. The molecular weight of gold is 196.97 g/mol. Since 30 mL of 30 mM solution of hydrogen tetrachloroaurate trihydrate were added, 0.9 mmol of gold atoms are present, and thus, 0.9 mmol × 196.97 g/mol (177.3 mg) of gold is present in solution. To obtain a concentration of about 1 mg/mL of gold, the total volume of the solution should be adjusted to 177 mL. The volume of the solution should be 80 mL at this point, and hence, 97 mL of DMAP solution (0.1 M) has to be added.
5. The DMAP-Au NPs can be kept at 4°C for several months without apparent degradation.
6. See ref. 23–25 for full characterization details.
7. Upon freeze-drying, VPGVG–Au NPs were obtained as a black/deep purple hygroscopic powder that could be readily redissolved in water at temperatures below the LCST to form clear red solutions. Samples of VPGVG–Au NPs were stable both in solution and in powder form (freeze-dried), and could be kept at room temperature under ambient atmosphere. The solutions remained clear red, and no precipitate formed even after several months.
8. Characterization of the VPGVG-AuNPs: ^{1}H NMR (DMSO-d$_6$+TFA for solubility, 400 MHz) δ 8.30 (t, J=5.8 Hz, 1H), 8.18 (t, J=6.0 Hz, 1H,), 8.08 (d, J=8.5 Hz, 1H), 7.61 (d, J=9.0 Hz, 1H), 4.30 (m, 2H), 4.18 (m, 1H), 3.80–3.40 (m, 6H), 2.10–1.90 (m, 4H), 1.90–1.75 (m, 2H), 0.90 (d, J=6.7 Hz, 3H), 0.87 (d, J=6.7 Hz, 6H), 0.82 (d, J=6.7 Hz, 3H). IR (solid): ν 3,284 (NH); 3,071, 2,964, 2,924, 2,872 (CH); 1,621 (C=O amide I); 1,529 (amide II); 1,442, 1,394, 1,311, 1,237, 1,200, 1,033 (unassigned). Transmission electron microscopy (TEM): An average diameter of 3.2 nm was obtained from the analysis of several micrographs where the diameter of at least 200 particles was measured. TGA (on 1.727 and 2.525 mg samples): weight loss of 30% between 250 and 500°C. The number of thiol-VPGVG ligands on the surface of each Au NP is evaluated to be approximately 211, assuming the presence of 1,289 gold atoms in the core of a nanoparticle with an average diameter of 3.2 nm (26).

9. Neutral solutions of VPGVG–Au NPs show no LCST behavior. This is anticipated since the end-group of the VPGVG ligand is a free carboxylic acid, which is expected to be largely deprotonated at pH around 7. At low pH, the carboxylic acid moieties are protonated, giving the nanoparticles a more hydrophobic character. Under these conditions, a clear hydrophilic–hydrophobic transition is observed as the solutions become turbid upon heating. Summary of the LCSTs: pH 2.1: 14°C, pH 2.8: 25°C, pH 3.0: 31°C, pH 3.3: 41°C, pH 3.6: no LCST. *See* ref. 19 for a more detailed description.

Acknowledgments

The authors thank the Netherlands Organisation for Scientific Research (NWO) and the Fonds québécois de la recherche sur la nature et les technologies (FQRNT) for their financial support. Christine Lavigueur is acknowledged for the design of Fig. 1.

References

1. Zhu M-Q, Wang L-Q, Exarhos GJ, Li ADQ (2004) Thermosensitive gold nanoparticles. J Am Chem Soc 126:2656–2657
2. Raula J, Shan J, Nuopponen M, Niskanen A, Jiang H, Kauppinen EI, Tenhu H (2003) Synthesis of gold nanoparticles grafted with a thermoresponsive polymer by surface-induced reversible-addition-fragmentation chain-transfer polymerization. Langmuir 19:3499–3504
3. Kim J-H, Lee TR (2004) Thermo- and pH-responsive hydrogel-coated gold nanoparticles. Chem Mater 16:3647–3651
4. Salmaso S, Caliceti P, Amendola V, Meneghetti M, Magnusson JP, Pasparakis G, Alexander C (2009) Cell up-take control of gold nanoparticles functionalized with a thermoresponsive polymer. J Mater Chem 19:1608–1615
5. Shen Y, Kuang M, Shen Z, Nieberle J, Duan H, Frey H (2008) Gold nanoparticles coated with a thermosensitive hyperbranched polyelectrolyte: towards smart temperature and pH nanosensors. Angew Chem Int Ed 47:2227–2230
6. Bakhtiari ABS, Hsiao D, Jin G, Gates BD, Branda NR (2009) An efficient method based on the photothermal effect for the release of molecules from metal nanoparticle surfaces. Angew Chem Int Ed 48:4166–4169
7. Schild HG (1992) Poly(N-isopropylacrylamide): experiment, theory and application. Prog Polym Sci 17:163–249
8. Tatham AS, Shewry PR (2002) Comparative structures and properties of elastic proteins. Philos Trans R Soc Lond B Biol Sci 357: 229–234
9. Urry DW (1984) Protein elasticity based on conformations of sequential polypeptides: the biological elastic fiber. J Protein Chem 3:403–436
10. Ayres L, Vos MRJ, Adams PJHM, Shklyarevskiy IO, van Hest JCM (2003) Elastin-based side-chain polymers synthesized by ATRP. Macromolecules 36:5967–5973
11. Ayres L, Koch K, Adams PHHM, van Hest JCM (2005) Stimulus responsive behavior of elastin-based side chain polymers. Macromolecules 38:1699–1704
12. Fernández-Trillo F, Duréault A, Bayley JPM, van Hest JCM, Thies JC, Michon T, Weberskirch R, Cameron NR (2007) Elastin-based side-chain polymers: improved synthesis via RAFT and stimulus responsive behavior. Macromolecules 40:6094–6099
13. Fernández-Trillo F, van Hest JC, Thies JC, Michon T, Weberskirche R, Cameron NR (2008) Fine-tuning the transition temperature of a stimuli-responsive polymer by a simple blending procedure. Chem Commun (Camb) 19:2230–2232
14. Reiersen H, Clarke AR, Rees AR (1998) Short elastin-like peptides exhibit the same temperature-induced structural transitions as elastin polymers: implications for protein engineer ing. J Mol Biol 283:255–264
15. Urry DW (1993) Molecular machines: how motion and other functions of living organisms

can result from reversible chemical changes. Angew Chem Int Ed 32:819–841

16. Nagarsekar A, Crissman J, Crissman M, Ferrari F, Cappello J, Ghandehari H (2002) Genetic synthesis and characterization of pH- and temperature-sensitive silk-elastin like protein block copolymers. J Biomed Mater Res 62:195–203
17. Huang H-C, Koria P, Parker SM, Selby L, Megeed Z, Rege K (2008) Optically responsive gold nanorod-polypeptide assemblies. Langmuir 24:14139–14144
18. Nath N, Chilkoti A (2001) Interfacial phase transition of an environmentally responsive elastin biopolymer adsorbed on functionalized gold nanoparticles studied by colloidal surface plasmon resonance. J Am Chem Soc 123:8197–8202
19. Lemieux V, Adams PHHM, van Hest JCM (2010) Elastin-based stimuli-responsive gold nanoparticles. Chem Commun 46:3071–3073
20. Atherton E, Sheppard RC (1989) Solid phase peptide synthesis. IRL, Oxford
21. Fields GB, Noble RL (1990) Solid phase peptide synthesis utilizing 9-fluorenylmethoxycarbonyl amino acids. Int J Pept Protein Res 35:161–214
22. Kaiser E, Colescot RL, Bossinge CD, Cook PI (1970) Color test for detection of free terminal amino groups in the solid-phase synthesis of peptides. Anal Biochem 34:595–598
23. Gittins DI, Caruso F (2001) Spontaneous phase transfer of nanoparticulate metals from organic to aqueous media. Angew Chem Int Ed 40:3001–3004
24. Rucareanu S, Gandubert VJ, Lennox RB (2006) 4-(*N*, *N*-Dimethylamino)pyridine-protected Au nanoparticles: versatile precursors for water- and organic-soluble gold nanoparticles. Chem Mater 18:4674–4680
25. Gandubert VJ, Lennox RB (2005) Assessment of 4-(dimethylamino)pyridine as a capping agent for gold nanoparticles. Langmuir 21:6532–6539
26. Hostetler MJ, Wingate JE, Zhong C-J, Harris JE, Vachet RW, Clark MR, Londono JD, Green SJ, Stokes JJ, Wignall GD, Glish GL, Porter MD, Evans ND, Murray RW (1998) Alkanethiolate gold cluster molecules with core diameters from 1.5 to 5.2 nm: core and monolayer properties as a function of core size. Langmuir 14:17–30

Index

Volkmar Weissig et al. (eds.), *Cellular and Subcellular Nanotechnology: Methods and Protocols*, Methods in Molecular Biology, vol. 991, DOI 10.1007/978-1-62703-336-7,

D

E

K

L

M

N

O

P

T

U

V

W

Y

Z

MIX
Papier aus verantwortungsvollen Quellen
Paper from responsible sources
FSC® C105338

If you have any concerns about our products,
you can contact us on
ProductSafety@springernature.com

In case Publisher is established outside the EU,
the EU authorized representative is:
Springer Nature Customer Service Center GmbH
Europaplatz 3, 69115 Heidelberg, Germany

Printed by Libri Plureos GmbH
in Hamburg, Germany